普通高等教育“十四五”规划教材

能源装备腐蚀与防护：理论与应用

（双语教材）

齐建涛　魏　午　刘国荣◎主编

中国石化出版社
·北京·

内容提要

本书根据能源装备腐蚀与防护等领域的研究成果，并结合国内外最新相关研究进展，针对化学腐蚀与电化学腐蚀理论、常见局部腐蚀（电偶腐蚀、点蚀、缝隙腐蚀、选择性腐蚀、晶间腐蚀、应力腐蚀开裂、腐蚀疲劳、磨损腐蚀）、环境腐蚀、防护技术、炼油装备防护技术进行了论述。本书为基于中英文、全英文授课编写的教材。

本书既可作为高等院校材料工程及相关专业的教材，也可以作为从事材料表面防腐处理与腐蚀测试等工程技术人员、科研人员和管理人员的参考资料。

图书在版编目（CIP）数据

能源装备腐蚀与防护：理论与应用：英汉对照 / 齐建涛，魏午，刘国荣主编．-- 北京：中国石化出版社，2025.4
普通高等教育“十四五”规划教材　双语教材
ISBN 978-7-5114-6968-7

Ⅰ.①能…　Ⅱ.①齐…②魏…③刘…　Ⅲ.①能源－装备－腐蚀－高等学校－教材－英、汉②能源－装备－防腐－高等学校－教材－英、汉　Ⅳ.①TK01

中国国家版本馆CIP数据核字（2023）第020034号

未经本社书面授权，本书任何部分不得被复制、抄袭，或者以任何形式或任何方式传播。版权所有，侵权必究。

中国石化出版社出版发行

地址：北京市东城区安定门外大街58号
邮编：100011　电话：(010)57512500
发行部电话：(010)57512575
http://www.sinopec-press.com
E-mail：press@sinopec.com
宝蕾元仁浩（天津）印刷有限公司印刷
全国各地新华书店经销

*

787毫米×1092毫米16开本18.75印张375千字
2025年4月第1版　2025年4月第1次印刷
定价：58.00元

Preface

Corrosion and protection is based on materials and environmental science, involving materials, machinery, physics, chemistry and other multidisciplinary compounds, infiltrated into an interdisciplinary discipline, in materials, machinery, chemical industry, energy, architecture, aerospace, transportation and many other fields of engineering technology has been widely used.

In recent decades, with the rapid development of energy equipment, people's long-term stable operation performance of equipment materials under high temperature, high pressure, high speed, high automation and harsh working conditions, especially the surface corrosion resistance of materials, has been continuously improved. Corrosion is the main problem to improve the service life and reliability of materials, and the materials used in energy equipment involve metallic materials (strength steel, light alloys and high-entropy alloys, etc.) and non-metallic materials (plastics, ceramics, etc.). In a sense, the surface of these industrial materials needs to be protected by high-quality material surfaces, and the internal microstructure of the materials is carefully designed. Anti-corrosion coating, electrochemical anti-corrosion technology, corrosion inhibitor, surface alloying design and microstructure control have become the most widely used corrosion protection technologies in engineering. Therefore, corrosion and protection technology has quickly become a discipline with great research and application value, many new anti-corrosion technologies continue to emerge, traditional anti-corrosion technology is constantly improved, perfected, crossed and compounded, thus greatly enriching and developing the original corrosion engineering.

"Energy Equipment Corrosion and Protection: Theory and Application" is a teaching textbook compiled based on teaching in Chinese and English. The main purpose of the textbook is to deeply discuss the keywords-corrosion and protection, the mechanism of action and application technology in it, based on the current situation and problems of chemical and other energy equipment, and is the planning textbook for the elective course of process equipment and control engineering and environmental protection equipment in the new energy college. The characteristics of this textbook are combined with the characteristics of professional "equipment" and "new energy" of the college, highlighting the symbols of the times of professional elective courses. In addition, with the corresponding knowledge of Chinese and English, the school's purpose of internationalization has been practiced, and the gap in English teaching materials has been filled.

The book invited professor Wei Wu of Beijing University of Technology, professor Liu Guorong of China University of Petroleum(East China), associate Dr. Liu Zuojia, general manager of Midland Corrosion and Protection Company, and engineer Li Tengjiao of Henan Boiler and Pressure Vessel Safety Testing Institute to participate, and strive to aim at the theoretical cornerstone and practical cases in the field of energy equipment corrosion and protection in academic and engineering practice, and innovate the old and become an indispensable reference book in the field of future energy equipment safety service.

The book is edited by Qi Jiantao, Wei Wu and Liu Guorong, and Qi Jiantao is responsible for the compilation planning and overall manuscript of the whole book; Li Tengjiao, Bian Guixue and Liu Zuojia served as deputy editors, and Sun Wentao, Feng Meng, Hu Shuangfei, Yu Deshui, Wang Yu, and Wang Ning participated in the compilation of this book; Liu Xiaohui and Lin An presided over the trial.

In the process of compiling this book, I draw on the research results of many experts and scholars, and I would like to express my heartfelt thanks! Thanks to Acknowledgements to the Shandong Provincial Key Laboratory of Advanced Aluminum-based Materials and Technology, and the project supported by the National Natural Science Foundation of China: "Multi-scale Characterization and Controllable Preparation of Functional Conversion Coating Materials on Non-alloyed Magnesium Surfaces (51701239)" the Ministry of Education's industry-university-research project Binzhou Weiqiao Passing Higher Technology Research Institute "Design and performance research of new green and efficient anti-corrosion coating on aluminum alloy surface (BINTECH-KJZX-20220831-35)" for the technical support for the publication of this book!

The topics studied in this book are relatively broad, and although we strive for perfection in the writing process, it is inevitable that there will be improprieties in the book due to the level of the author. Experts and readers are welcome to comment on it.

前　言

腐蚀与防护是建立在材料和环境科学基础之上，涉及材料、机械、物理、化学等多学科相互复合、渗透而成的一门交叉学科，是在材料、机械、化工、能源、建筑、宇航、交通等诸多领域得到广泛应用的工程技术之一。

近几十年来，随着能源装备的飞速发展，人们对装备材料在高温、高压、高速、高度自动化和恶劣工况下长期稳定运转的性能，尤其是材料表面耐腐蚀性能的要求不断提高。腐蚀是提高材料使用寿命和可靠性的主要问题，而能源装备使用材料涉及金属材料（强度钢、轻合金及高熵合金等）和非金属材料（塑料、陶瓷等）。在一定意义上，这些工业材料表面需要进行高质量的材料表面防护，并且其材料内部微观结构进行了精细设计。防腐涂层、电化学防腐技术、缓蚀剂、表面合金化设计和微观组织控制等成为工程应用最为广泛的腐蚀与防护技术。因而，腐蚀与防护技术迅速成为一门极具研究和应用价值的学科，许多新的防腐技术不断涌现，传统的防腐技术不断得到改进、完善、交叉和复合，从而大大丰富和发展了原有的腐蚀工程学。

《能源装备腐蚀与防护：理论与应用》是基于中英文、全英文授课而编写的教材。教材基于化工等能源装备的现状及问题，深入探讨关键词——腐蚀与防护，其中的作用机理及应用技术，是新能源学院过程装备与控制工程及环保设备专业选修课规划教材。本教材的特点是结合专业“装备”和学院“新能源”特色，凸显专业选修课的时代符号。另外，配合中英文对应知识实践了学校国际化办学宗旨，弥补了英文教材的空白。

本教材编写邀请了北京工业大学魏午副教授、中国石油大学（华东）刘国荣教授、英国米德兰腐蚀与防护公司总经理刘佐嘉博士及河南省锅炉压力容器安全检测研究院李腾蛟工程师参与，力求在学术和工程实践中，瞄准能源装备腐蚀及防护领域的理论基石和实践案例，推陈出新，成为未来能源装备安全使役领域不可缺少的参考书。

本书由齐建涛、魏午和刘国荣担任主编，齐建涛负责全书的编写规划及统稿；由李腾蛟、卞贵学、刘佐嘉担任副主编，孙文涛、冯孟、胡爽飞、于得水、王瑜、王宁参加了本书的编写工作；由刘小辉和林安担任主审。

在本书编写过程中，借鉴了许多专家和学者的研究成果，在此表示衷心的感谢！同时，感谢山东省先进铝基材料与技术重点实验室、国家自然科学基金资助项目“镁合金表面功能化转化膜材料的多尺度表征及可控制备研究（51701239）”和教育部产学研项目滨州魏桥国科高等技术研究院“铝合金表面新型绿色高效防腐涂层设计及性能研究（BINTECH-KJZX-20220831-35）”对本书出版的技术支撑！

本书研究的课题涉及的内容比较宽泛，尽管在写作过程中力求完美，但由于作者水平有限，书中难免有不妥之处，敬请各位专家和读者批评斧正。

目　录

Chapter 1　Energy Equipment and Corrosion　能源装备腐蚀与防护　/ 001

1.1　Introduction　简介 ………………………………………………… 001
1.2　Energy equipment and corrosion　国内能源装备与腐蚀现状 ……………… 003
1.3　A brief history of the development of corrosion science and technology
腐蚀科学技术发展简史 ………………………………………………… 008
1.4　Material corrosion classification and velocity characterization methods
材料腐蚀分类及速度表征方法 ………………………………………… 012
1.5　References　参考文献 ………………………………………………… 014

Chapter 2　Chemical Corrosion　化学腐蚀　/ 015

2.1　Oxidation of metals　金属的氧化现象 ……………………………… 015
2.2　Thermal dynamic criterion　金属氧化的热力学判据 ………………… 022
2.3　Formation and properties　金属氧化膜的形成与性质…………………… 025
2.4　Oxidation kinetics of metals　金属氧化动力学 ……………………… 042
2.5　Common chemical corrosion　常见的化学腐蚀 ……………………… 049
2.6　References　参考文献 ………………………………………………… 058

Chapter 3　Electrode Potential and Pourbaix Diagram　电极电位与 Pourbaix 图　/ 059

3.1　Electrode system reacts with electrode　电极系统与电极反应 …………… 059
3.2　Electrode potential and its determination equilibrium potential
电极电位与平衡电位 ………………………………………………… 061

3.3 Thermodynamic criterion 热力学判据 ······ 065
3.4 Potential-pH figure 电位 –pH 图 ······ 066
3.5 References 参考文献 ······ 078

Chapter 4 Electrochemical Corrosion 电化学腐蚀 / 079

4.1 Corroded battery 腐蚀电池 ······ 079
4.2 Polarization and depolarization 极化与去极化 ······ 090
4.3 Evans polarization diagram Evans 极化图 ······ 097
4.4 Passivation of metals 金属的钝化 ······ 108
4.5 Hydrogen evolution corrosion and oxygen absorption corrosion
析氢腐蚀与吸氧腐蚀 ······ 118
4.6 Influencing factors of electrochemical corrosion 电化学腐蚀影响因素 ······ 135
4.7 References 参考文献 ······ 138

Chapter 5 Localized Corrosion 局部腐蚀 / 139

5.1 The corrosion morphology 腐蚀形态 ······ 139
5.2 Galvanic corrosion 电偶腐蚀 ······ 142
5.3 Pitting 点蚀 ······ 145
5.4 Crevice corrosion 缝隙腐蚀 ······ 151
5.5 Selective corrosion 选择性腐蚀 ······ 157
5.6 Intergranular corrosion 晶间腐蚀 ······ 161
5.7 Stress corrosion cracking 应力腐蚀开裂 ······ 165
5.8 Corrosion fatigue 腐蚀疲劳 ······ 170
5.9 Friction corrosion 磨损腐蚀 ······ 173

Chapter 6 Natural Environment Corrosion 自然环境腐蚀 / 175

6.1 Atmospheric corrosion 大气腐蚀 ······ 175
6.2 Seawater corrosion 海水腐蚀 ······ 198

6.3 Soil corrosion 土壤腐蚀 …… 211

Chapter 7 Corrosion Protection Techniques 腐蚀防护技术 / 223

7.1 Design for corrosion and protection 腐蚀与防护设计 …… 223
7.2 Electrochemical protection 电化学防护 …… 226
7.3 Metal protection layer 金属保护层 …… 234
7.4 Non-metallic protective layers 非金属保护层 …… 241
7.5 Corrosion inhibitors 缓蚀剂 …… 248
7.6 References 参考文献 …… 256

Chapter 8 Application of Corrosion and Protection in Refinery Equipment
炼油装备腐蚀防护案例 / 258

8.1 Principles and significance of formulating anticorrosion strategies
防腐策略制定原则与意义 …… 258
8.2 Corrosion and protection of atmospheric-vacuum distillation units
常减压装置的腐蚀与防护 …… 269
8.3 Corrosion and protection of catalytic cracking units
催化裂化装置的腐蚀与防护 …… 278
8.4 Corrosion and protection of delayed coking units
延迟焦化装置的腐蚀与防护 …… 284
8.5 References 参考文献 …… 289

Energy Equipment and Corrosion 能源装备腐蚀与防护 1

1.1 Introduction

Energy equipment is the carrier of energy technology and an important and core part of the equipment manufacturing industry. At present, China's energy technology equipment manufacturing is facing the challenges of energy development and structural adjustment, and the contradictions such as weak independent innovation ability, lack of some key core technologies, relative excess of traditional products and insufficient supporting capacity of key parts and components are still prominent, and urgently need to be transformed and upgraded.

National Development and Reform Commission, Ministry of Industry and Information Technology, and National Energy Administration issued the "Made in China 2025—Energy Equipment Implementation Plan", which identified energy equipment development

1.1 简介

能源装备是能源技术的载体，是装备制造业的重要和核心部分。目前，我国能源技术装备制造面临能源发展和结构调整的挑战，自主创新能力较弱、部分关键核心技术缺失、传统产品相对过剩和关键零部件配套能力不足等矛盾仍然突出，亟待转型升级。

国家发改委、工信部、国家能源局印发《中国制造2025——能源装备实施方案》，确定了15个领域的能源装备发展任务。其中包括煤炭绿色智

tasks in 15 fields. These include green and intelligent coal mining and washing equipment, oil and gas storage, transportation and transportation equipment, deepwater and unconventional oil and gas exploration and development equipment, clean and efficient coal-fired power generation equipment, hydropower equipment, wind power equipment, solar power generation equipment, fuel electric pool, geothermal energy equipment, marine energy equipment, gas turbines, energy storage equipment, advanced power grid equipment and coal deep processing equipment [1].

能采掘洗选装备、油气储运和输送装备、深水和非常规油气勘探开发装备、清洁高效燃煤发电装备、水电装备、风电装备、太阳能发电装备、燃料电池、地热能装备、海洋能装备、燃气轮机、储能装备、先进电网装备和煤炭深加工装备[1]。

At present, high-end manufacturing in developed countries such as Europe and the United States has returned, and Germany and the United States have successively proposed the concept of Industry 4.0 and industrial Internet. Under the situation of promoting the green and low-carbon development of energy and structural transformation, traditional energy technology and equipment urgently need to be innovated and upgraded, and a number of new energy technology and equipment industries are budding.

当前，欧美等发达国家高端制造回流，德国、美国相继提出工业4.0和工业互联网概念。在推动能源绿色低碳发展和结构转型大形势下，传统能源技术装备亟须革新和提升水平，一批新能源技术装备产业正在萌芽。

Figure 1-1 shows different energy equipment, among which, Figure 1-1(a) shows that the “Offshore Oil 981” drilling platform built by CNOOC can operate at a water depth of 3,000 meters and drill to a depth of 12,000 meters. Figure 1-1(b) shows the “H_2 Kusatsu Farm” hydrogen station built by the Panasonic Group in Osaka, Japan, and plans to commercialize hydrogen fuel cells in Japan around April 2021 [2]. Figure 1-1(c) shows the expansion bolts of the medium-load coupling of a heavy gas turbrine, and the required materials cannot be produced in China at present.

图1–1所示为不同能源装备。其中，图1–1（a）为中国海油建造的“海洋石油981”钻井平台，可在水深3000m作业，钻进深度达到12000m。图1–1（b）为松下集团在日本大阪建造的“H_2 Kusatsu Farm”氢气站，并计划于2021年4月左右在日本国内实现氢燃料电池的商业化[2]。图1–1（c）为重型燃气轮机中负荷联轴器膨胀螺栓，所需材料目前国内无法生产。

(a)CNOOC's "Offshore Oil 981" platform
中海油建造的"海洋石油981"钻井平台

(b)"H_2 Kusatsu Farm" hydrogen station
"H_2 Kusatsu Farm"氢气站

(c)Load coupling expansion bolts
燃气轮机中负荷联轴器膨胀螺栓

Figure 1-1 Different energy equipment
图 1-1 不同能源装备

1.2 Energy equipment and corrosion

China has become a world leader in the innovation and development of clean energy equipment. Taking power technology as an example, China's installed power generation capacity in 2022 is about 2.47 billion kilowatts, making it the world's largest power country. China's leading position is mainly reflected in the following three aspects: Firstly, electric power technology and equipment have fully entered the "million era", and the single-unit capacity million-kilowatt power generation and million-volt AC-DC transmission technology and equipment continuously breaking world records. Secondly, renewable energy equipment came later. As of the end of August this year, China's installed wind power capacity is about 340 million kilowatts, and solar power generation capacity is about 350 million kilowatts. The output and installed scale of wind turbines rank first in the world. The wind power industry chain is complete, forming a complete wind power equipment manufacturing system, and 7 wind power machine manufacturing enterprises are among the top ten in the world. The photovoltaic industry occupies a leading position globally, with polysilicon, silicon wafers,

1.2 国内能源装备与腐蚀现状

中国已成为世界清洁能源装备创新发展的领军者。以电力技术为例，我国2022年发电装机容量约24.7亿kW，是名副其实的世界第一电力大国。我国的领军者地位主要体现在以下三个方面：一是电力技术与装备全面进入"百万时代"，单机容量百万千瓦级发电和百万伏交直流输电技术与装备不断刷新世界纪录。二是可再生能源装备后来居上。截至2022年8月底，我国风电装机容量约3.4亿kW；太阳能发电装机容量约3.5亿kW。风电机组产量、装机规模稳居世界第一。风电产业链完整，形成了完备的风电设备制造体系，7家风电整机制造企业入列全球前10。光伏产业占据全球主导地位，多晶硅、硅片、电池片

cells and modules accounting for 76%, 96%, 83% and 76% of global production, respectively. Thirdly, UHV transmission technology and equipment have set world records in a row. China has independently developed, designed, manufactured and built a series of UHV transmission projects such as Jindongnan - Nanyang - Jingmen, Yunnan - Guangdong and Changji - Guquan with the highest voltage level, the longest transmission distance and the largest transmission capacity in the world, and has fully mastered the transmission technology of 1000 kV AC and ± 1100 kV DC and below, and the technical level, equipment scale and test capacity of transmission equipment are leading the world.

和组件分别占全球产量的76%、96%、83%和76%。三是特高压输电技术装备连创世界纪录。我国自主研发、设计、制造并建设了世界上电压等级最高、输送距离最长、输送容量最大的晋东南—南阳—荆门、云南—广东、昌吉—古泉等系列特高压输电工程，全面掌握了1000kV交流、± 1100kV直流及以下等级的输电技术，输电装备技术水平、设备规模、试验能力等均领先世界。

The photovoltaic industry is a combination of semiconductor technology and new energy demand derived industry, and is a technology-supported frontier industry, with broad development prospects, which is characterized by the development of silicon materials in photoelectric conversion applications, including high-purity polysilicon raw material production, solar cell production, solar cell module production, and related production equipment manufacturing, etc. As one of the national strategic emerging industries, under the dual role of industrial policy guidance and market demand, the photovoltaic industry has achieved rapid development and has gradually been recognized by the global market, and has the world's most complete photovoltaic industry chain, becoming one of the few industries in China that can participate in international competition and gain leading advantages.

光伏产业是半导体技术与新能源需求相结合而衍生的产业，是以技术为支撑的前沿产业，发展前景广泛，其特点是硅材料在光电转换应用方面的开发，包括高纯多晶硅原材料生产、太阳能电池生产、太阳能电池组件生产、相关生产设备的制造等。光伏产业作为国家战略性新兴产业之一，在产业政策引导和市场需求驱动的双重作用下，全国光伏产业实现了快速发展，并逐步得到全球市场的认可，已具备全球最完整的光伏产业链，成为我国为数不多可参与国际竞争并取得领先优势的产业。

The entire photovoltaic industry chain (Figure 1-2) can be summarized as upstream raw material extraction (collection of crystalline silicon and processing and

整个光伏产业链（图1–2）可概括为上游原材料提炼（晶体硅的采集和硅锭、硅棒、硅

manufacturing of silicon ingots, silicon rods and silicon wafers), midstream photovoltaic product manufacturing (photovoltaic cells and photovoltaic modules), and downstream photovoltaic power station system integration and operation (centralized power station and distributed photovoltaic). The photovoltaic module segment can be regarded as the end product of the midstream manufacturing process, and at the same time a unit directly used by the downstream power generation system. The industrial link is located between photovoltaic cells and photovoltaic power generation systems, playing a role in connecting the upper and lower, because it is not a technology-intensive industry itself, so China has less say in the photovoltaic module link.

片的加工制造），中游光伏产品制造（光伏电池和光伏组件），下游光伏电站系统的集成和运营（集中式电站和分布式光伏）。光伏组件环节可视作中游制造环节的终端产品，同时是下游发电系统直接使用的单元。产业环节位于光伏电池和光伏发电系统之间，起到承上启下的作用，由于其本身不属于技术密集型产业，因而我国在光伏组件环节中话语权较小。

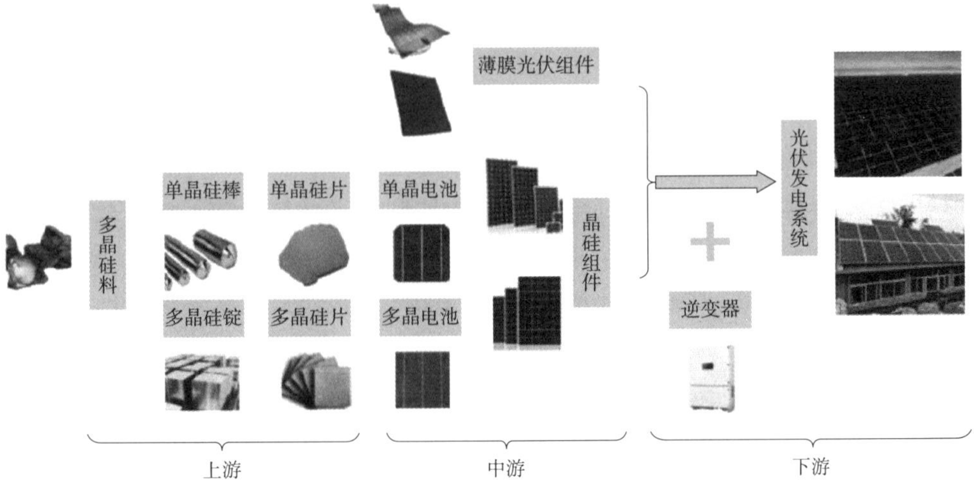

Figure 1–2 Panorama of the photovoltaic industry chain (China Photovoltaic Association "2021 China Photovoltaic Industry Development Roadmap")

图 1–2 光伏产业链全景（中国光伏协会《2021 年中国光伏产业发展路线图》）

Photovoltaic modules, commonly known as solar panels, are the core part of the solar power generation system, mainly composed of cells, welding strips (divided into bus strips and

光伏组件俗称太阳能电池板，是太阳能发电系统中的核心部分，主要由电池片、焊带（分为汇流条与互联条）、钢化

interconnection strips), tempered glass, EVA, backplane, aluminum alloy, silicone, junction box and other core parts, in which the cell interconnection determines the electrical performance of the module, in addition to the cell, generally known as photovoltaic auxiliary materials. At present, the standard number of solar cells for photovoltaic modules is 60 or 72, corresponding to 10 or 12 copper wires as a bus bar to connect them, 6 groups are interconnected into one photovoltaic module. The service life of photovoltaic module products is at least 25 years, it is necessary to ensure its environmental tolerance, and has certain mechanical properties, so after the cell is interconnected, in general, tempered glass, EVA, cell, backplane in order from top to bottom, laminated together, the backplane and tempered glass cell and EVA are encapsulated inside.

玻璃、EVA、背板、铝合金、硅胶、接线盒等核心部分组成，其中电池片互联决定了组件的电性能，除电池片以外的部分，一般被称为光伏辅材。目前，光伏组件的标准电池片数量为60片或72片，对应以10条或12条铜线作为汇流条将其连接，6组互联为一个光伏组件。光伏组件产品的使用寿命至少为25年，需要保证它的环境耐受性，并且具备一定的机械性能，因此在电池片互联后，一般而言，钢化玻璃、EVA、电池片、背板按照从上到下的顺序，经过层压的方式封装在一起，背板与钢化玻璃电池片和EVA封装在内部。

It is worth noting that on October 31, 2022, SPIC's Shandong Peninsula South No.3 Offshore Wind Farm Deep Offshore Floating Photovoltaic 500kV Passed Verification and Acceptance 500 kilowatts passed the demonstration acceptance, becoming the world's first deep sea wind and solar floating photovoltaic demonstration project, which realizes wind power and photovoltaic power generation in the same field. The successful implementation of the project has verified the wind and wave resistance marine environmental weather resistance of the floating body, anchoring, power generation components, the weather resistance of the marine environment and the technical feasibility of the wind and light grid-connected on the same site. In the future, marine photovoltaic equipment and its anti-corrosion technology will become a strategic highland for new energy equipment.

值得注意的是，2022年10月31日，国家电投山东半岛南3号海上风电场深远海漂浮式光伏500kV通过实证验收，成为全球首个深远海风光同场漂浮式光伏实证项目。该项目成功实施，验证了浮体、锚固、发电组件的抗风浪能力、海洋环境耐候性以及风光同场并网的技术可行性。未来海洋光伏装备及其防腐技术成为新能源装备战略高地。

In addition, the petroleum equipment manufacturing industry is a basic and strategic industry that supports the development of the petroleum industry, an important part of the national economy, and plays a pivotal role. Petroleum equipment involves a wide range, mainly including drilling equipment, oil production equipment, refining equipment, oil pipes, power equipment, logging equipment, etc. In the past ten years, the petroleum equipment manufacturing business has made remarkable achievements in the research and development of major equipment such as platform construction and drilling equipment and tools, development and operation equipment, petroleum steel pipe and pipeline equipment, power equipment, marine equipment and refining equipment [3].

此外，石油装备制造业是支撑石油工业发展的基础性、战略性产业，是国民经济的重要组成部分，具有举足轻重的地位。石油装备涉及的范围较广，主要包括钻井设备、采油装备、炼化装备、石油管材、动力装备、测井装备等。近十年来，石油装备制造业务在研发条件、平台建设和钻井装备与工具、开发与作业装备、石油钢管与管道装备、动力设备、海洋装备和炼化设备等重大装备研发等方面取得了显著的成绩[3]。

The corrosion or aging behavior of materials used in energy equipment is closely related to its service environment and determines the reliability and life of engineering structures. Energy equipment materials operation terrain is complex, climate is diverse, not only by high temperature, high humidity, high salt and other harsh environment, but also by the equipment operation brought by vibration, fatigue, airflow, current and other factors, multi-factor coupling effect will not only lead to engineering materials and structural corrosion is extremely serious, but also its geographical distribution law is complex. It is difficult to simply identify and predict, must be through full scientific corrosion investigation and on-site observation to obtain full systematic understanding. Therefore, it is important to carry out material corrosion big data monitoring and prediction technology for the operation and maintenance safety of facilities and equipment. Measurement technology, in the equipment operation

能源装备用材料腐蚀或老化行为与其服役环境密切相关，并决定了工程结构的可靠性和寿命。能源装备用材料运行地形复杂、气候多样，不但受到高温、高湿、高盐等严酷环境的影响，还受到装备运行时带来的振动、疲劳、气流、电流等因素的影响，多因素耦合作用不仅会导致工程材料和结构腐蚀异常严重，而且其地域分布规律复杂，难以简单判别和预测，必须经过充分科学的腐蚀调查与现场观测才能获得充分系统的认知。因此，开展面向设施装备运维安全的材料腐蚀大数据监测和预测技术，在装备运行重点区段的路基、轨道、桥梁、隧道等

of key sections of the roadbed, track, bridges, tunnels and other locations, and the system layout material service big data sensor conducts real-time dynamic monitoring and high-throughput data collection on the service failure state of key facilities and equipment such as corrosion and wear, fatigue vibration, displacement deformation and the service environment status of key facilities and equipment, forming an Internet of Things for intelligent monitoring of service damage of engineering facilities and equipment. Study the corrosion failure mechanism and environmental adaptability law of engineering equipment materials, clarify the corrosion and performance degradation laws of metals, concrete, polymers and coated materials under complex service conditions and the relationship between them and the main influencing factors of geology and climatic environment, establish a corrosion database and shared service platform for materials for engineering equipment to guide the design and application of corrosion-resistant materials and protection technologies in the construction of major national engineering construction. It is of great significance to promote the long-life and safe operation of infrastructure and major projects.

位置，系统布局材料服役大数据传感器对关键设施与装备腐蚀磨损、疲劳振动、位移变形等服役失效状态及其所处服役环境状态进行实时动态监测和高通量数据采集，形成工程设施与装备服役损伤智能监测物联网，研究工程装备材料腐蚀失效机理和环境适应性规律，厘清复杂服役条件下金属、混凝土、高分子及涂镀层材料的腐蚀、老化、磨损和疲劳等破坏机理及性能退变规律及其与地质、气候环境主要影响因素之间的关系，建立工程装备用材料腐蚀数据库和共享服务平台，对指导国家重大工程建设中耐腐蚀材料和防护技术的设计与应用，促进基础设施和重大工程长寿命安全运行等系列工作，具有重要意义。

1.3 A brief history of the development of corrosion science and technology

1.3 腐蚀科学技术发展简史

The discipline of corrosion and protection is an applied science that has always been inseparable from human economic activity. Corrosion science and protection technology of materials is a comprehensive marginal discipline involving multiple disciplines, and its theory and practice are closely related to metallurgy,

腐蚀与防护学科是一门应用科学，始终与人类经济活动密不可分。材料的腐蚀科学与防护技术是一个涉及多门学科的综合性边缘学科，其理论和实践与金属学、冶金学、物

metallurgy, physics, materials science, electrochemistry, chemistry, physical chemistry, engineering mechanics, fracture mechanics, fluid mechanics, chemical engineering, mechanical engineering, microbiology, surface science, electricity, and computer science. Therefore, as an independent discipline, the discipline of corrosion and protection can be traced back to middle of the 18th century, when the research and elaboration of corrosion phenomena and protection technologies have appeared successively. In 1748, Jlomohocob explained the phenomenon of metal oxidation. In 1790, Keir described the passivation of iron in nitric acid. In 1800, Volta discovered the principle of galvanic batteries. In 1801, Wollaton proposed the theory of galvanic corrosion. In 1824, Davy used iron as a sacrificial anode to provide cathodic protection from the bottom of a British Navy copper ship. In 1833, Davy's assistant, the British scientist Faraday, proposed Faraday's law of electrolysis. In 1929, the British metallurgical scientist, U. R. Evans, established a corrosion polarization map and promoted the quantitative study of the electrochemical nature of corrosion. In addition, Belgian scientist M. Pourbaix has established a potential-pH diagram based on thermodynamics, which can be used to clearly determine the thermodynamic tendency of a material to corrosion in environmental systems. That same year, Brenner and Riddell proposed electroless nickel plating. In 1950, The American scientist Uhlig proposed an autocatalytic mechanism model of pitting corrosion and compiled a corrosion science manual to query the corrosion characteristics of various metals. In 1970, Epellboin first studied corrosion processes using electrochemical impedance spectroscopy. From the

理学、材料学、电化学、化学、物理化学、工程力学、断裂力学、流体力学、化学工程学、机械工程学、微生物学、表面科学、电学、计算机科学密切相关。因此，作为独立学科的腐蚀与防护学科最早可追溯到18世纪中叶，当时陆续出现对腐蚀现象和防护技术的研究及阐述。其中，Jlomohocob在1748年解释了金属氧化现象。1790年，Keir描述了铁在硝酸中的钝化现象。随后1800年，Volta发现了原电池的原理。1801年，Wollaton提出了电化学腐蚀理论。1824年，Davy采用铁作为牺牲阳极为英国海军铜船底提供阴极保护。1833年，Davy的助手英国科学家法拉第提出了法拉第电解定律。1929年，英国冶金科学家U. R. Evans建立了腐蚀极化图，并推动了腐蚀电化学本质的定量化研究。此外，同时期，比利时科学家M. Pourbaix基于热力学建立了电位-pH图，可参考该图，清晰判断一种材料在环境体系中的腐蚀热力学倾向。同年，Brenner和Riddell提出化学镀镍技术。1950年，美国科学家Uhlig提出了点蚀的自催化机理模型，并编著了《腐蚀手册》，可查询各种金属的腐蚀特

middle and second half of the 20th century, corrosion science and protection technology have developed rapidly, and the applied science of corrosion science-corrosion engineering and corrosion prevention system engineering has been formed. In addition, some advanced surface analysis instruments and techniques, such as Auger electron spectroscopy (AES), X-ray electron spectroscopy (XPS), secondary ion mass spectrometry (SIMS), scanning electron microscopy SEM, transmission electron microscopy (TEM), atomic force microscopy (AFM), scanning vibration microscopy (SVET) have been applied in corrosion research, further revealing the microscopic mechanism and nature of many corrosion processes. In the 60s and 70s of the 20th century, electron beam, ion beam, laser beam and plasma technology were successively introduced surface technology methods, and promoted the development of modern material surface anti-corrosion technology and surface engineering. In recent years, nanoscience and additive manufacturing have promoted new materials surface corrosion prevention technologies.

性。1970年，Epellboin首次用电化学阻抗谱研究腐蚀过程。从20世纪中下叶起，腐蚀科学和防护技术得到了迅速发展，腐蚀科学的应用科学——腐蚀工程学和防腐蚀系统工程学形成。此外，一些先进的表面分析仪器和技术，如俄歇电子能谱（AES），X射线光电子能谱（XPS），二次离子质谱（SIMS），扫描电子显微镜（SEM），透射电子显微镜（TEM），原子力显微镜（AFM），扫描振动显微镜（SVET）在腐蚀研究中得到应用，进一步揭示了许多腐蚀过程的微观机制和本质。20世纪60—70年代，电子束、离子束、激光束和等离子技术等先后引入表面技术方法，并推动了现代材料表面防腐技术和表面工程学发展。近年来，纳米科学和增材制造等促进了新型材料表面防腐蚀技术。

China’s corrosion science started late, but developed rapidly. After the founding of People’s Republic of China, especially in the 20 years of reform and opening up, with the increase in the demand for China’s economic construction and national defense construction, corrosion and protection technology has become a key application technology for equipment safety service, and many achievements have reached or approached the international advanced level, and have also achieved greater economic and social benefits.

我国腐蚀科学起步较晚，但发展很快。中华人民共和国成立以后，尤其是改革开放20年内，随着我国经济建设和国防建设需求提升，腐蚀与防护技术成为装备安全服役的关键性应用技术，也有不少成果达到或接近国际先进水平，也取得了较大的经济效益和社会效益。

China's earliest corrosion subject magazine is Wuhan Institute of Materials Protection in the early 60s of the last century "Material Protection", and later in 1979 with the establishment of the Chinese Society of Corrosion and Protection, "Journal of the Chinese Society of Corrosion and Protection", "Corrosion Protection and Protection Technology", "Corrosion and Protection" and "Corrosion Protection News" were successively established. Research institutions continue to expand, and the Chinese Academy of Sciences has established a professional Institute of Metal Corrosion and Protection in Shenyang, of which Mr. Li Xun played an important role as founder. The State Education Commission has established a material failure open laboratory in the University of Science and Technology Beijing, the Chinese Academy of Sciences has established an open laboratory for corrosion science, and the State Development Planning Commission approved the establishment of the State Key Laboratory of Metal Corrosion and Protection in 1993, unprecedented strength of specialised research in the discipline. In terms of talents, researchers Shi Changxu, Xiao Jimei and Cao Chunan, who are engaged in research in this discipline, have been successively elected as academicians of the Chinese Academy of Sciences, and researchers Ke Wei and Hou Baorong have been elected as academicians of the Chinese Academy of Engineering. The University of Science and Technology Beijing, the Shanghai Institute of Metallurgy of the Chinese Academy of Sciences and the Shenyang Institute of Metal Corrosion and Protection have cultivated a large number of academic leaders and scientific and technological backbone. In recent years, with the development of the aerospace industry, Professor

我国最早的腐蚀学科杂志是武汉材料保护研究所于20世纪60年代初创刊的《材料保护》，后于1979年，随着中国腐蚀与防护学会成立，《中国腐蚀与防护学会学报》《腐蚀科学与防护技术》《腐蚀与防护》和《腐蚀防护报》等相继创刊。研究机构不断扩大，中国科学院在沈阳建立了专业性的金属腐蚀与防护研究所，其中李薰先生作为创始人起到了重要的推动作用。国家教委在北京科技大学建立了材料失效开放实验室，中国科学院成立了腐蚀科学开放实验室，国家计划委员会于1993年批准成立了金属腐蚀与防护国家重点实验室等，该学科专业研究力量空前壮大。在人才方面突出的一点是，从事本学科研究的师昌绪、肖纪美、曹楚南研究员先后被评选为中国科学院院士，柯伟研究员和侯保荣研究员被选为中国工程院院士。北京科技大学、中科院上海冶金所和沈阳金属腐蚀与防护研究所培养了大批学术带头人和科技骨干。近年来，随着航空航天行业发展，北京航空航天大学的宫声凯教授和重庆大学的潘复生教授也入选中国工程院院士。除了高等院校和科研机

Gong Shengkai of Beihang University and Professor Pan Fusheng of Chongqing University have also elected academicians of the Chinese Academy of Engineering. In addition to colleges and universities and scientific research institutions, enterprises and institutions with industry characteristics has also established corrosion and protection research institutes. Sinopec Safety Engineering Research Institute (Qingdao), Qingdao Marine Corrosion Research Institute Co., Ltd., General Iron and Steel Research Institute, etc. have continuously practiced the professional technology of corrosion and protection disciplines in relevant national standard formulation and engineering applications, providing a safety guarantee for China's national production and economic development.

构外，具有行业特色的企事业单位也相继成立了腐蚀与防护研究所。中国石化安全工程研究院（青岛）、钢铁研究总院青岛海洋腐蚀研究所有限公司等在相关国家标准制定和工程应用中不断践行腐蚀与防护学科的专业技术，为我国国民生产和经济发展提供了安全保障。

1.4 Material corrosion classification and velocity characterization methods

Material corrosion refers to the linearity of material deterioration, deterioration and fracture caused by chemical or electrochemical action of materials in a corrosive environment. According to the environment in which the material is located, the corrosion classification method can be divided into: natural environment corrosion (atmospheric corrosion, soil corrosion, seawater corrosion, etc.), industrial environment corrosion (acid and alkali salt solution corrosion, industrial water corrosion, molten salt corrosion, nuclear industry corrosion, etc.) and biological environment corrosion (tooth decay, etc.). With aerospace, deep-sea and polar exploration, material corrosion in related special environments has also attracted more and more

1.4 材料腐蚀分类及速度表征方法

材料腐蚀是指材料在腐蚀环境中发生化学或者电化学作用导致材料出现材料变质、性能恶化和断裂的现象。腐蚀分类方法按照材料所处环境可分为：自然环境腐蚀（大气腐蚀、土壤腐蚀、海水腐蚀等）、工业环境腐蚀（酸碱盐溶液腐蚀、工业水腐蚀、熔盐腐蚀、核工业腐蚀等）和生物环境腐蚀（蛀牙等）。随着航空航天和深海、极地探索，相关特殊环境下的材料腐蚀也越来越受到关注。

attention.

According to the corrosion reaction mechanism, material corrosion can be divided into, chemical corrosion (high temperature oxidation, etc.), electrochemical corrosion (electrochemical action, coupling of electrochemical and mechanical action, coupling of electrochemical and biological action).

按照腐蚀反应机理，材料腐蚀可分为：化学腐蚀（高温氧化等）、电化学腐蚀（电化学作用、电化学与机械作用耦合、电化学与生物作用耦合）。

Material corrosion can be divided into uniform corrosion (also known as total corrosion) and local corrosion (pitting corrosion, crevice corrosion, galvanic corrosion, intergranular corrosion, stress corrosion cracking, etc.) according to the corrosion forms. Compared with uniform corrosion, local corrosion is the key to corrosion failure of materials. It is worth noting that local corrosion is difficult to predict and detect, which has become a problem in anti-corrosion technology. At present, the scheme of reserved corrosion margin is adopted in the anti-corrosion design of engineering materials, and in view of the design of corrosion margin, the corrosion rate detection of uniform corrosion has become the key.

材料腐蚀按照腐蚀形态可分为：均匀腐蚀（也称为全面腐蚀）和局部腐蚀（点蚀、缝隙腐蚀、电偶腐蚀、晶间腐蚀、应力腐蚀开裂等）。相较于均匀腐蚀，局部腐蚀是材料腐蚀失效的关键。值得注意的是，局部腐蚀因难以预测和检测，而成为防腐技术的难题。目前，工程材料防腐设计方面采用预留腐蚀裕量的方案，鉴于腐蚀裕量的设计，均匀腐蚀的腐蚀速度检测成为关键。

For uniform corrosion, gravimetric methods (gain-in-weight or loss-in-weight methods), depth methods, and current density methods are often used to characterize the average rate of corrosion. In addition, neutral salt spray accelerated corrosion test, acid salt spray accelerated corrosion test is also widely used in material corrosion speed test and protection technology corrosion resistance evaluation, the current domestic standard is GB/T 10125—2021 “artificial atmosphere corrosion test”. Compared with the laboratory test corrosion rate, outdoor environmental corrosion evaluation is more in line with engineering applications, such as the six-month

对于均匀腐蚀，通常采用重量法（增重法或失重法）、深度法和电流密度法来表征腐蚀的平均速度。除此之外，中性盐雾加速腐蚀测试、酸性盐雾加速腐蚀测试也被广泛用于材料腐蚀速度测试和防护技术耐蚀性能评价，目前国内采用的标准是GB/T 10125—2021《人造气氛腐蚀试验盐雾试验》。相较于实验室检测腐蚀速度，室外环境腐蚀评价更

test of seaside exposure GB/T 1766—2008 “Grading method for coating aging of pigments and varnishes”.

符合工程应用，如海边暴露6个月测试采用GB/T 1766—2008《色漆和清漆涂层老化的评级方法》。

1.5 References

[1] Made in China 2025-Energy Equipment Implementation Plan, National Development and Reform Commission, Ministry of Industry and Information Technology and National Energy Administration

[2] Panasonic hydrogen refueling station officially operated in Japan [J]. Auto & Parts, 2019, 23:23.

[3] XIONG Q R, XU X F, LI Y H, et al. Development status and standardization of green manufacturing of petroleum equipment [J]. China Standardization, 2022, 12:75-80, 84.

1.5 参考文献

[1] 国家发展改革委，工业和信息化部，国家能源局. 中国制造2025——能源装备实施方案.[2016-6-20].

[2] 松下加氢站日本正式运营[J]. 汽车与配件，2019（23）：23.

[3] 熊庆人，许晓锋，李炎华，等. 石油装备绿色制造发展现状及其标准化[J]. 中国标准化，2022（12）：75-80，84.

Chemical Corrosion 2
化学腐蚀

2.1 Oxidation of metals

2.1.1 Chemical corrosion of metals

Refers to the metal and the surrounding medium direct chemical reaction caused by the phenomenon of metamorphism and damage. In the process of corrosion, electron transfer is directly carried out between metal and medium, without the generation of corrosion micro current. This type of corrosion is a purely chemical REDOX reaction in which metal atoms with valence electrons interact directly with molecules of the reactant, such as oxygen. Thus, the transformation of the metal into an ionic state and the reduction of the oxidizer component in the medium occur simultaneously and at the same location. The most important form of chemical corrosion is gaseous corrosion, such as the oxidation process of metals or the chemical interaction of metals

2.1 金属的氧化现象

2.1.1 金属的化学腐蚀

金属的化学腐蚀是指金属与周围介质直接发生化学反应而引起的变质和损坏现象。在腐蚀过程中，电子传递在金属与介质间直接进行，没有腐蚀微电流的产生。这类腐蚀过程是一种氧化还原的纯化学反应，带有价电子的金属原子直接与反应物（如氧）的分子相互作用。因此，金属转变为离子状态和介质中氧化剂组分还原是在同时、同一位置发生的。最重要的化学腐蚀形式是气体腐蚀，如金属的氧化过程或金属在高温下与

with SO_2, water vapor, etc at high temperature.

The corrosion products of chemical corrosion form a surface film on the metal surface, and the property of the surface film determines the chemical corrosion rate. If the integrity, strength and plasticity of the film are good, if the expansion coefficient of the film is close to the metal, and the affinity between the film and the metal is strong, it is beneficial to protect the metal and reduce the corrosion rate. Chemical corrosion can be divided into:

(1) Corrosion in dry gas

Corrosion in dry gases usually refers to the corrosion of metals under the action of high temperature gases. For example, thick oxide sheet is generated during steel rolling and gas turbine blade is corrodes under working condition.

(2) Corrosion in non-electrolyte solutions

Corrosion in non-electrolyte solutions refers to the corrosion of metals in a non-conducting liquid. For example, the corrosion of Al in CCl_4, $CHCl_3$ or CH_3CH_2OH, and the corrosion of magnesium and titanium in CH_3OH. In fact, pure chemical corrosion is very rare, and electrochemical corrosion is more common[1].

2.1.2 Oxidation of metals

Metal oxidation(Figure 2-1) has narrow sense of oxidation and broad sense of oxidation, in which narrow sense of metal oxidation refers to the process of metal and oxidation synthesis into oxide, is the process of metal elements lose electrons oxygen elements gain electrons, and eventually generate metal oxide. Metal oxidation

SO_2、水蒸气等的化学作用。

化学腐蚀的腐蚀产物在金属表面形成表面膜，表面膜的性质决定了化学腐蚀的速度。如果膜的完整性、强度、塑性都较好，若在膜的膨胀系数与金属接近、膜与金属的亲和力较强等情况下，则有利于保护金属、降低腐蚀速度。化学腐蚀可分为：

（1）在干燥气体中的腐蚀

在干燥气体中的腐蚀通常指金属在高温气体作用下的腐蚀。例如，轧钢时生成厚的氧化铁皮、燃气轮机叶片在工作状态下的腐蚀。

（2）在非电解质溶液中的腐蚀

在非电解质溶液中的腐蚀是指金属在不导电的液体中发生的腐蚀。例如，Al在CCl_4、$CHCl_3$或CH_3CH_2OH中的腐蚀，镁和钛在CH_3OH中的腐蚀等。实际上单纯的化学腐蚀是很少见的，更常见的是电化学腐蚀[1]。

2.1.2 金属的氧化

金属的氧化（图2-1）有狭义的氧化和广义的氧化，其中狭义的金属氧化指金属与氧化合成为氧化物的过程，是金属元素失电子、氧元素得电子的过程，最终生成金属氧化物。

in a broad sense refers to the process in which a metal loses electrons and thus increases its positive valence. The oxidation of metals depends on thermodynamic conditions (see reduction-oxidation reaction, oxygen potential diagram).

广义的金属氧化指金属失去电子从而正原子价升高的过程。金属的氧化能否发生，取决于热力学条件（见还原–氧化反应、氧势图）。

Oxidation in a narrow sense: oxidation in a narrow sense refers to the process of forming metal oxides by combining metal with oxidation in the environmental medium.

狭义氧化：是指金属与环境介质中的氧化合成为金属氧化物的过程。

$$M+\frac{n}{2}O_2 = MO_n \quad (2\text{–}1)$$

Generalized oxidation: the process by which a metal reacts with a medium and loses electrons.

广义氧化：是指金属与介质作用失去电子的过程。

$$xM+yN = M_xN_y \quad (2\text{–}2)$$

Figure 2–1 Oxidation of waste metals

图 2–1 废弃金属的氧化

(1) Oxygen position (oxygen potential)

（1）氧位（氧势）

The relative chemical potential of oxygen in the system is called oxygen potential, which can be used to compare the strength of the oxidation atmosphere. The oxygen potential of oxide can be used to measure the stability of oxide. The smaller the oxygen potential is, the more stable the oxide is. Conversely, the more unstable the oxide is.

体系中氧气的相对化学位称为氧位（氧势），可用来比较氧化气氛的强弱。氧化物的氧势可用来衡量氧化物的稳定性，氧势越小，氧化物越稳定；反之，氧化物越不稳定。

$$\pi_0 = \Delta\mu_{O_2} - \mu_{O_2} - \mu_{O_2}^{\theta} = RT\ln P_{O_2} \quad (2\text{–}3)$$

At a certain temperature T, the oxide formation reaction:

一定温度 T 下，氧化物形成反应：

$$\frac{2x}{y}M(s)+O_2 = \frac{2}{y}M_xO_y(s) \tag{2-4}$$

When equilibrium is reached $RT\ln P_{O_2}$, which balances oxygen in the gas phase, is the oxygen position of the oxide. The oxygen potential of oxide is expressed as follows:

达到平衡时，平衡气相中氧气的 $RT\ln P_{O_2}$ 即是氧化物的氧位。氧化物的氧势表示符号及关系式如下：

$$\pi_{O(M_xO_y)} = RT\ln P_{O_2} = -RT\ln K_{反应} = \Delta G^{\theta}_{反应} \tag{2-5}$$

$$\pi_{O(M_\xi O_\psi)} = \Delta\Gamma^{\theta}_{反应} = \frac{\psi}{2}\Delta_{\varphi}\Gamma^{\theta}_{O(M_\xi O_\psi)} = A + BT \tag{2-6}$$

$$\Delta G^{\circ}=RT\ln P_{O_2} \tag{2-7}$$

(2) Oxygen potential (oxygen position) diagram

Oxygen potential diagram mainly introduces the relationship between the oxygen potential (oxygen potential) of oxides and temperature[2]. The main components are abscissa and ordinate; Oxygen potential line cluster of oxide; Three scale line; H, C, O at three. Where the ordinate is oxygen position $RT\ln P_{O_2}=\Delta G^{\theta}$, the unit is $kJ/molO_2$ (or $kCal/molO_2$), the maximum ordinate is 0; The abscissa is the temperature, in K (or ℃); Around them are the scale coordinates of $\frac{P_{CO}}{P_{CO_2}}$、$\frac{P_{M_2}}{P_{MO}}$、P_{O_2}, The leftmost vertical line is the line T=0 K (absolute zero). The standard free energy changes for the formation of some oxides are shown in Figure 2-2.

（2）氧势（氧位）图

氧势图主要介绍了氧化物的氧势（氧位）与温度之间的关系[2]。主要构成为横坐标、纵坐标；氧化物的氧势线簇；三标尺线；H、C、O三点。其中纵坐标为氧位 $RT\ln P_{O_2}=\Delta G^{\theta}$，单位为 $kJ/molO_2$（或 $kCal/molO_2$），纵坐标最大值为0；横坐标为温度，单位为K（或℃）；周围依次是 $\frac{P_{CO}}{P_{CO_2}}$、$\frac{P_{M_2}}{P_{MO}}$、P_{O_2} 的刻度坐标；最左侧的垂直线为T=0K直线（绝对零度线）。图2-2所示为一些氧化物标准自由能与温度的关系。

(3) Trend of oxygen potential line

① The oxygen potential of the condensed oxide is basically upward sloping.

② The CO oxygen potential line is downward sloping.

（3）氧势线的走向

① 凝聚相似的氧化物氧势基本上向上倾斜。

② CO氧势线向下倾斜。

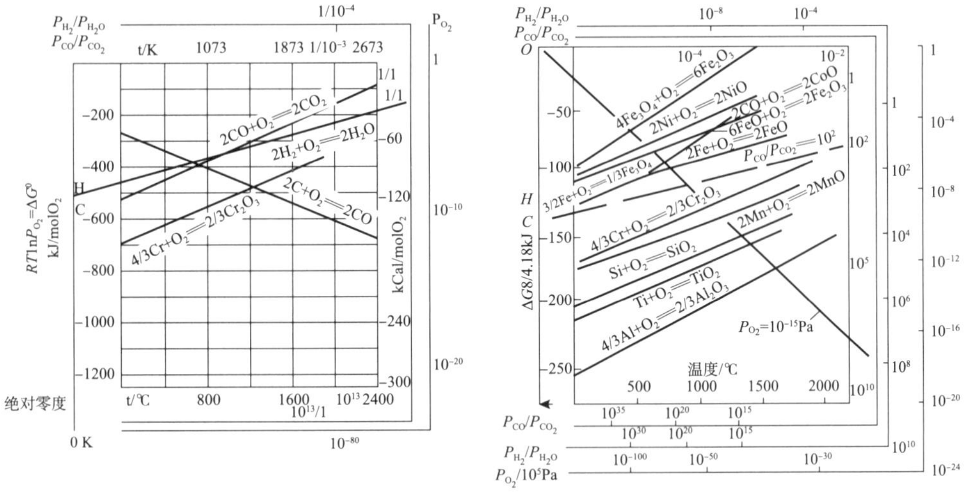

Figure 2-2 Diagram of standard free energy and temperature for some oxides

图 2-2 一些氧化物标准自由能与温度的关系

① and ② reasons: the trend and slope of oxygen potential line are determined by the state of the reaction ($-\Delta S^\theta$), and the entropy change is of the same sign as the change in gas volume before and after the reaction. If the gas volume change of $2M(s)+O_2 = 2MO(s)$ reaction is less than 0, then the reaction $-\Delta S^\theta>0$. For the reaction $-\Delta S^\theta>0$, the change of gas volume is greater than 0, and the reaction $-\Delta S^\theta$ is less than 0. Therefore, the oxygen potential of the oxide in the condensed phase is inclined upward, and the CO oxygen potential line is inclined downward.

①与②原因：氧化物氧势线走向与斜率决定于生成反应的熵变（$-\Delta S^\theta$），而熵变与反应的前后气体体积变化同号，$2M(s)+O_2 = 2MO(s)$反应的气体体积变化小于0，则其反应的$-\Delta S^\theta>0$。$2C+O_2 = 2CO$反应的气体体积变化大于0，则其反应的$-\Delta S^\theta<0$。因此，凝聚相似的氧化物氧势线向上倾斜，CO氧势线向下倾斜。

③ $2C+O_2 = 2CO_2$, the oxygen potential line is horizontal.

③ $2C+O_2 = 2CO_2$氧势线呈水平状。

Reason: The direction and slope of the oxide oxygen potential line is determined by the entropy change ($-\Delta S^\theta$) of the generation reaction. and the state of the gas volume before and after the reaction is the same as $2C+O_2 = 2CO_2$, the state of the gas volume before and after the reaction is 0, $-\Delta S^\theta=0$, therefore, $2C+O_2 = 2CO_2$ oxygen potential line is horizontal.

原因：氧化物氧势线走向与斜率决定于生成反应的熵变（$-\Delta S^\theta$），而熵变与反应的前后气体体积变化同号，$2C+O_2 = 2CO_2$反应前后气体体积变化为0，$-\Delta S^\theta=0$，因此，$2C+O_2 = 2CO_2$氧势线呈水平状。

④ Some of the oxygen potential lines have shifted

④一些氧势线发生了转折

Reason: The trend and slope of oxide oxygen potential line depend on the state of formation reaction ($-\Delta S^\theta$). When some substances undergo a phase change in an oxide generation reaction, the entropy change of the reaction is subsequently altered. If the reactants phase change (such as melting, evaporation, etc.), the total ΔS^θ of the reaction decreases, $-\Delta S^\theta$ increases, and the slope of the oxide oxygen potential line increases. If the total ΔS^θ of the reaction increases and $-\Delta S^\theta$ decreases, the slope of the oxide oxygen potential line decreases.

原因：氧化物氧势线走向与斜率决定于生成反应的熵变（$-\Delta S^\theta$），氧化物生成反应中当一些物质发生相变时，反应的熵变随之发生改变。若反应物发生相变（如熔化、蒸发等），反应的总ΔS^θ降低，$-\Delta S^\theta$增大，氧化物氧势线斜率增大。若产物发生相变时，反应的总ΔS^θ增大，$-\Delta S^\theta$减小，氧化物氧势线的斜率降低。

(4) Application of oxygen potential diagram

（4）氧势图的应用

① Determine the stability of oxide under certain conditions.

① 判定一定条件下氧化物的稳定性；

② To determine the relative stability of different oxides under certain conditions; (①, ② Basis: the smaller the oxygen potential of the oxide, the better its stability, and vice versa, the worse its stability)

② 判定一定条件下不同氧化物之间的相对稳定性；（①② 依据：氧化物氧势越小，稳定性越好，反之，稳定性越差）

③ Determination of reducing agent for reduction reaction; (According to: the oxygen potential of the oxide product of the reducing agent is less than the oxygen potential of the reduced oxide, or the reaction $\Delta G^\theta<0$)

③确定还原反应的还原剂；（依据：还原剂的氧化物产物氧势小于被还原氧化物的氧势，或反应的$\Delta G^\theta<0$）

④ Calculation:

④计算：

The standard Gibbs energy ΔG^θ for oxide formation at a certain temperature;

氧化物在一定温度时的标准生成吉布斯能ΔG^θ；

Decomposition pressure of oxide at a certain temperature;

氧化物在一定温度下的分解压力；

ΔG of the oxidation reaction of elements at a certain temperature and partial pressure of oxygen.

元素的氧化反应在一定温度及氧分压下的ΔG。

2.1.3 Medium of metal chemical corrosion

2.1.3 金属化学腐蚀的介质

Chemical corrosion is the corrosion process in

化学腐蚀是金属与周围介

which the metal and the surrounding medium only have chemical interaction and no current is generated. The main media in the corrosion process are dry gas, high temperature gas, other oxidants and non-electrolyte solution.

质只发生化学作用而没有电流产生的腐蚀过程，腐蚀过程中的主要介质有干燥气体、高温气体、其他氧化剂、非电解质溶液。

(1) Corrosion of metals in dry gases

（1）金属在干燥气体中的腐蚀

The surface corrosion of metal in a completely dry environment is called gas corrosion, which is characterized by slow speed and little harm.

金属在完全干燥环境中产生的表面腐蚀称为气体腐蚀，其特点为速度慢、危害小。

(2) Corrosion of metals in hot gases

（2）金属在高温气体中的腐蚀

Serious harm, ethylene cracking furnace furnace tube, ammonia industry reformer, waste heat boiler and so on these equipment are in the environment of high temperature gas operation, often cause high temperature oxidation corrosion.

危害严重，如乙烯裂解炉炉管、合成氨工业中的转化炉、废热锅炉等设备均处在高温气体的环境中运行，常引起高温氧化腐蚀。

(3) Chemical corrosion caused by other oxidants

（3）其他氧化剂引起的化学腐蚀

Sulfur, halogens, or other atoms and groups.

硫、卤素或其他原子及原子团。

(4) Corrosion of metals in non-electrolyte solutions

（4）金属在非电解质溶液中的腐蚀

Liquid corrosion is the corrosion of metals in organic solvents. For example: Al in CCl_4, Mg and Ti in methanol corrosion, sulfur gas pipeline surface.

液体腐蚀是金属在有机溶剂中的腐蚀现象。如Al在CCl_4、Mg和Ti在甲醇中的腐蚀，含硫的煤气管道表面等。

2.1.4 Characteristics of metal chemical corrosion

2.1.4 金属化学腐蚀的特点

Chemically corrosive media are non-conductive and non-ionizing substances. Chemical corrosion is the result of direct pure chemical reaction, the gain and loss of electrons in the same part at the same moment, no

化学腐蚀介质是不导电、不电离的物质。化学腐蚀是直接的纯化学反应的结果，电子的得失在同一部位于同一瞬间

corrosion current generation. Chemical corrosion has nothing to do with potential and potential change and can not be controlled by electrochemical protection.

完成，没有腐蚀电流产生。化学腐蚀与电位及电位变化无关，不能用电化学保护的方法予以控制。

2.2 Thermal dynamic criterion

2.2 金属氧化的热力学判据

2.2.1 Free-energy change of metal oxidation process

Criterion for spontaneous chemical reaction Gibbs free energy:

Material changes in nature can be roughly divided into two categories, one is the process that can be spontaneous, the other is the process that cannot be spontaneous. For any chemical reaction, if the reaction can be spontaneous, it must be accompanied by a reduction in energy, otherwise it cannot be spontaneous, and metal corrosion is such a spontaneous process. The standard free energy changes for the formation of some oxides are shown in Table 2-1.

Metal corrosion and most chemical reactions are generally carried out in an open system at constant temperature and pressure. In this case, the Cibbs free energy criterion is usually used to judge the direction of the reaction. Under isothermal conditions:

$\Delta G<0$ occurs spontaneously;

$\Delta G=0$ reaction balance;

$\Delta G>0$ reaction cannot be spontaneous;

G—Gibbs free energy, short for free energy, also

2.2.1 金属氧化过程的自由能变化

化学反应能否自发进行的判据——吉布斯自由能：

将自然界的物质变化大体分为两类：一类是能够自发进行的过程，另一类是不能自发进行的过程。对于任何化学反应，如果反应能够自发进行，必须伴随着能量的降低，否则不能自发进行，金属腐蚀就是这样一种自发过程。表2-1所示为一些氧化物形成的标准自由能变化。

金属腐蚀和大多数化学反应一般在恒温恒压的敞开体系条件下进行。这种情况下，通常用吉布斯自由能判据来判断反应的方向。在等温条件下：

$\Delta G<0$　反应自发进行；

$\Delta G=0$　反应平衡；

$\Delta G>0$　反应不能自发进行；

G——吉布斯自由能，简

enthalpy, Ji function, isothermal isobaric, etc.

称自由能，亦称自由焓、吉布斯函数等。

Take the oxidation of metals in oxygen:

以金属在氧气中的氧化为例：

$$M+O_2 = MO_2 \quad (2\text{-}8)$$

According to the Van't Hoff isothermal equation:

根据范特霍夫等温方程式：

$$\Delta G = \Delta G^0 + RT\ln\frac{\alpha_{MO_2}}{\alpha_M \alpha_{O_2}} \quad (2\text{-}9)$$

Where R——general gas constant, R =8.314.

式中 R——通用气体常数，取R=8.314。

α——activity, solid substance, $\alpha_M = \alpha_{MO_2} = 1$; In the gaseous phase, alpha $\alpha_{O_2} = P_{O_2}$.

α——活度，固态物质，$\alpha_M = \alpha_{MO_2} = 1$；气态物质，$\alpha_{O_2} = P_{O_2}$。

ΔG^0——standard free energy of formation of matter; Stable elemental, take ΔG^0=0.

ΔG^0——物质的标准生成自由能；稳定单质，取ΔG^0=0。

$$\Delta G^0 = \Delta G^0_{MO_2} - \Delta G^0_M - \Delta G^0_{O_2} = \Delta G^0_{MO_2} \quad (2\text{-}10)$$

$$\Delta G = \Delta G^0 + RT\ln\frac{1}{P_{O_2}} = \Delta G^0 - RT\ln P_{O_2} \quad (2\text{-}11)$$

Table 2-1 Standard free energy changes of some oxides formation (ΔG^0)

表 2-1 一些氧化物形成的标准自由能变化 (ΔG^0)

oxide 氧化物	ΔG^0/(kcal/mol)	ΔG^0/(kg/mol)
Ag_2O	−5.1	−21.3
Cu_2O	−69.2	−289.2
PbO	−90	−376.2
NiO	−102.8	−430
FeO	−109.2*	−455.6
ZnO	−152.4	−631
Cr_2O_2	−157.8*	−660
SiO_2	−197.1	−824
Al_2O_3	−252	−1060
MgO	−273	−1140

Note: data marked with * are measured at 500K, and the rest data are measured at 300K.

注：带 * 号的数据为 500K 温度下测得的，其余数据为 300K 温度下测得的。

Two important conclusions of metal oxidation process:

(1) Under normal atmospheric conditions, most metal materials have the tendency of spontaneous oxidation, that is, metal oxides are stable, and the metal itself is unstable.

(2) The ΔG of oxidation reaction changes from negative to positive with the increase of temperature, That is, the tendency of spontaneous oxidation of metals decreases with increasing temperature. (This is the thermodynamic reason why metal smelting takes place at high temperatures.)

Decompression of metal oxides:

金属氧化过程的两个重要结论如下：

（1）常态大气条件下，绝大多数金属材料都有自发氧化的倾向，即金属氧化物是稳定的，而金属本身是不稳定的。

（2）氧化反应的ΔG随着温度升高呈由负向正变化的趋势，即金属自发氧化的趋势随温度上升而减小。（这正是金属冶炼要在高温下进行的热力学依据）

金属氧化物的分解压：

$$M+O_2 \longrightarrow MO_2 \quad (2\text{–}12)$$

When the reaction equilibrium is reached, the partial pressure of oxygen in the system is called the partial decompression of metal oxide, namely P_{O_2}:

达到反应平衡时，体系中氧气的分压称为金属氧化物的分解压，即为P_{O_2}：

$$K=\frac{\alpha_{MO_2}^2}{\alpha_M P_{O_2}}=\frac{1}{P_{O_2}} \quad (2\text{–}13)$$

At this point, the equilibrium state

此时，平衡态

$$\Delta G^0 = RT\ln P_{O_2} = -RT\ln K \quad (2\text{–}14)$$

$$\Delta G = \Delta G^0 - RT\ln P_{O_2} = 0 \quad (2\text{–}15)$$

2.2.2 Criteria for metal oxidation

2.2.2 金属氧化的判据

So the change of gibbs free energy of the whole system is:

所以整个体系的吉布斯自由能变化为：

$$\Delta G = -RT\ln K - RT\ln P_{O_2} = -RT\ln\frac{1}{P_{O_2}} - RT\ln P_{O_2} = -RT\ln\frac{1}{P_{O_2}} \quad (2\text{–}16)$$

(1) When oxygen partial pressure is higher than oxide partial pressure, $\Delta G<0$, metal oxidation occurs spontaneously.

(2) When the partial pressure of oxygen is lower than the partial decompression of metal oxide, $\Delta G>0$, the

（1）氧分压高于氧化物的分解压时，$\Delta G<0$，金属氧化自发进行。

（2）氧分压低于金属氧化物的分解压时，$\Delta G>0$，金属

metal oxidation reaction cannot proceed spontaneously, and the reverse process of oxidation reaction—decomposition reaction will occur spontaneously.

氧化反应不能自发进行，此时会自发进行氧化反应的逆过程——分解反应。

According to the above conclusions, the partial decompression of most metal oxides at room temperature is far less than the partial pressure of oxygen in the atmosphere, so most metals have a tendency of spontaneous oxidation at room temperature. With the increase of temperature, the decompression of metal oxides increases, and the thermodynamic trend of spontaneous metal oxidation decreases; Adjusting and controlling the oxygen partial pressure of the system can affect or even change the thermodynamic tendency of spontaneous oxidation of metals.

根据以上有如下推论：常温下绝大多数金属氧化物的分解压远小于大气中的氧分压，因此常温下大多数金属都有自发氧化的趋势；随着温度升高，金属氧化物的分解压增大，金属自发氧化的热力学趋势减小。调节和控制体系氧分压能够影响或改变金属的自发氧化热力学倾向。

2.3 Formation and properties

2.3 金属氧化膜的形成与性质

2.3.1 Formation of oxidation film

Under certain conditions, only the surface of the metal is oxidized, and the oxide can form a continuous layer on the surface of the metal, called the oxide film, which can protect the metal inside the oxidation. Such as iron passivation film for γ-Fe_2O_3, Fe_3O_4, aluminum passivation film for non-porous γ-Al_2O_3 and so on. Some reducing anions, such as Cl^-, have great damage to the oxide film. There are two types of stress in oxide film: growth stress generated when oxide film grows at constant temperature and thermal stress generated when temperature changes due to different thermal expansion coefficient of metal and oxide.

2.3.1 氧化膜的形成

金属在一定条件下，只有其表面被氧化，氧化物能在金属表面形成连续的一层，称为氧化膜，它能保护里面的金属再被氧化。如铁钝化膜为γ-Fe_2O_3、Fe_3O_4，铝钝化膜为无孔的γ-Al_2O_3等。一些还原性阴离子，如Cl^-对氧化膜破坏作用较大。氧化膜中存在两种类型的应力：氧化膜恒温生长时产生的生长应力和温度变化时由于金属与氧化物的热膨胀系数不同而产生的热应力。

Film on the metal surface: metal high temperature oxidation on the metal surface to form a layer of oxide film, usually called oxide or rust oxide film thickness is generally 10^{-10}~10^{-9} m. According to the thickness of the film, the metal film can be divided into three categories:

金属表面的膜：金属高温氧化在金属表面形成一层氧化膜，通常称为氧化皮或锈皮。氧化膜厚度一般为10^{-10}~10^{-9}m。按照膜的厚度，可以把金属膜分为以下三类：

(1) The thick film is visible to the naked eye, the film thickness >500 nm, such as The high temperature oxidation of Fe in 900℃ air, its thickness can be up to 600 μm.

（1）厚膜是肉眼可见的，膜厚>500nm，如Fe在900℃空气中的高温氧化，其厚度可达到600μm。

(2) Film of medium thickness, which can be revealed by interfering color on metal surface, thickness of 40~500 μm.

（2）中等厚度的膜，可通过金属表面的干扰色而显现，厚度为40~500μm。

(3) Thin films, which are invisible and can only be detected by testing methods, are less than 40 nm thick. For example, the film thickness of Fe and Cu formed in dry air at normal temperature is 1~3 nm, and the film thickness on Al is about 10 nm.

（3）薄膜，它是不可见的，只能通过测试手段检查出来，其膜厚<40nm。如常温下Fe和Cu在干燥空气中形成的薄膜厚度为1~3nm，在Al上的膜厚约为10nm。

The formation process of oxide film is as follows:

氧化膜形成过程如下：

$$O_2 \xrightarrow{\text{吸附、分解}} O \xrightarrow{+2e^-} O^{2-} \xrightarrow{M^{2+}} MO \longrightarrow \text{完整氧化膜}$$

Factors affecting the formation of oxide film:

影响氧化膜形成的因素如下：

In the process of oxide film formation, the main influencing factors are oxygen concentration and temperature. The effect of oxygen concentration on oxide film is as follows: increasing oxygen concentration can increase the nucleation rate of oxide; The influence of temperature on oxide film is reflected in that increasing temperature can not only improve the nucleation rate of oxide but also increase the growth rate of oxide core, thus accelerating the process of forming continuous oxide film [3]. A schematic diagram of the formation of

在氧化膜形成的过程中，主要的影响因素有氧浓度和温度。氧浓度对氧化膜的影响表现在：提高氧浓度可以提高氧化物的生核速度；温度对氧化膜的影响表现在：提高温度不仅可以提高氧化物的生核率，而且可以提高氧化物核心的长大速度，从而加速形成连续氧化膜的进程[3]。图2-3所示为

the surtave oxide film is shown in Figure 2-3.

表面氧化膜的形成示意。

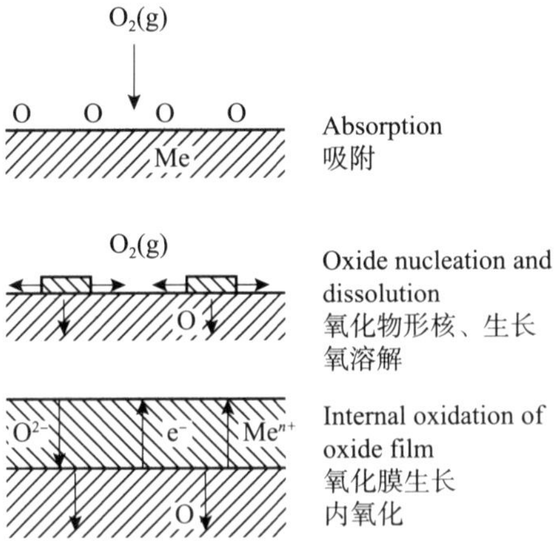

Figure 2-3 Formation of oxide film

图 2-3 氧化膜的形成示意

2.3.2 Oxidation mechanism of metals

2.3.2 金属的氧化机理

After metal contacts with oxidizing medium, oxidation product film is formed on its interface. According to the oxidution realtion equation:

金属与氧化性介质接触后，其界面上形成氧化产物膜，根据氧化反应方程式：

$$M(固)+\frac{1}{2}O_2(气)=\!=\!=MO(固) \quad (2-17)$$

It can be seen that the metal oxide MO can be divided into the following two reactants. Once the dense and thin oxide film is formed, the growth of the film only depends on the diffusion process through the oxide. A model of metal oxidation at high temperature is shown below. A model of the metal oxidation process at high temperatures is shown in Figure 2-4.

可以看出：金属氧化物MO可分为以下两个反应物，一旦形成致密而薄的氧化膜之后，膜的成长主要取决于经由氧化物的扩散过程。金属高温氧化过程模型如图2-4所示。

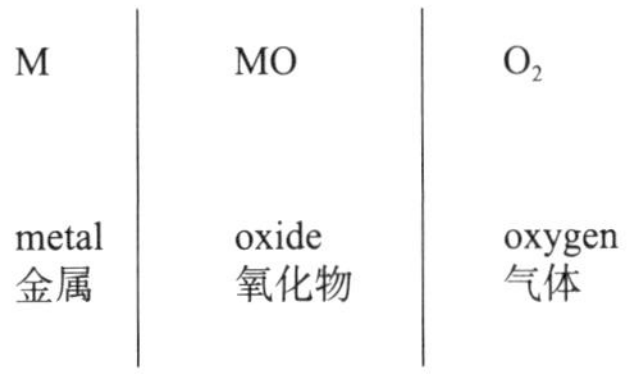

Figure 2-4 Formation of MO metal oxide, including metal (M), metal oxide (MO) and oxygen (O_2)

图 2-4 金属氧化物 MO 形成

(1) The metal ions involved in the reaction have a high diffusion gradient B_i, namely $B_{M^{2+}} \gg B_{O^{2-}}$.

（1）参与反应的金属离子具有较高的扩散梯度B_i，即$B_{M^{2+}} \gg B_{O^{2-}}$。

Reaction process a: Metal ions diffuse outwards in one direction and react on the oxide-gas interface, so that the growth region of the membrane is on the outer surface of the membrane. Such as the oxidation of copper.

反应过程a：金属离子单向向外扩散，在氧化物-气体界面反应，因而膜的生长区域在膜的外表面，如铜的氧化过程。

(2) Oxygen ions have a high diffusion gradient, that is, $B_{O^{2-}} \gg B_{M^{2+}}$.

（2）氧离子具有较高的扩散梯度，即$B_{O^{2-}} \gg B_{M^{2+}}$。

Reaction process b: Oxygen diffuses unidirectionally inward and is reacted on the metal-oxide interface, so the growth region of the film is at the metal-membrane interface. The oxidation of metals such as titanium and zirconium. A schematic diagram of the growth process of the oxide film is shown in Figure 2-5.

反应过程b：氧单向向内扩散，在金属-氧化物界面反应，因而膜的生长区域在金属与膜界面处，如钛、锆等金属的氧化过程。图2-5所示为氧化膜成长历程示意。

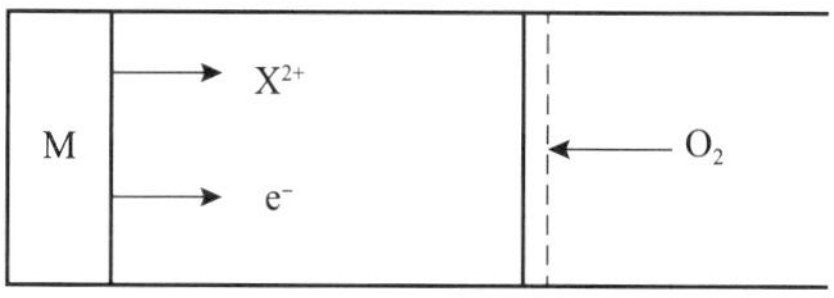

(a)Oxide/gas interface
氧化物/气体界面生成

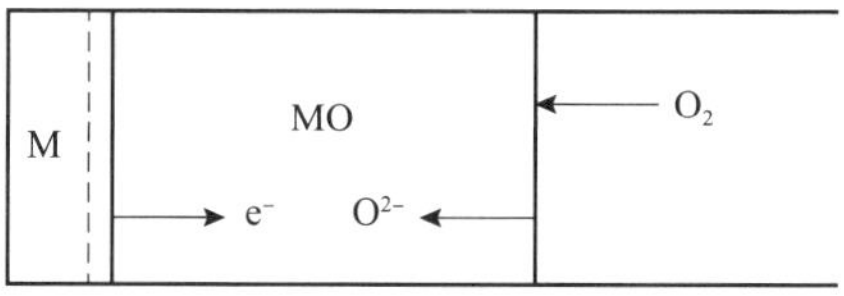

(b)Metal/oxide interface is formed
金属/氧化物界面生成

Figure 2-5 Schematic diagram of oxide film growth process

图2-5 氧化膜成长历程示意

In the process of oxidation film growth, there are three transport forms of reactants. Figure 2.6 describes the relationship between the location of film formation and diffusion direction, including metal ion outward diffusion (Cu) [Figure 2-6(a)], oxygen ion inward diffusion (Ti) [Figure 2-6(b)], and bidirectional diffusion (Co) [Figure 2-6(c)].

在氧化膜生长过程中，反应物传输形式有三种，图2-6所示为膜生成的位置与扩散方向的关系，其中介绍了金属离子向外扩散（Cu）[图2-6（a）]，氧离子向内扩散（Ti）[图2-6（b）]，双向扩散（Co）[图2-6（c）]。

Several ways of reactant transport pathways within the oxide film exist depending on the metal system and oxidation temperature:

反应物质在氧化膜内传输途径根据金属体系和氧化温度的不同存在以下几种方式：

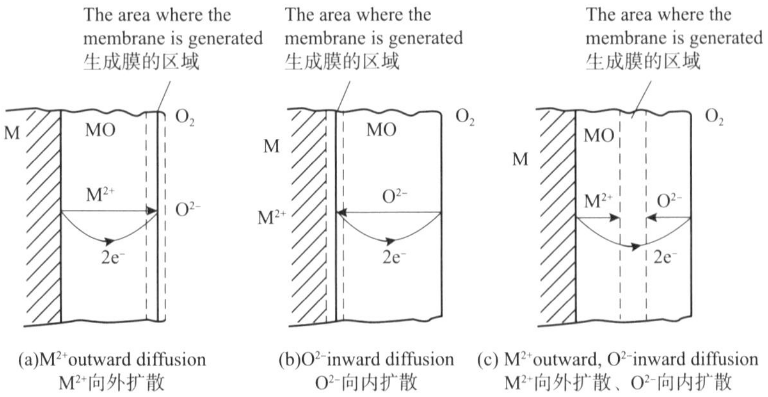

Figure 2-6 The relationship between the position of film formation and the direction of diffusionn
图 2-6 膜生成的位置与扩散方向的关系

① Diffusion through the lattice. It is common in the case of high temperature, dense oxide film, and high concentration of vacancy defects inside the oxide film. By measuring the oxidation rate, the diffusion coefficient of the reacting substance can be calculated directly, such as the oxidation of cobalt.

① 通过晶格扩散。常见于温度较高，氧化膜致密，而且氧化膜内部存在高浓度的空位缺陷的情况下，通过测量氧化速度，可直接计算出反应物质的扩散系数，如钴的氧化。

② Diffusion through grain boundaries. At lower temperatures, grain boundary diffusion is more important because the activation energy of grain boundary diffusion is smaller than lattice diffusion, and the grain size of oxides at low temperatures is smaller and the grain boundary area is larger, such as the oxidation of nickel, chromium and aluminum.

② 通过晶界扩散。在较低的温度下，由于晶界扩散的激活能小于晶格扩散，而且低温下氧化物的晶粒尺寸较小，晶界面积大，因此晶界扩散显得更加重要，如镍、铬、铝的氧化。

③ Simultaneous diffusion through the lattice and grain boundaries. Such as titanium, error, give in the medium temperature region (400~600℃) under prolonged oxidation conditions.

③ 同时通过晶格扩散和晶界扩散。如钛、锗、铬在中温区域（400~600℃）长时间氧化条件下。

The oxide film may also form and grow in another form due to the presence of grain boundary diffusion. When metal ions diffuse outward in one direction, it corresponds to the migration of metal ion vacancies to the metal-oxide film interface. If the oxide film is too thick to maintain contact with the metal matrix by

由于存在晶界扩散，氧化膜还可能以另一种形式形成和生长。当金属离子单向向外扩散时，相当于金属离子空位向金属－氧化膜界面迁移。如果氧化膜太厚而不能通过变形来

deformation, these vacancies coalesce and eventually form holes at the metal-oxide film interface. If the diffusion of metal ions through the grain boundaries of the oxide film is faster than the lattice diffusion, the grain boundary region acts as a microscopic channel connecting the hole to the external environment. This channel will allow the migration of molecular oxygen to the metal and produce oxidation on the surface of the hole, forming an internal porous oxide layer.

维持与金属基体的接触，这些空位凝聚，最后在金属－氧化膜界面形成孔洞。若金属离子通过氧化膜的晶界扩散速度大于晶格扩散，则晶界区域起到连接孔洞与外部环境的显微通道作用。这种通道将允许分子氧向金属迁移，并在孔洞表面产生氧化，形成内部多孔的氧化层。

2.3.3 Electrochemical model of metal oxidation

2.3.3 金属氧化的电化学模型

General metals are alloys, doped with, for example, carbon and other non-metallic substances, these substances and the metal to form a microcell, the metal's electron conduction to the carbon to accelerate the corrosion of the metal, which actually belongs to electrochemical corrosion, so metal oxidation is a chemical corrosion, can be explained by electrochemical models. Thus, metal oxidation is a form of chemical corrosion that can be explained by electrochemical models, as shown in Figure 2-7.

一般的金属都是合金，掺杂有如碳等非金属物质，这些物质可以和金属构成微电池，将金属的电子导到碳上加速金属腐蚀，这实际上属于电化学腐蚀，因此金属氧化属于化学腐蚀，可以用电化学模型来解释，如图2-7所示。

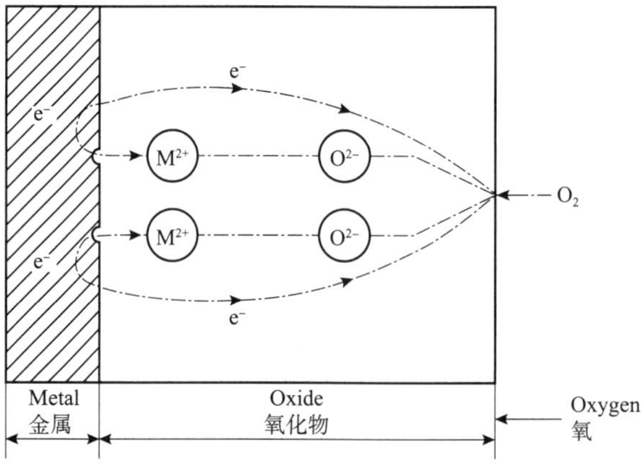

Figure 2-7 An electrochemical model of metal oxidation
图 2-7 金属氧化的电化学模型

Anodic reaction at the M / MO interface:

在M/MO界面进行阳极反应：

$$M \longrightarrow M^{2+} + 2e^- \quad (2\text{–}18)$$

Cathodic reaction at the MO / O_2 interface:

在MO/O_2界面进行阴极反应：

$$\frac{1}{2}O_2 + 2e^- \longrightarrow O^{2-} \quad (2\text{–}19)$$

2.3.4 Protective effect of metal oxide film

2.3.4 金属氧化膜的保护作用

The structure and properties of oxide films vary greatly in their protective capacity. At a certain temperature, different metal oxides may have different substances due to the metallic state.

由于金属氧化膜的结构和性质各异，其保护能力有很大差别。一定温度下，不同金属氧化物可能有不同物态。

For example: Cr、Mo、V in 1000℃ air are oxidized, its oxide state is different:

Cr、Mo、V在1000℃空气中均被氧化，其氧化物状态则各不相同：

$$2Cr + \frac{3}{2}O_2 \longrightarrow Cr_2O_3(\text{固态}) \quad (2\text{–}20)$$

$$Mo + \frac{3}{2}O_2 \longrightarrow MoO_3(\text{气态，450℃以上开始挥发}) \quad (2\text{–}21)$$

$$2V + \frac{3}{2}O_2 \longrightarrow V_2O_5(\text{液态，熔点 658℃}) \quad (2\text{–}22)$$

It can be seen that only Cr,O is solid and protective among these three oxides at 1000℃, while the oxides of Mo and V are not protective.

可见，在1000℃下这三种氧化物中只有Cr、O为固态，有保护性，而Mo和V的氧化物则无保护性。

It has been proved that not all solid oxide films are protective, only those with dense structure and complete coverage of the metal surface are protective. Therefore, the protective properties of oxide films depend on the following factors: the metal oxide film should be dense and complete, and can effectively isolate the metal from the oxygen in the medium; the metal oxide itself is

实践证明，并非所有的固态氧化膜都有保护性，只有那些组织结构致密，能完整覆盖金属表面的氧化膜才有保护性。因此，氧化膜的保护性取决于以下因素：金属氧化膜应当是致密、完整，能有效隔离

stable, refractory, non-volatile, and not easily damaged by the action of the medium; the metal oxide is well bonded to the substrate, has a similar coefficient of thermal expansion, and will not peel off by itself or by external action; the oxide has sufficient strength and plasticity to withstand certain stress, strain and friction. Enough to withstand a certain amount of stress, strain and friction.

金属与介质中的氧接触；金属氧化物本身是稳定、难溶、不挥发，且不易与介质作用发生破坏；金属氧化物与基体结合良好，有相近的热膨胀系数，不会自行或受外界作用而剥离脱落；氧化物有足够的强度、塑性，足以经受一定的应力、应变和摩擦作用。

2.3.5 Stress and stress relaxation in oxide film

The oxidation kinetic curves of metals often deviate from the theoretical equation and lose their protective properties due to the occurrence of cracking and peeling of the oxide film. This is due to the presence of stress in the oxide film, and stress and relaxation in the oxide film are important factors in determining the oxidation kinetics. There are two main sources of stress in the oxide film one is growth stress, which is generated in the constant growth of the oxide film, and the other is thermal stress, which is generated due to the different thermal expansion and contraction of the substrate and the oxide film. Growth stresses are generated for the following reasons:

(1) PB Ratio. Proposed by Pilling and Bedworth, defined as the ratio of the volume of metal oxide films generated by oxidation V_{MO} to the volume of metal eliminated by generating these oxide films V_M, the necessary condition for the integrity of oxide films is that the volume of metal oxide is greater than the volume of metal consumed by oxidation, i.e., the Pilling-Bedworth principle:

2.3.5 氧化膜中的应力与应力松弛

金属的氧化动力学曲线常常由于发生氧化膜的开裂与剥落而偏离理论公式，并失去保护性。这是由于氧化膜中存在应力，氧化膜中的应力与应力松弛是决定氧化动力学的重要因素。氧化膜中的应力有两种主要来源：一种是生长应力，是在氧化膜恒温生长中产生的；另一种是热应力，是由于基体和氧化膜的热膨胀和收缩不同而产生的。生长应力产生原因如下：

（1）PB比。由Pilling和Bedworth提出，定义为氧化生成的金属氧化膜的体积V_{MO}与生成这些氧化膜所消耗金属的体积V_M之比，氧化膜完整性的必要条件是金属氧化物的体积要大于氧化消耗金属的体积，即Pilling–Bedworth原理：

$$\varnothing = \frac{V_{MO}}{V_M} = \frac{M/ZD}{m/\rho} = \frac{M\rho}{ZmD} > 1 \qquad (2-23)$$

Among them

M——molar mass of metal oxide;

m——molar mass of the metal;

ρ——density of the metal;

Z——number of metal atoms in the metal oxide;

D——density of metal oxide.

It should be noted that PB ratio greater than 1 is a necessary condition for the integrity of the oxide film but not a sufficient condition; The PB ratio can not quantitatively reflect the quality of the protective performance of the oxide film, and the protection performance is better in the range of 1~3.

The PB ratios of some metal oxide films are given in Table 2-2. When the PB ratio > 1, compressive stress is applied in the oxide film, which is also necessary to generate protective oxide films; when the PB ratio < 1, tensile stress is applied in the oxide film, which cannot generate protective oxide films. For example, with PB ratio <1 for alkali metals and alkaline earth metals, the generated oxide film undergoes cracking and shows a linear oxidation pattern. However, the growth stress caused by PB ratio is also related to the growth mechanism of the oxide film. The general rule is that growth stress can be generated by PB ratio when the new oxide is generated at the metal/oxide film interface, and it is not possible to generate growth stress due to PB ratio if the new oxide is generated at the oxide film/gas phase interface. Therefore, PB ratio is one of the necessary conditions for growth stress generation, but not sufficient condition.

式中 M——金属氧化物的摩尔质量；

m——金属的摩尔质量；

ρ——金属的密度；

Z——金属氧化物中金属原子个数；

D——金属氧化物的密度。

需要注意的是，PB比大于1是氧化膜完整的必要条件而非充分条件；PB比不能定量反映氧化膜保护性能的好坏，在1~3范围内，保护性能较好。

表2-2中给出一些金属氧化膜的PB比，当PB比>1时，氧化膜中受到压应力，也是生成保护性氧化膜的必要条件；当PB比<1时，氧化膜中受到拉应力，无法生成保护性氧化膜。例如，碱金属和碱土金属的PB比<1，生成的氧化膜发生开裂，呈线性氧化规律。但是由PB比引起的生长应力还与氧化膜的生长机制有关，一般规律是当新的氧化物在金属/氧化膜界面生成时，由PB比可以产生生长应力。若新的氧化物在氧化膜/气相界面生成时，则不可能因PB比产生生长应力。因此，PB比是生长应力产生的必要条件之一，但不是充分条件。

Table 2-2 PB ratios of some metal oxide films

表 2-2 某些金属氧化膜的 PB 比

Metal oxide film 金属氧化膜	PB ratio PB 比	Metal oxide film 金属氧化膜	PB ratio PB 比	Metal oxide film 金属氧化膜	PB ratio PB 比	Metal oxide film 金属氧化膜	PB ratio PB 比
MoO_3	3.4	Co_3O_4	1.99	NiO	1.52	MgO	0.99
WO_3	3.4	TiO_2	1.95	ZrO_2	1.51	BaO	0.74
V_2O_5	3.18	MnO	1.79	PdO_2	1.40	CaO	0.65
Nb_2O_3	2.68	FeO	1.77	SnO_2	1.32	SrO	0.65
Sb_2O_5	2.35	Cu_2O	1.68	ThO_2	1.32	Na_2O	0.58
Ta_2O_5	2.33	ZnO	1.62	HgO	1.31	Li_2O	0.57
Bi_2O_5	2.27	PdO	1.60	Al_2O_3	1.28	Cs_2O	0.46
Si_2O_5	2.27	BeO	1.59	CdO	1.21	K_2O	0.45
Cr_2O_3	1.99	Ag_2O	1.59	Ce_2O_3	1.16	RbO	0.45

(2) Epitaxial stress. The oxide formed at the beginning of oxidation maintains an epitaxial relationship with the matrix lattice. This epitaxial constraint leads to stresses due to the different parameters of the intrinsic lattice of the metal and the oxide. When the oxide film thickens, the oxide film regains its intrinsic lattice and the epitaxial stress disappears.

（2）外延应力。氧化初期形成的氧化物与基体晶格保持外延关系，由于金属和氧化物固有晶格的参数不同，这种外延约束导致应力的产生。当氧化膜变厚时，氧化膜恢复固有晶格，外延应力消失。

(3) Changes in alloy or oxide film composition. When an alloy is oxidized by vacancy selection, changes in the alloy surface composition cause changes in the alloy lattice resulting in stresses, and similarly changes in the oxide film composition can also cause stresses. Oxygen dissolution in the metal can also lead to stress generation.

（3）合金或氧化膜成分的变化。合金发生选择氧化时，合金表面成分变化引起合金晶格变化导致产生应力，同样氧化膜成分变化也能产生应力。氧在金属中溶解也可导致应力产生。

(4) Point Defect Stress. The presence of point defect gradients in the oxide film causes stresses due to changes in lattice parameters. Likewise, stresses are generated in the matrix due to outward migration of cations that create

（4）点缺陷应力。氧化膜中存在点缺陷梯度引起晶格参数变化而产生应力。同样，由于阳离子向外迁移，沿金属截面产生空位梯度，引起空位周

vacancy gradients along the metal cross section, causing shrinkage around the vacancies.

围收缩，在基体中产生应力。

(5) The generation of new oxides within the oxide film can create compressive stresses within the oxide film.

（5）新氧化物在氧化膜内生成可在氧化膜内产生压应力。

(6) Recrystallization stresses. Recrystallization in oxide films is one of the causes of stresses, but it can also eliminate growth stresses.

（6）再结晶应力。氧化膜中再结晶是应力的起因之一，但也可以消除生长应力。

(7) Stress caused by the aggregate shape of the form. The oxide film formed at the convex surface is susceptible to tensile stress; the oxide film generated at the concave surface is susceptible to compressive stress.

（7）试样的几何形状引起的应力在凸表面处形成的氧化膜易产生拉应力，在凹表面生成的氧化膜易产生压应力。

2.3.6 Semiconductor properties of metal oxide films

2.3.6 金属氧化膜的半导体性质

1. Oxide types and diffusion mechanisms in the membrane

1.氧化物类型和膜中扩散机制

Metal oxides are ionic crystals composed of metal ions and oxygen ions. In case of perfect crystals, the movement of ions is difficult. However, as long as there are defects in the crystal such as vacancies, interstitial atoms, dislocations, etc., ions can diffuse through the lattice defects. The existence of lattice defects and the form of ion movement through them are currently inferred from their electrical conductivity. Based on their conductivity magnitudes, general compounds can be classified as ionic conductors and semiconductors.

金属氧化物是由金属离子和氧离子组成的离子晶体。若是完善晶体，离子移动是难以进行的。但是晶体中只要有缺陷，如空位、间隙原子、位错等，离子就能通过晶格缺陷进行扩散。晶格缺陷的存在和通过它所进行的离子移动形式，目前主要通过其电导率大小来推测。根据其电导率大小，可将一般化合物分为离子导体和半导体。

Most metal oxides have an ion-metal bond composition and are semiconductors whose composition does not strictly conform to stoichiometric ratios, with the presence of a stable excess of cations or anions. The distribution of ions in a semiconductor is shown in Figure 2-8.

大多数金属氧化物由离子–金属键组成，属于半导体，其成分并不严格符合化学计量比，存在稳定过量的阳离子或者阴离子。图2–8所示为半导体中离子分布。

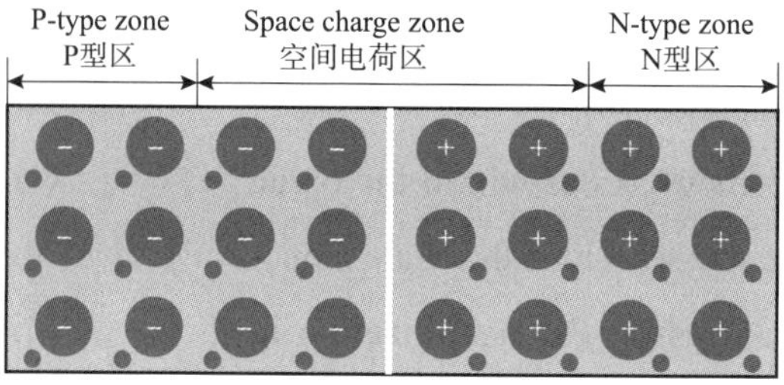

Figure 2-8 Distribution of ions in semiconductors

图 2-8 半导体中离子分布

(1) Metal ion excess type oxide (N type)

（1）金属离子过剩型氧化物（N型）

For example, ZnO、CdO、BeO、RaO、V_2O_5、PbO_2、MoO_2、WO_3、BaS、TiO_2、CdS、Cr_2S_3、TiS_2, etc., all belong to this type of oxide, which is characterized by excess metal ions in the lattice interstices. The oxide as a whole is electrically neutral, so the corresponding electrons exist in the gap and are in motion, and the process conducts electricity mainly through the movement of negatively charged free electrons. The diagram shows a schematic representation of a ZnO semiconductor. A metal ion excess oxide tilm is shown in Figure 2-9, a schematic of a ZnO semiconductor metal ion excess semiconductor is shown in Figure 2-10, indicating a ZnO semiconductor.

例如ZnO、CdO、BeO、RaO、V_2O_5、PbO_2、MoO_2、WO_3、BaS、TiO_2、CdS、Cr_2S_3、TiS_2等，均属于这种类型的氧化物。这种类型的氧化物的特点是，过剩的金属离子处于晶格间隙。氧化物作为整体是电中性的，所以间隙中存在相应的电子并在运动着，过程中主要是通过带负电荷的自由电子的移动导电。图2-9所示为金属离子过剩型氧化膜；图2-10所示为ZnO金属离子过剩型半导体示意。

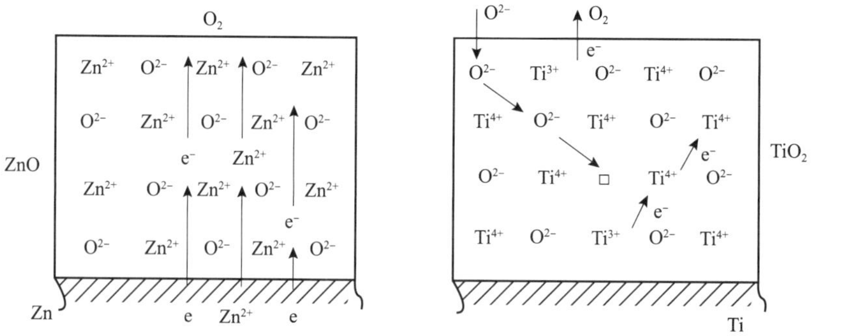

(a)The presence of gap cations and free electrons 存在间隙阳离子和自由电子

(b)The presence of anion vacancies and free electrons 存在阴离子空位和自由电子

Figure 2-9 Excess metal ion oxide film

e: Free electron, i: Gap; TI^{3+}: Free electrons; □: O^{2-} vacancy

图 2-9 金属离子过剩型氧化膜

e: 自由电子 , i: 间隙 ; Ti^{3+}: 自由电子 ; □: O^{2-} 空位

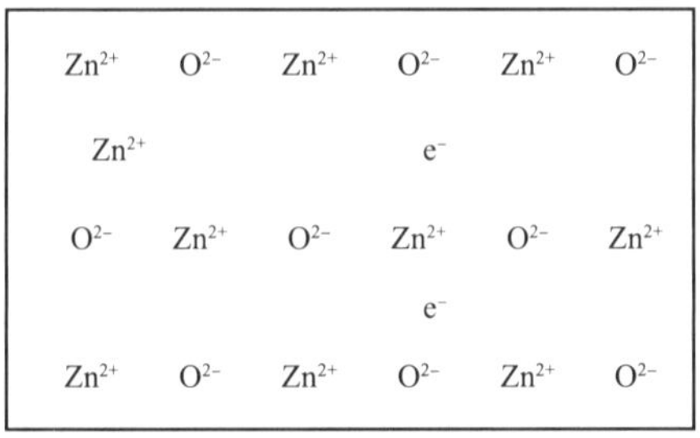

Figure 2-10 Schematic diagram of ZnO metal ion excess type semiconductor

图 2-10 ZnO 金属离子过剩型半导体示意

During oxidation, interstitial ions (in Zn^{2+}) and interstitial electrons (e^-) migrate through the membrane to the ZnO/O_2 interface and absorb O_2 to produce ZnO and absorbs O_2 to produce ZnO.

氧化时，间隙离子（Zn^{2+}中）和间隙电子（e^-）通过膜向ZnO/O_2界面迁移，并吸收O_2而生成ZnO。

$$\frac{1}{2}O_2 + Zn^{2+} + 2e^- \xlongequal{\quad} ZnO \tag{2-24}$$

(2) Metal ion-deficient oxide (P type)

（2）金属离子不足型氧化物（P型）

Conducting electricity mainly through the movement of electron holes (i.e. positive vacancies), such as Cu_2O、NiO、FeO、Ag_2O、MnO、CoO、Bi_2O_3、FeS、Cu_2S、Ag_2S、SnS, etc. are metal ion-deficient oxides, or oxygen excess oxides. Since the radius of oxygen ions is larger than that of metal ions, the excess oxygen ions cannot be in the lattice gap position, but occupy the junction position.

主要是通过电子空位（正空）的运动导电，如Cu_2O、NiO、FeO、Ag_2O、MnO、CoO、Bi_2O_3、FeS、Cu_2S、Ag_2S、SnS等是金属离子不足型氧化物，或氧过剩型氧化物。由于氧离子半径比金属离子大，过剩的氧离子不能在晶格间隙位置，而是占据着节点位置。

Figure 2-11 reprosent a schematic diagram of a metal ion deficient oxide film.

图2-11所示为金属离子不足型氧化膜示意。

The lattice of NiO in the Figure 2-11 represents a Ni^{2+} hole and Ni^{3+} represents an electron hole, also called a cation. In order to maintain the electrical neutrality of the lattice, a vacancy, if any, in NiO is present in the ratio of two Ni^{3+}. In order to form O^{2-} electrons must be provided; in order to generate a

图2-11中NiO的晶格中表示一个Ni^{2+}空位，Ni^{3+}表示一个电子空位，也称阳穴。为维持晶格的电中性，在NiO中，如果有一个空位，就以两个Ni^{3+}的比例存在。为形成O^{2-}必须

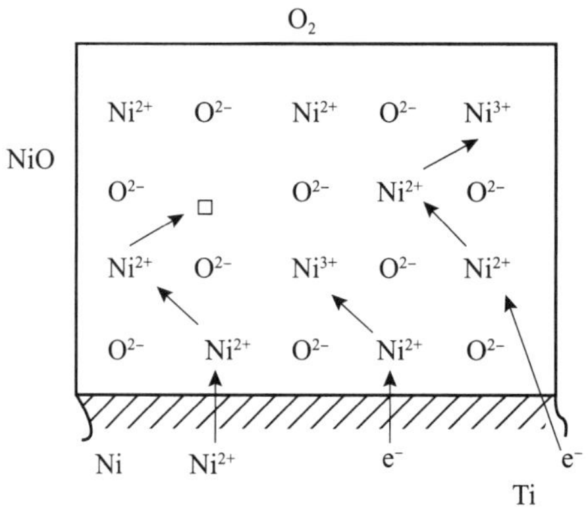

Figure 2-11 NiO Metal ion deficient oxide film (presence of cation vacancies and electron vacancies)

Ni^{3+}: Positive hole = □e^- (electronic vacancy); □: Ni^{2+} vacancy

图 2-11 NiO 金属离子不足型氧化膜（存在阳离子空位和电子空位）

Ni^{3+}: 正孔 =□e^-(电子空位); □:Ni^{2+} 空位

new nickel oxide layer, Ni^{2+} must be provided, and both electrons and Ni^{2+} are provided by the crystal. The generation of new NiO leads to the presence of cation vacancies (□Ni^{2+}) and electron vacancies (□e^-) within the crystal. The electron hole can be thought of as Ni^{3+} losing an electron at the Ni^{2+} position to become Ni^{3+}, i.e., the electron hole is positively charged when it loses an electron, while the cation vacancy (□Ni^{2+}) is negatively charged. In the high temperature oxidation, NiO/O_2 interface o, and the role of electrons to generate O_2, and then with the role of Ni^{2+} to generate new NiO, the reaction is as follows:

提供电子；为生成新的氧化镍层，必须提供Ni^{2+}，而电子和Ni^{2+}都是晶体提供的。生成新的NiO时，导致晶体内出现阳离子空位（□Ni^{2+}）和电子空位（□e^-）。电子空位可认为是Ni^{3+}在Ni^{2+}位置上失去一个电子变成Ni^{3+}，亦即失去电子空位荷正电，而阳离子空位（□Ni^{2+}）则荷负电。在高温氧化时，NiO/O_2界面的O与电子作用生成O_2，再与Ni^{2+}作用生成新的NiO，其反应如下：

$$\frac{1}{2}O_2 = NiO + Ni^{2+} + 2e^- \quad (2-25)$$

In n-type semiconductors, the oxide interstitial metal ions and electrons migrate to the outer surface and contact oxygen at the oxide-oxygen interface to generate a new oxide layer; in p-type semiconductor oxides, oxygen ions, cation vacancies and electron holes migrate

在N型半导体中，氧化物间隙金属离子和电子向外表面迁移，在氧化物－氧界面与氧接触生成新氧化层；在P型半导体氧化物中，氧离

inward, while metal ions and electrons migrate outward and form a new oxide layer inside the lattice.

子、阳离子空位和电子空位都向内迁移，金属离子和电子则向外迁移，并在晶格内部形成新的氧化物层。

(3) Semiconductor: PN junction

（3）半导体：PN结

An N-type semiconductor and a P-type semiconductor are formed on both sides of an intrinsic semiconductor by diffusing different impurities. In this case, the following physical processes will be formed on the bonding surfaces of the N-type and P-type semiconductors.

在一块本征半导体两侧通过扩散不同的杂质，分别形成N型半导体和P型半导体。此时将在N型半导体和P型半导体的结合面形成以下物理过程。

1) Diffusion of most carriers

1）多数载流子的扩散

When a P-type semiconductor and an N-type semiconductor are combined organically, there is a difference in the concentration of holes and electrons at their interface because there are more holes on the P-zone side and more electrons on the N-zone side. So the holes in the P region will diffuse to the N region and be compounded by electrons in the N region. The electrons in the N region will also diffuse to the P region and be compounded by holes in the P region. As the diffusion movement continues, the positive and negative ions revealed on both sides of the interface gradually increase, and the space charge area widens, so that the internal electric field is continuously enhanced, and the drift movement is then enhanced.

P型半导体和N型半导体有机结合在一起时，因为P区一侧空位多，N区一侧电子多，所以在它们的界面处存在空位和电子的浓度差。于是，P区中的空位会向N区扩散，并在N区被电子复合。而N区中的电子也会向P区扩散，并在P区被空位复合。随着扩散运动的不断进行，界面两侧显露出的正、负离子逐渐增多，空间电荷区展宽，使内电场不断增强，于是漂移运动随之增强。

2) Drift of a few carriers

2）少数载体的漂移

Under the action of the internal electric field, the lesser free electrons in the P region drift toward the N region, while the lesser holes in the N region drift toward the P region, weakening the internal electric field.

在内电场的作用下，P区中的少子自由电子向N区漂移，而N区中的少子空位向P区漂移，使内电场削弱。

3) Dynamic equilibrium of diffusion and drift

3）扩散与漂移的动态平衡

When the internal electric field reaches a certain

当内电场达到一定值时，

value, the diffusion motion of the multiplets and the drift motion of the fewts reach dynamic equilibrium, at which point the net carrier number through the interface is zero, although diffusion and drift are still ongoing. The space charge region no longer changes, and this space charge region, is called a PN junction.

The space charge region has no carriers to stay, so it is called the depletion layer, also called the blocking layer or potential barrier layer. When there is no external electric field, the PN junction internal carrier movement, but no directional current formation.

In practice, if the doping concentration of the P and N regions are the same, the depletion region is symmetrical relative to the interface, called symmetrical junction. If the doping concentration on one side is large (heavy doping) and the doping concentration on the other side is small (light doping), it is called an asymmetric junction. It is denoted by P+N or PN+ (the + sign indicates the heavily doped region). At this point the depletion zone mainly extends to the lightly doped side. The ion flow in the depletion region is shown in Figure 2-12.

多子的扩散运动与少子的漂移运动达到动态平衡时，这时，虽然扩散和漂移仍在不断进行，但通过界面的净载流子数为0。空间电荷区不再变化，这个空间电荷区就称为PN结。

空间电荷区无载流子停留，故称为耗尽层，又称阻挡层或势垒层。无外电场作用时，PN结内部虽有载流子运动，但无定向电流形成。

实际中，如果P区和N区的掺杂浓度相同，则耗尽区相对界面对称，称为对称结。如果一边掺杂浓度大（重掺杂），一边掺杂浓度小（轻掺杂），则称为不对称结。用P+N或PN+表示（+号表示重掺杂区）。这时耗尽区主要伸向轻掺杂区一边。图2-12所示为耗尽区离子流动。

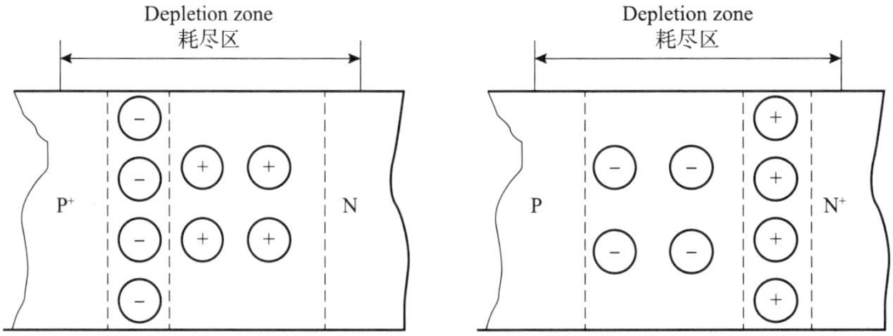

Figure 2-12 Ion flow in the depletion region

图2-12 耗尽区离子流动

The formation process of PN junction: In a piece of intrinsic semiconductor on both sides by diffusion of

PN结的形成过程：在一块本征半导体两侧通过扩散不同

different impurities, N-type semiconductor and P-type semiconductor are formed. At this time, the following physical processes are formed on the bonding surfaces of the N-type and P-type semiconductors:

的杂质，分别形成N型半导体和P型半导体。此时将在N型半导体和P型半导体的结合面形成以下物理过程：

Due to the concentration difference → diffusion movement of the multiplet → by the impurity ions to form a space charge region → space charge region to form an internal electric field at the same time so that the internal electric field to promote the drift of the minority, the internal electric field to prevent the multiplet diffusion.

因浓度差→多子的扩散运动→由杂质离子形成空间电荷区→空间电荷区形成内电场同时使内电场促使少子漂移、内电场阻止多子扩散。

Finally, the diffusion of the multiplets and the drift of the fewtons reach a dynamic equilibrium, forming a PN junction.

最后多子扩散和少子漂移达到动态平衡，形成PN结。

2. Hauff's atomic valence law—alloying changes the antioxidant properties

2. 哈菲原子价规律——合金化改变抗氧化性能

Hauff experimentally summarized an atomic valence law that describes the effect of alloying elements on the lattice defects, electronic and ionic conductivity, and oxidation rate of oxide films.

哈菲通过实验总结出一个原子价规律，它描述了合金元素对氧化膜晶格缺陷、电子和离子导电性以及氧化速率的影响。

N-type oxides: When a higher valence alloying element is added, such as Al to Zn, Zn^{2+} is replaced by Al^{3+}, the number of interstitial ions must decrease, the tendency to move must weaken, and the oxidation rate of the metal becomes slower. If lower-valent alloying elements are added, the number of interstitial ions increases, the diffusion rate increases, and the oxidation rate of the metal is accelerated.

N型氧化物：加入更高价的合金元素，Zn中加入Al，Zn^{2+}被Al^{3+}取代，间隙离子的数量必定减少，移动趋势必定减弱，金属的氧化速度变慢。若加入较低价的合金元素，间隙离子数量增多，扩散速度增大，金属的氧化速度加快。

P-type oxides: By adding higher-valent alloying elements, more cationic vacancies and electron vacancies will appear in the dot matrix, prompting the metal to oxidize more quickly. On the contrary, if low-valent alloying elements are added, such as Li in Ni, the

P型氧化物：加入更高价的合金元素，点阵中将出现更多的阳离子空位和电子空位，促使金属更快氧化。相反，若加入低价的合金元素，如Ni中加

cation vacancies and electron holes in the dot matrix are reduced accordingly, the diffusion rate of cations decreases, and the oxidation rate slows down.

入Li，则点阵中阳离子空穴和电子空位相应地减少，阳离子的扩散速度减小，氧化速度减慢。

2.4 Oxidation kinetics of metals

2.4 金属氧化动力学

There are three main factors affecting the rate of metal oxidation, the effect of temperature, the effect of oxygen pressure and the effect of gas medium. Among them, an increase in temperature causes a significant increase in the rate of metal oxidation; the partial pressure of oxygen in the gas medium has a different effect on metal oxidation due to the different types of oxide films; the effect of different gas media on the oxidation of metals and alloys varies greatly.

影响金属氧化速度的因素主要有三大方面，温度的影响、氧压的影响以及气体介质的影响。其中，温度升高会使金属氧化的速度显著增大；由于氧化膜的类型不同，气体介质中的氧分压对金属氧化的影响不同；不同的气体介质对金属和合金的氧化影响有很大差别。

1. Weight representation method

The size of metal corrosion can be used before and after the corrosion of the specimen mass change to assess, that is, the use of unit time unit area of metal weight change to indicate the rate of oxidation. Customarily referred to as “weight loss” or “weight gain method”.

(1) weight loss method - the oxidation of the weight of the metal per unit of time on the unit area of the reduction to express the rate of oxidation, unit g / (m^2·h), applicable to the case of oxide easy to flake.

1.质量表示法

金属腐蚀程度的大小可用腐蚀前后试样质量的变化来评定，即利用单位时间内单位面积上金属质量的变化来表示氧化速度。习惯上称为“失重法”“增重法”。

（1）失重法

用氧化后单位时间内单位面积上金属质量的减少来表示氧化速度，单位g/（m^2·h），适用于氧化物易剥落的情况。

$$K^{-}=\frac{g_0-g_1}{S\tau} \tag{2-26}$$

g_0——weight of the sample before oxidation (g);

式中　g_0——样品氧化前的质量，g；

g_1——weight of the sample after oxidation (g);
S——surface area of the sample (m^2);
τ——oxidation time (h).

g_1——样品氧化后的质量，g；
S——样品的表面积，m^2；
τ——氧化时间，h。

(2) weight gain method - the oxidation rate is expressed by the increase in the weight of the metal per unit area per unit time after oxidation, in $g/(m^2 \cdot h)$. Applicable to uniform corrosion that the oxide is not easy to flake or easy to collect.

（2）增重法

用氧化后单位时间内单位面积上金属质量的增加来表示氧化速度，单位g/（$m^2 \cdot h$）。适用于均匀腐蚀即氧化物不易剥落或易于收集的情况。

$$\Delta G^0 = RT \ln P_{O_2} = -RT \ln K \quad (2\text{–}27)$$

g_0——weight of the sample before oxidation (g);
g_2——weight of the sample after oxidation (g);
S——surface area of the sample (m^2);
τ——oxidation time (h).

式中 g_0——样品氧化前的质量，g；
g_2——样品氧化后的质量，g；
S——样品的表面积，m^2；
τ——氧化时间，h。

2. Kinetic curves of metal oxidation

(1) Oxidation depth method

Engineering, the depth of corrosion of the material or the degree of corrosion thinning of the components are directly affected by the life of the material components, so the measurement of its more practical significance. In the evaluation of different densities of metal corrosion degree, more suitable for the use of this method.

The formula for converting the rate of metal weight loss corrosion to corrosion depth is, in mm/a.

2.金属氧化的动力学曲线

（1）氧化深度法

工程上，材料的腐蚀深度或构件腐蚀变薄的程度均直接影响材料部件的寿命，因此对其测量更具有实际意义。在评定不同密度金属的腐蚀程度时，更适合采用氧化深度法。

将金属失重腐蚀速率换算为腐蚀深度的公式如下（单位：mm/a）。

$$K = \frac{K^-}{\rho} \times \frac{24 \times 365}{1000} = 8.76 \times \frac{K^-}{\rho} \quad (2\text{–}28)$$

According to the different depth of corrosion per year, the corrosion resistance of metals can be classified

根据每年腐蚀深度的不同，可将金属耐蚀性按十级标

according to the ten standard and three standards. According to the depth of the corrosion rate units, foreign literature is also commonly used ipy (inch / year inchs pey year abbreviation) and mpy (mil/year mils per year abbreviation). 1 mil = 10^{-3} in (inch), 1 in = 25.4 mm. the schedule lists the conversion relationship between different corrosion rate units. The conversion relationships between different corrosion rate units are listed in Table 2-3 and Table 2-4.

准和三级标准分类。按深度表示腐蚀速度，国外文献上还常用ipy（英寸/年inchs pey year的缩写）和mpy（密耳/年mils per year的缩写）。1mil=10^{-3}in（英寸），1in=25.4mm。表2-3、表2-4中列出了不同腐蚀速率单位间的换算关系。

Table 2-3 Ten levels of metal corrosion criteria

表 2-3 金属腐蚀性的十级标准

Corrosion resistance classification 耐蚀性分类		Corrosion resistance grade 耐腐蚀性等级	Corrosion rate/（mm/a） 腐蚀速率 /（mm/a）
Ⅰ	Total corrosion protection 完全耐蚀	1	<0.001
Ⅱ	Very corrosion protection 很耐蚀	2	0.001~0.005
		3	0.005~0.01
Ⅲ	Corrosion protection 耐蚀	4	0.01~0.05
		5	0.05~0.1
Ⅳ	Still corrosion resistant 尚耐蚀	6	0.1~0.5
		7	0.5~1.0
Ⅴ	Under-corrosion resistance 欠耐蚀	8	1.0~5.0
		9	5.0~10.0
Ⅵ	Incorrosion resistant 不耐蚀	10	>10.0

Table 2-4 Grade Ⅲ standards for metal corrosion

表 2-4 金属腐蚀性的三级标准

Corrosion resistance classification 耐蚀性分类	Corrosion resistance grade 耐蚀性等级	Corrosion rate/（mm/a） 腐蚀速率 /（mm/a）
durable 耐用	1	<0.01
available 可用	2	0.1~1.0
not available 不可用	3	>1.0

(2) Oxide film thickness method

Through the metal oxidation in a unit of time after the formation of the thickness of the oxide film to determine the corrosion rate, mm/h.

（2）氧化膜厚度法

通过单位时间内金属氧化后形成的氧化膜厚度来判断腐蚀速率（单位：mm/h）。

$$y=\frac{K^{+}M_{O_2}}{M_{MO_x}D}\times 10^{-3} \quad (2\text{-}29)$$

M_{MO_x} ——relative molecular weight of the oxide;

M_{O_2} ——relative molecular weight of oxygen;

D——density of the metal oxide.

式中 M_{MO_x} ——氧化物的相对分子量；

M_{O_2} ——氧的相对分子量；

D——金属氧化物的密度。

The most basic method of studying oxidation kinetics: determination of the thermostatic kinetic curve of the oxidation process (*W-t* curve). The speed limiting link in the oxidation process, the protective nature of the oxide film, the speed constant of the reaction, the energy change of the process. Typical metal oxidation kinetic curves are linear, parabolic, cubic, logarithmic and antilogarithmic laws. A theermostatic kinetic curve for the oxidation process is shown in Figure 2-13.

研究氧化动力学最基本的方法：图2-13所示为氧化过程的恒温动力学曲线（*W-t*曲线）。在氧化过程的速度限制性环节、氧化膜的保护性、反应的速度常数、过程的能量变化。典型的金属氧化动力学曲线包括直线规律、抛物线规律、立方规律、对数与反对数规律。

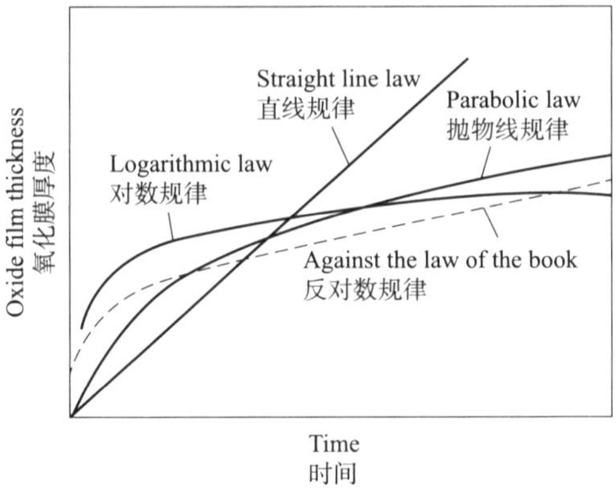

Figure 2-13 Thermostatic kinetic curve of the oxidation process (*W-t* curve)

图2-13 氧化过程的恒温动力学曲线（*W-t* 曲线）

1) Line laws of oxidation kinetics

When the metal is oxidized, if a protective oxide film cannot be generated, or if the product of the gas or liquid phase is formed during the reaction and leaves the metal surface, the oxidation rate is directly determined by the chemical reaction of the oxide formation, and

1）氧化动力学的直线规律

金属氧化时，若不能生成保护性氧化膜，或在反应期间形成气相或液相产物而脱离金属表面，则氧化速率直接由形成氧化物的化学反应所决定，

the oxidation rate is constant, i.e., the oxidation rate = the reaction rate of the oxide formation oxidation rate is constant, in accordance with the linear law.

因而氧化速率恒定不变，即氧化速度=形成氧化物的反应速度，氧化速度恒定不变，符合直线规律。

$$\Delta G = -RT\ \ln K - RT\ \ln P_{O_2} = -RT\ \ln\frac{1}{P_{O_2}} - RT\ \ln P_{O_2} = -RT\ \ln\frac{P_{O_2}}{P_{O_2}} \quad (2\text{-}30)$$

That is, the film thickness (or weight gain) is linearly related to the oxidation time t:

亦即膜厚（或增重）与氧化时间t呈直线关系：

$$y = k_L t + C \quad (2\text{-}31)$$

Where: k_L——Linear velocity constant of oxidation;

C——integration constant.

式中　k_L——氧化的线性速度常数；

C——积分常数。

Metals such as Mg, Mo and V follow this linear law when they oxidize at high temperatures.

Mg、Mo、V等金属高温氧化时遵循这一直线规律。

2) Parabola law of oxidation kinetics

2）氧化动力学的抛物线规律

The oxidation kinetic curve of most metals and alloys is parabolic. The oxidation rate is inversely proportional to the film thickness because a denser oxide film is formed on the metal surface when oxidizing over a wide temperature range:

多数金属和合金的氧化动力学曲线为抛物线。因为在较宽的温度范围内氧化时，金属表面形成较致密的氧化膜，氧化速率与膜厚成反比：

$$O_2 \xrightarrow{\text{吸附、分解}} O \xrightarrow{+2e} O^{2-} \xrightarrow{M^{2+}} MO \longrightarrow \text{Complete oxide film} \quad (2\text{-}32)$$

Or the integral gets to a parabolic equation:

或积分的抛物线方程：

$$y^2 = k_p^{t+C} \quad (2\text{-}33)$$

where: k_p——parabolic (Parabola) rate constant;

C——integral constant, which reflects the deviation from the parabolic regularity in the initial phase of oxidation.

式中　k_p——抛物线速率常数；

C——积分常数，它反映了氧化初始阶段对抛物线规律的偏离。

The parabolic law of metal oxidation is mainly due to the protective nature of the oxide film and the fact that the oxidation reaction is mainly controlled by the diffusion of metal ions and oxygen in the film. In

金属氧化的抛物线规律，主要因为氧化膜具有保护性，氧化反应主要受金属离子和氧在膜中的扩散控制。实际上，

fact, the oxidation of many metals deviates from the square parabolic law, so the equation is often written as a general equation:

许多金属的氧化偏离平方抛物线规律，因此常将公式写成通式：

$$y^n=k_pt+C \quad (2\text{–}34)$$

At n<2, the diffusion block of oxidation is not completely proportional to the growth of the membrane thickness, and the stress, hole and crystal boundary may cause the diffusion deviation from the parabola, or the oxidation process may be controlled by the diffusion and the surface oxidation reaction speed.

当n<2时，氧化的扩散阻滞并非完全与膜厚的增长成正比，应力、孔洞和晶界可能是扩散偏离抛物线的原因；也可能是氧化过程受扩散和表面氧化反应速度共同控制的。

At n>2, the diffusion arrest is more severe than the arrest produced by membrane thickening, i. e., there are other factors that inhibit the diffusion process.The doping of alloy oxides and the formation of a dense barrier layer are all possible causes.

当n>2时，扩散阻滞作用比膜增厚所产生的阻滞更严重，即还有其他因素抑制扩散过程。合金氧化物的掺杂，致密阻挡层的形成都是可能的原因。

3) Cube rules

3）立方规律

In a certain temperature range, the oxidation of certain metals obeys the cubic law, i.e.:

在一定温度范围内，某些金属的氧化服从立方规律，即：

$$y^3=k_pt+C \quad (2\text{–}35)$$

k_p is the cubic rate constant and C is a constant. For example, the constant temperature oxidation of copper at 100~300℃ and each air pressure, and zirconium at 600~900℃ and 1×10^5Pa in oxygen are all cubic laws.

式中，k_p为立方速率常数，C为常数。如铜在100~300℃及各气压下，锆在600~900℃、1×10^5Pa氧中的恒温氧化均属立方规律。

4)Logarithmic and antilogarithmic laws of oxidation kinetics

4）氧化动力学的对数与反对数规律

Some metals obey the logarithmic or antilogarithmic law when oxidized at low or room temperature. The logarithmic law is related by the equation:

有些金属在低温或室温氧化时服从对数或反对数规律。对数规律的关系式为：

$$\frac{dy}{dt}=A\,e^{-By} \quad (2\text{–}36)$$

The law against counting is:

反对数规律的关系式为：

$$\frac{dy}{dt}=Ae^{By} \tag{2-37}$$

where $y=k_1\cdot\lg(k_2\cdot t+k_3)$ and $1/y=k_4-k_5\cdot\lg t$, where A, B, and k_1、k_2、k_3、k_4 and k_5 are all constants.

式中，$y=k_1\cdot\lg(k_2\cdot t+k_3)$，$1/y=k_4-k_5\cdot\lg t$；$A$、$B$以及$k_1$、$k_2$、$k_3$、$k_4$和$k_5$皆为常数。

Both of these laws occur when the oxide film is quite thin, and it indicates that the oxidation process is far more hindered than in the parabolic relationship. For example, the oxidation of Mg, Al, Cu at room temperature; Zn, Fe, Ni, etc. at 100~200℃ are logarithmic laws. The oxidation of Al and Ta, etc. at 100~200℃ obeys the logarithmic law. In fact it is difficult to distinguish these two laws.

这两种规律都是在氧化膜相当薄时才出现，表明氧化过程受到的阻滞远比抛物线关系中的大。例如室温下的Mg、Al、Cu；100~200℃下的Zn、Fe、Ni等金属的氧化为对数规律。100~200℃下的Al和Ta等金属的氧化服从反对数规律。实际上很难将这两种规律区分。

5) Multi-layer oxide film (copper oxide)

5）多层氧化膜（铜氧化）

At lower oxygen partial pressures, only Cu_2O is produced and the parabolic velocity constant is a function of oxygen partial pressure; at higher oxygen partial pressures, Cu_2O and CuO double-layer oxide film, kinetics is independent of oxygen partial pressure; Cu_2O grows fast, oxidation kinetics is determined by Cu_2O and the Cu_2O layer is always at Cu_2O/CuO equilibrium oxygen partial pressure conditions.

如图2-14所示，在较低氧分压下，只生成Cu_2O，抛物线速度常数是氧分压的函数；在较高氧分压下，Cu_2O和CuO双层氧化膜，动力学与氧分压无关。Cu_2O生长快，氧化动力学由Cu_2O决定，且Cu_2O层始终处于Cu_2O/CuO平衡氧分压条件下。

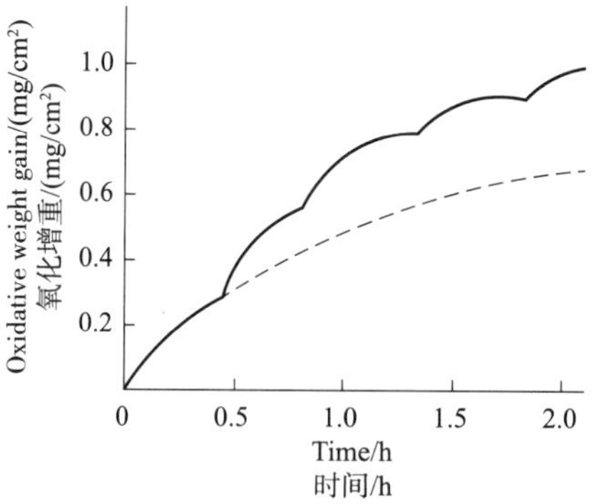

Figure 2-14 Copper oxidation kinetics at 500℃, and dashed line indicates the situation when the oxide film does not break down

图2-14 铜在500℃时的氧化曲线（虚线表示氧化膜不发生破坏时的情况）

2.5 Common chemical corrosion

2.5.1 Gas corrosion at high temperat

Steel materials in the high temperature gas environment is susceptible to corrosion, common forms of corrosion are mainly oxidation, decarburization, hydrogen corrosion, internal oxidation. As shown in Figure 2-15, oxide composition from the inside to the outside of FeO, Fe_3O_4 and Fe_2O_3.

2.5 常见的化学腐蚀

2.5.1 高温气体腐蚀

钢铁材料在高温气体环境中很容易受蚀，常见的腐蚀形式主要有氧化、脱碳、氢蚀、内氧化。图2-15所示为钢在570℃以上高温氧化时氧化物的形成；氧化物组成由内到外为FeO、Fe_3O_4和Fe_2O_3。

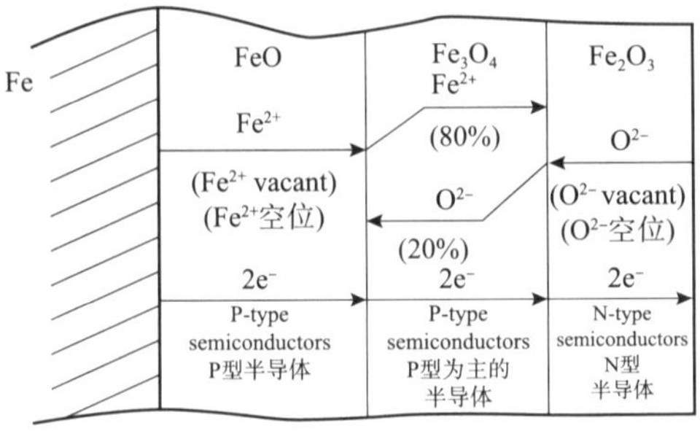

Figure 2-15 Formation of oxides when steel is oxidized at high temperatures above 570℃

图2-15 钢在570℃以上高温氧化时氧化物的形成

(1) Decarbonization of steel

Steel in the occurrence of oxidation at the same time, gas O_2, CO_2, H_2, H_2O, etc. may also be with the steel surface Fe_3C (or graphite), so that Fe_3C reduction into Fe, while generating carbon gas CO_2, CO, CH_4, etc., the result of the material surface carbon content is reduced, this chemical corrosion is called decarburization. As shown in Figure 2-16, decarburization reaction generated gas products, not only to destroy the integrity of the surface film and bonding, accelerating the oxidation process, but also seriously deteriorate the mechanical properties of the material, affecting the normal use of the material. The chemical equation for decarburization of steel is as follows:

（1）钢铁的脱碳

钢铁在发生氧化的同时，气体中O_2、CO_2、H_2、H_2O等还可能与钢铁表面的Fe_3C（或石墨）作用，使Fe_3C还原成Fe，同时生成含碳气体CO_2、CO、CH_4等，结果使材料表层含碳量降低，这种化学腐蚀称为脱碳。脱碳形貌如图2-16所示；脱碳反应生成的气体产物，不仅破坏表面膜的完整性和结合力，加速了氧化过程，而且还严重恶化材料的力学性能，影响材料的正常使用。钢铁脱碳的化学方程式如下：

$$Fe_3C+\frac{1}{2}O_2 \longrightarrow 3Fe+CO \quad (2-38)$$

$$Fe_3C+\frac{1}{2}CO_2 \longrightarrow 3Fe+2CO \quad (2-39)$$

$$Fe_3C+H_2O \longrightarrow 3Fe+CO+H_2 \quad (2-40)$$

$$Fe_3C+H_2 \longrightarrow 3Fe+CH_4 \quad (2-41)$$

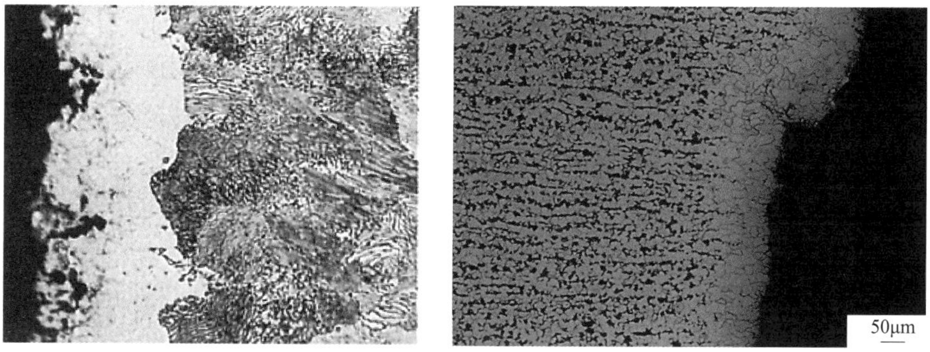

(a) First decarburization topography
第一种脱碳形貌图

(b) Second decarburization topography
第二种脱碳形貌图

Figure 2-16 Decarburization of steel
图 2-16 钢铁的脱碳形貌

(2) Hydrogen corrosion

Hydrogen corrosion is a partial decomposition of molecular hydrogen into atomic hydrogen or ionic hydrogen under high temperature and pressure conditions, and diffusion into the steel through the metal lattice and grain boundaries, diffusion of hydrogen invading the steel and unstable carbide chemical reaction, the generation of methane bubbles (it contains the nucleation process and growth of methane), and in the intergranular cavities and non-metallic inclusions parts of the aggregation, and methane in the steel diffusion capacity is very small, aggregation in the grain boundaries The original microscopic pores (or sub-microscopic pores) within the formation of local high pressure, resulting in stress concentration, so that the

（2）氢腐蚀

氢腐蚀是在高温高压条件下，分子氢发生部分分解而变成原子氢或离子氢，并通过金属晶格和晶界向钢中扩散，扩散侵入钢中的氢与不稳定的碳化物发生化学反应，生成甲烷气泡（它包含甲烷的成核过程和成长），并在晶间空穴和非金属夹杂部位聚集，而甲烷在钢中的扩散能力很小，聚集在晶界原有的微观孔隙（或亚微观孔隙）内，形成局部高压，造成应力集中，使晶界变宽，并发展成为裂纹，开始时

grain boundaries widen and develop into cracks, which are very small at the beginning, but at a later stage, numerous cracks are connected, causing a decrease in the strength, ductility and toughness of the steel with simultaneous intergranular fracture. Since this embrittlement phenomenon is the result of a chemical reaction, he has an irreversible nature, also known as permanent embrittlement phenomenon.

很微小，但到后期，无数裂纹相连，引起钢的强度、延性和韧性下降及同时发生晶间断裂。由于这种脆化现象是发生化学反应的结果，所以它具有不可逆的性质，也称永久脆化现象。

Hydrogen corrosion occurs in equipment operating at high temperatures and pressures in hydrogen gas in two forms: surface decarburization (as shown in Figure 2-17) and internal decarburization(as shown in Figure 2-18). Surface decarburization does not produce cracks, which is similar to the decarburization of steel exposed to some gases such as air, oxygen or carbon dioxide, etc. The effect of surface decarburization is generally very clear, with a local decrease in the strength and hardness of the steel and an increase in ductility. Internal decarburization is due to the reaction of hydrogen diffusion into the steel to generate methane, which cannot diffuse outside the steel and collects near grain boundaries or inclusions. A very high local stress is formed, causing the steel to crack, craze or bulge, and its mechanical properties become apparent.

在高温高压氢气中操作的设备所发生的氢腐蚀有两种形式：一是表面脱碳，如图2–17所示；二是内部脱碳，如图2–18所示。表面脱碳不产生裂纹，这与钢材暴露在空气、氧气或二氧化碳等一些气体中所产生的脱碳相似。表面脱碳的影响一般很清楚，其钢材的强度和硬度局部有所下降而延性有所提高。内部脱碳是由于氢扩散侵入钢中发生反应生成甲烷，而甲烷又不能扩散到钢外，就聚集于晶界或夹杂物附近。形成了很高的局部应力，使钢产生龟裂、裂纹或鼓包，其力学性能更加显著。

Simply put, hydrogen corrosion is the phenomenon of steel becoming brittle or even breaking by the action of high temperature and pressure hydrogen, and the chemical equation of the process is as follows:

简单来说，氢腐蚀就是钢材受高温高压的氢气作用变脆甚至破裂的现象，其过程的化学方程式如下：

$$Fe_3C+2H_2 \longrightarrow Fe+CH_4 \quad (2\text{–}42)$$

$$\begin{bmatrix} Fe_3C \rightarrow 3Fe+C \\ C+2H_2 \rightarrow CH_4 \end{bmatrix} \quad (2\text{–}43)$$

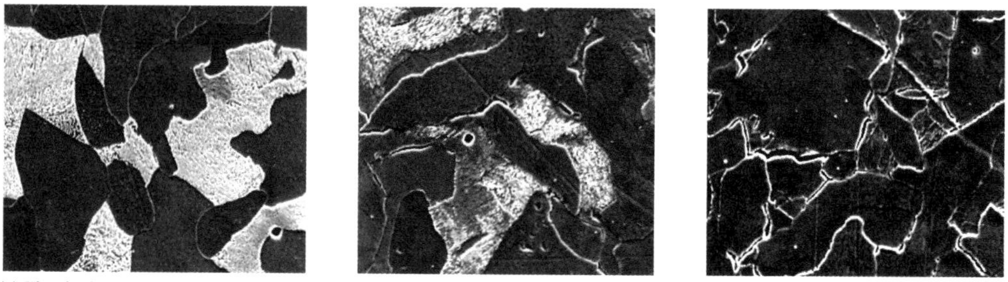

(a) First hydrogen corrosion topography 第一种氢腐蚀形貌图 (b)Second hydrogen corrosion topography 第二种氢腐蚀形貌图 (c)Third hydrogen corrosion topography 第三种氢腐蚀形貌图

Figure 2-17 Hydrogen corrosion surface

图 2-17 氢腐蚀表面

Nelson line is a combination of temperature and pressure conditions that indicate the onset of hydrogen corrosion of steel, and Nelson line diagrams are widely used around the world as a reference to select steel for hydrogen corrosion resistance.

纳尔逊曲线是表示钢材发生氢腐蚀的起始温度和压力的组合条件，目前世界各国广泛应用纳尔逊曲线图作参考来选用抗氢腐蚀的钢材。

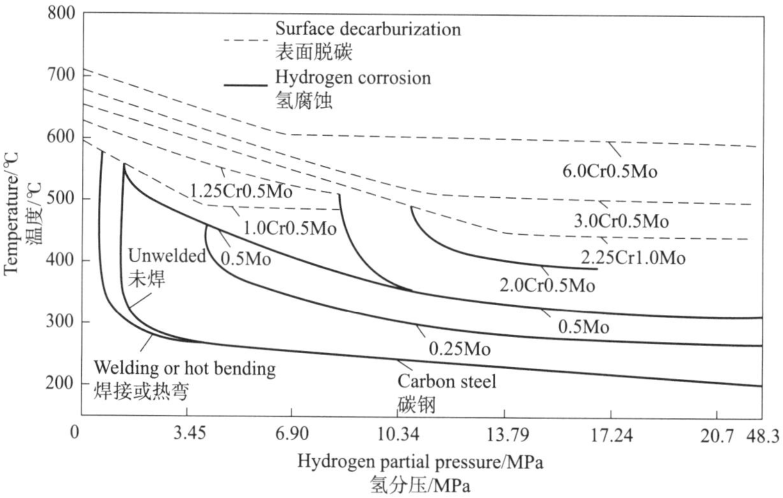

Figure 2-18 Hydrogen corrosion curve of various steel products (Nelson-line diagram)

图 2-18 各种钢材的氢腐蚀曲线（纳尔逊曲线图）

(3) High temperature vulcanization

（3）高温硫化

The contact between metal and sulfur-containing gas (S vapor, SO_2 or H_2S, etc.), which reacts to produce sulfide and causes continuous corrosion of the metal, is called sulfide.

金属与含硫气体（S蒸气、SO_2或H_2S等）接触，反应生成硫化物，使金属不断腐蚀的现象称为硫化。

Sulphide has the following special properties:

硫化物具有以下特殊的性能：

① The PB ratio of metal sulfide is greater, thus generating sulfide film with greater internal stress, easy to make the film rupture.

①金属硫化物的PB比更大，因而生成硫化物膜有更大的内应力，易于使膜破裂。

② The concentration of lattice defects in metal sulfides is much higher than that of the corresponding oxides, making it easier for various ions to diffuse in the film.

②金属硫化物的晶格缺陷浓度比相应的氧化物高得多，使各种离子在膜中扩散更容易。

③ The melting point of metal sulfide is much lower than the corresponding oxide, especially the eutectic point of metal-metal sulfide is lower, easy to appear in the liquid phase and lose the protective properties, intensifying the corrosion process.

③金属硫化物的熔点比对应的氧化物低得多，特别是金属–金属硫化物的共晶点更低，容易出现液相而失去保护性，加剧腐蚀进程。

④ Sulfur has a strong destructive ability to metal oxides.

④硫对金属氧化物有很强的破坏能力。

(4) Sulfuric Acid Dew Point Corrosion of Flue Gas

（4）烟气的硫酸露点腐蚀

Flue gas dew point corrosion is due to the combustion of sulfur elements in the fuel to generate SO_2, SO_3, when the outer surface temperature of the heat exchanger surface is lower than the flue gas dew point temperature, sulfuric acid mist dew droplets will be formed on the heat exchanger surface, resulting in heat exchanger surface corrosion.

烟气的硫酸露点腐蚀是由于燃料中硫元素在燃烧时生成SO_2、SO_3，当换热面的外表面温度低于烟气露点温度时，在换热面上就会形成硫酸雾露珠，导致换热面腐蚀。

When the fuel is burned, the hydrogen (H_2) and oxygen (O_2) are combined to produce water vapor (H_2O), and most of the burners use steam atomization, thus making the flue gas in the furnace with a large amount of water vapor. In addition, the sulfur (S) in the fuel is burned to produce sulfur dioxide (SO_2), of which a small amount of SO_2 is further oxidized to sulfur trioxide (SO_3) sulfur trioxide combined with water vapor in the flue gas to produce sulfuric acid (H_2SO_4). The dew point of the flue gas containing sulfuric acid vapor is greatly increased, and when the wall temperature of the heated

燃料在燃烧时，其中的氢（H_2）和氧（O_2）化合生成水蒸气（H_2O），而燃烧器大部分又采用蒸汽雾化，因而使炉子中的烟气带有大量的水蒸气。另外，燃料中的硫（S）在燃烧后生成SO_2，其中少量的SO_2进一步又氧化成SO_3，SO_3与烟气中的水蒸气结合生成硫酸（H_2SO_4）。含有硫酸蒸汽的烟气露点大为升高，当受热面

surface is lower than the dew point, the vapor containing sulfuric acid will condense into a liquid containing sulfuric acid on the heated surface, which will cause serious corrosion of the heated surface. Because it is in the lower temperature of the heated surface corrosion, so called low-temperature corrosion. As only in the heated surface after the condensation of this corrosion, so also known as the dew point corrosion. As shown in Figure 2-19, dew point temperature of the high and low in addition to the amount of sulfur in the fuel, but also with the excess air coefficient and the amount of sulfur trioxide generation and other factors. The higher the furnace chamber temperature, the less excess air, the smaller the share of sulfur generated in the combustion of SO_2 is oxidized to SO_3, the lower the dew point temperature. The chemical equation of the corrosion process is as follows:

的壁温低于露点时，含有硫酸的蒸汽就会在受热面凝结成含有硫酸的液体，对受热面产生严重腐蚀。因为它是在温度较低的受热面发生的腐蚀，故称为低温腐蚀。由于只有在受热面结露后才发生这种腐蚀，所以又称露点腐蚀。如图2-19所示，钢铁烟气的硫酸露点腐蚀，露点温度的高低除与燃料中的含硫量有关外，还与过剩空气系数和SO_3的生成量等因素有关。炉膛温度越高过剩空气越少，则燃烧中的硫生成的SO_2被氧化成SO_3的份额就越小，露点温度越低。腐蚀过程的化学方程式如下：

$$S+O_2 \longrightarrow SO_2\uparrow \tag{2-44}$$

$$2SO_2+O_2 \longrightarrow 2SO_3\uparrow \tag{2-45}$$

$$SO_3\uparrow+H_2O\uparrow \xrightarrow{400℃} H_2SO_4\uparrow \tag{2-46}$$

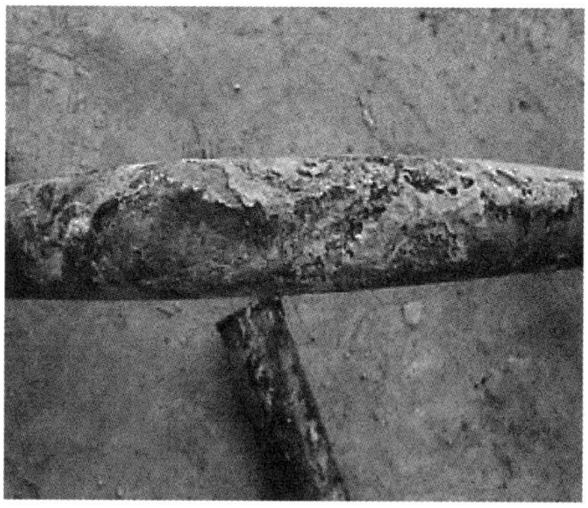

Figure 2-19 Sulfuric acid dew point corrosion of steel flue gas

图2-19 钢铁烟气的硫酸露点腐蚀

Flue gas dew point corrosion of sulfuric acid mainly occurs in the low temperature parts of industrial boilers,

烟气的硫酸露点腐蚀主要发生在工业锅炉的低温部

such as convection section and chimney. In order to reduce flue gas corrosion, the industry mainly uses low sulfur fuel, increase the exhaust temperature, choose dew point corrosion resistant steel such as ND steel (09CrCuSb) and other measures.

位，如对流段和烟囱处等。为减少烟气腐蚀，工业上主要采用低硫燃料、提高排烟温度、选择耐露点腐蚀钢材如ND钢（09CrCuSb）等措施。

2.5.2 High temperature corrosion of liquid metals

2.5.2 液态金属的高温腐蚀

Metal corrosion occurs in liquid metals because of their good thermal conductivity, high stability and resistance to irradiation, therefore, low melting point metals such as sodium, sodium-potassium, lithium, bismuth and aluminum are used as coolants in nuclear energy reactors in nuclear engineering. Fast neutron breeder reactors use sodium as a cooling medium.

Liquid metal corrosion forms can be divided into two categories: one is the metal material and the liquid metal or impurities contained therein, especially oxygen, carbon, nitrogen, hydrogen and other non-metals to form an alloy or compound. Such as soft steel in the oxygen-containing liquid sodium, the generation of very loose composite oxide, easy to fall off, so that the corrosion process is accelerated; the other category is a solid metal dissolved in liquid metal. As shown in Figure 2-20, the size of its solubility and the metal material alloying elements in the liquid metal solubility, dissolution rate, solute atomic diffusion rate and the concentration of non-metallic inclusions in the liquid metal, temperature related. Generally, the solubility increases with increasing temperature. When the liquid metal in a static isothermal system is saturated with solute, the dissolution reaction stops. However, in practice, often non-isothermal dynamic systems. Such as in the

液态金属中发生的金属腐蚀是由于液态金属具有良好的热传导性、高稳定性和抗辐照性，因此核工程上用钠、钠-钾、锂、铋、铝等低熔点金属作为核能反应堆的冷却剂。快中子增殖反应堆以钠作为冷却介质。

液态金属腐蚀形式可分为两类：一类是金属材料与液态金属或其中所含的杂质特别是氧、碳、氮、氢等非金属形成合金或化合物。如软钢在含氧液态钠中，生成很疏松的复合氧化物，易脱落，使腐蚀过程加速；另一类是固态金属溶解于液态金属中。图2-20所示为液态金属的高温腐蚀，其溶解度大小与金属材料的合金元素在液态金属中的溶解度、溶解速度、溶质原子的扩散速度以及液态金属中非金属夹杂物的浓度、温度有关。随着温度升高，溶解度增大。当静态等温系统中的液态金属被溶质饱和时，溶解反应就停止。但在

heat exchange device, high temperature zone solute metal gradually reach the equilibrium concentration at that temperature, when entering the low temperature zone, due to the temperature drop solubility becomes smaller, excess solute metal constantly precipitation, resulting in the high temperature zone of the metal material suffered corrosion; low temperature zone due to metal precipitation may cause pipeline blockage. This phenomenon is called the temperature difference mass migration. Even in the isothermal system, when more than two kinds of dissimilar metal materials and liquid metal contact, due to some common elements in these dissimilar metals in the value of activity is different, these elements from the activity of the large part of the dissolution, in the activity of the small part of the precipitation.

实践中，常常是非等温动态系统。如在热交换装置中，高温区溶质金属逐步达到在该温度下的平衡浓度，当进入低温区时，由于温度下降溶解度变小，过剩的溶质金属不断析出，致使处于高温区的金属材料遭受到腐蚀；低温区由于金属析出可能造成管路堵塞，这种现象称为温差质量迁移。即使在等温系统中，当两种以上的异种金属材料与液态金属相接触时，由于某些共同的组成元素在这些异种金属中活度值不同，这些元素从活度大的部位溶解，在活度小的部位析出。

(a)Lead-bismuth alloy ingot casting
铅铋合金铸锭

(b)Tester
试验装置

(c)Corrosion morphology of 316L in liquid lead-bismuth
316L在液态铅铋中的腐蚀形貌

Figure 2-20 High-temperature corrosion of liquid metal
图 2-20 液态金属的高温腐蚀

The main measures to prevent or slow down the corrosion of liquid metal are: to reduce as much as possible the content of oxygen, carbon, nitrogen, hydrogen and other non-metallic impurities in liquid metal; reasonable selection of materials.

防止或减缓液态金属腐蚀的主要措施是：尽可能地降低液态金属中氧、碳、氮、氢等非金属杂质含量；合理选材。

2.5.3 Corrosion of melting salt

2.5.3 熔盐腐蚀

Metal corrosion of metallic materials occurs in molten salts. Figure 2-21 shows the topography of

熔盐腐蚀是指金属材料在熔盐中发生的金属腐蚀。图

molten salt chemical corrosion of steel. Molten salts are widely used in industry because of their good thermal stability, fluidity, high thermal conductivity and high electrical conductivity. For example, they are often used as high-temperature carriers and electrolytes in the chemical industry. In some nuclear reactors, they are used as solvents and coolants for nuclear fuel. In all devices where molten salts are applied, they can be corrosive to metal materials.

2-21为钢材的熔盐化学腐蚀。由于熔盐具有良好的热稳定性、流动性、高热导性和高电导性，因此，在工业上得到广泛应用。例如，在化学工业中常用作高温载体和电解液。在某些核反应堆中用作核燃料的溶剂和冷却剂。凡应用熔盐的装置中，均可对金属材料产生腐蚀。

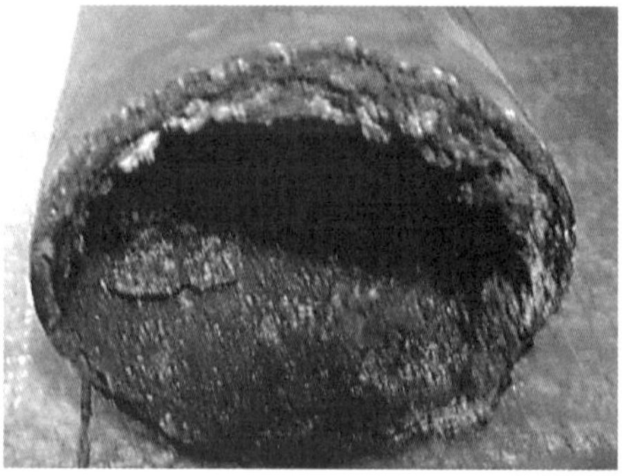

Figure 2-21 Chemical corrosion of steel by molten salt
图2-21 钢材的熔盐化学腐蚀

Molten salt corrosion forms are divided into two categories: one is the metal is oxidized into metal ions, with the same electrochemical corrosion process with aqueous corrosion, which is the main form of molten salt corrosion, the potential difference between the cathode and anode is the driving force of the corrosion reaction, and the rate of migration of the oxidizer controls the overall corrosion reaction rate; the other is the metal state dissolved in molten salt, not accompanied by oxidation, such as lead immersed in lead chloride molten salt generated by the corrosion. When the metal-molten salt system there is a temperature gradient, will produce temperature mass migration corrosion, in the high temperature parts of the metal is dissolved, in the

熔盐腐蚀形式分为两类：一类是金属被氧化成金属离子，具有与水溶液腐蚀相同的电化学腐蚀过程，这是熔盐腐蚀的主要形式，阴极、阳极间的电位差是腐蚀反应的推动力，而氧化剂的迁移速度控制着整个腐蚀的反应速度；另一类是以金属态溶解于熔盐中，不伴随氧化作用，如铅浸入氯化铅熔盐中产生的腐蚀。当金属-熔盐体系中存在温度梯度时，会产生温差质量迁移腐蚀，处于高温部位的金属被溶

low temperature parts of the metal precipitated from the molten salt. So repeated, can lead to localized corrosion in the high-temperature parts, low-temperature parts produce blockage. Even under isothermal conditions, as most of the practical metal materials are alloys, when it comes into contact with the molten salt, the more active components of the alloy is often selective corrosion.

解，处于低温部位的金属从熔盐中析出。如此往复，可导致高温部位发生局部腐蚀，低温部位产生堵塞。即使处于等温条件下，由于实用金属材料大部分是合金，当其与熔盐接触时，合金中比较活泼的组分也常发生选择性腐蚀。

The main measures to prevent or slow down molten salt corrosion are: removing air and water in the molten salt or slowing down the rate of diffusion and migration of oxygen and water in the molten salt; reasonable selection of materials; electrochemical protection.

防止或减缓熔盐腐蚀的主要措施是：除去熔盐中的空气和水分或减慢氧与水分在熔盐中的扩散迁移速度；合理选材；电化学保护。

2.6 References

[1] MA Y Y. Research on corrosion and protection of chemical Equipment [J]. Henan Science and Technology, 2013, 24: 52-53.

[2] WANG X B, FAN S H, YIN M Z. Construction principle and application of the standard free energy variation and temperature relationship diagram of oxidation reaction [J]. Journal of Fuyang Normal University (Natural Science edition), 2004, 21(3): 26-30.

[3] ZHAO X P, XU K Y, JIANG W Y, et al. Effects of metal oxide films on ash wear properties at different temperatures [J]. Journal of East China University of Technology, 1996, 18(1): 73-79.

2.6 参考文献

[1] 马媛媛 . 化工设备的腐蚀与防护研究 [J]. 河南科技，2013（24）：52–53.

[2] 汪晓波，范少华，殷梅珍 . 氧化反应标准自由能变与温度关系图的构建原理及其应用 [J]. 阜阳师范学院学报（自然科学版），2004，21（3）：26–30.

[3] 赵宪萍，徐开义，蒋伟元，等 . 不同温度下金属氧化膜对灰磨损性能的影响 [J]. 华东工业大学学报，1996，18（1）：73–79.

Electrode Potential and Pourbaix Diagram

电极电位与 Pourbaix 图

3

3.1 Electrode system reacts with electrode

An electrode is a system composed of an electronic conductor and an ionic conductor. Such as Cu||$CuSO_4$ said copper electrode, Zn||$ZnSO_4$ said zinc electrode. Usually, electrodes can be divided into three categories: single electrode, metal corrosion electrode, reversible and irreversible electrode. As shown in Figure 3-1, the zinc electrode is on the left and the copper electrode is on the right.

3.1.1 Single electrode

A single electrode is an electrode where only a single electrode reaction occurs at the metal-solution interface. It can be divided into metal electrode, gas electrode and REDOX electrode.

3.1 电极系统与电极反应

电极是电子导体和离子导体组成的体系，如Cu||$CuSO_4$称为铜电极，Zn||$ZnSO_4$称为锌电极。通常可以将电极分为三大类：单电极、金属腐蚀电极、可逆与不可逆电极。如图3-1所示，左侧为锌电极，右侧为铜电极。

3.1.1 单电极

单电极是在金属-溶液界面只发生单一的电极反应的电极。又可分为金属电极、气体电极和氧化-还原电极。

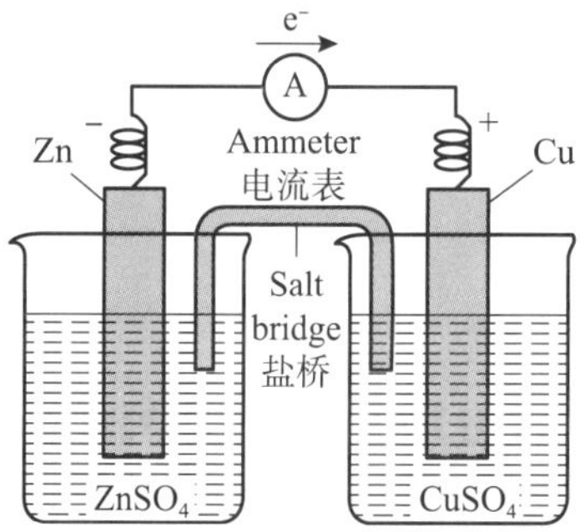

Figure 3-1 Copper and zinc galvanic cell
图 3-1 铜锌原电池

Metal electrode: an electrode in which a metal is formed in an electrolyte solution containing the metal ion and an electrode reaction occurs. The electrode reaction is as follows:

金属电极：金属在含有该金属离子的电解质溶液中构成的而且发生电极反应的电极。电极反应如下：

$$M \longleftrightarrow M^{n+} + ne^- \quad (3\text{-}1)$$

Gas electrode: an electrode consisting of metal in a solution containing gas and gas ions and in which the gas participates in the electrode reaction. The electrode reaction is as follows:

气体电极：金属在含有气体和气体离子的溶液中构成的而且发生气体参与电极反应的电极。电极反应如下：

$$H_2 \longleftrightarrow 2H^+ + 2e^- \quad (3\text{-}2)$$

$$O_2 + 2H_2O + 4e^- \longleftrightarrow 4OH^- \quad (3\text{-}3)$$

$$Cl_2 + 2e^- \longleftrightarrow 2Cl^- \quad (3\text{-}4)$$

REDOX electrode: The electrode with only electron exchange at the electrode interface, and no metal ion translocation between phases is called REDOX electrode. The electrode reaction is as follows:

氧化-还原电极：电极界面只有电子交换，而不发生金属离子在相间迁跃的电极称为氧化-还原电极。电极反应如下：

$$Fe^{3+} + e^- \longleftrightarrow Fe^{2+} \quad (3\text{-}5)$$

3.1.2 Metal etching electrode

3.1.2 金属腐蚀电极

Metal etching electrodes are electrodes where more than one electrode reaction occurs simultaneously on one electrode. Including double electrodes and multiple electrodes.

金属腐蚀电极是在一个电极上同时发生多于一个电极反应的电极，包括二重电极和多重电极。

3.1.3 Reversible and irreversible electrodes

3.1.3 可逆与不可逆电极

The electrode reaction of the reversible electrode is as follows:

可逆电极的电极反应如下。

Reduction reaction:

还原反应：

$$Cu^{2+} \cdot mH_2O + 2e^- \xrightarrow{J_1} Cu + mH_2O \qquad (3\text{-}6)$$

Oxidation reaction:

氧化反应：

$$Cu + mH_2O \xrightarrow{J_2} Cu^{2+} \cdot mH_2O + 2e^- \qquad (3\text{-}7)$$

There are irreversible electrodes in addition to reversible electrodes.

除了可逆电极外还有不可逆电极。

3.2 Electrode potential and its determination equilibrium potential

3.2 电极电位与平衡电位

3.2.1 Electrode potential and its determination

3.2.1 电极电位及其确定

Electrode potential is the potential difference between two sides of an electrode.

电极电位是指电极两侧的电位差。

Ignoring the liquid-boundary potential and metal-contact potential, the electromotive force of the cell is equal to the electrode potential of the positive electrode minus the electrode potential of the negative electrode, which is equal to the difference between the electrode potential of the cathode (where the reduction reaction occurs) and the anode (where the oxidation reaction occurs):

在忽略液界电位和金属接触电位的情况下，电池的电势等于正极的电极电位减去负极的电极电位，亦即等于阴极（发生还原反应）的电极电位 ϕ_C 与阳极（发生氧化反应）的电极电位 ϕ_A 之差：

电极电位与平衡电位

$$\phi = \phi_C - \phi_A \qquad (3\text{-}8)$$

In a corrosion cell, the metal anode dissolves (corrodes), and its potential is ϕ_A; The reduction reaction occurs at the cathode, and the potential is ϕ_C, and the electrochemical criterion of the corrosion tendency of the metal is:

腐蚀电池中，金属阳极发生溶解（腐蚀），其电位为 ϕ_A；而腐蚀剂在阴极发生还原反应，其电位为 ϕ_C，金属腐蚀倾向的电化学判据为：

$$\phi_A < \phi_C \text{ 则电位为 } \phi_A \text{ 的金属自发进行腐蚀} \quad (3\text{-}9)$$

$$\phi_A = \phi_C \text{ 平衡状态} \quad (3\text{-}10)$$

$$\phi_A > \phi_C \text{ 则电位为 } \phi_A \text{ 的金属不会自发进行腐蚀} \quad (3\text{-}11)$$

There are two methods to determine electrode potential: one is the measurement method, the other is the calculation method.

电极电位的确定有两种方法：一种是测量法，另一种是计算法。

Measurement method: The potential of the double electric layer at the interface between electrode and solution is called absolute electrode potential, which directly reflects the thermodynamic and kinetic characteristics of the electrode process, but the absolute electrode potential is measured with the method of the original one. In the actual study, the electromotive force of the galvanic cell composed of electrode potential is measured, and the electrode of the reference object used for measuring electrode potential is called the reference electrode. The reference electrode needs to meet the following conditions: ① the electrode reaction is reversible, the potential is stable and does not change with time; ② The exchange current density is large, not polarized or difficult to polarize; ③ The solution in the reference electrode is not impermeable to the corrosive medium, and the potential of the solution interface is small; ④ Low temperature coefficient.

测量法：电极和溶液界面双电层的电位称为绝对电极电位，它直接反映了电极过程的热力学和动力学特征，但绝对电极电位是无法测量的。在实际研究中，测量电极电位组成原电池的电势，而测量电极电位所用参考对象的电极称为参考电极。参考电极需要满足以下条件：①电极反应是可逆的，电位稳定而不随时间变化；②交换电流密度大，不极化或难极化；③参比电极内溶液与腐蚀介质互不渗污，溶液界面的电位小；④温度系数低。

Then the electromotive force of the battery is:

则该电池的电势为：

$$E = \varphi_{待测} - \varphi_{参比} \quad (3\text{-}12)$$

Calculation method: there are two main factors

计算法：影响电极电位的

affecting the electrode potential. One is the intrinsic factor of the electrode, that is, the thermodynamic process of the electrode (the nature of the substance) is the determinant factor affecting the electrode potential value. The other is external factors, that is, the concentration of the solution and the pH value of the solution have an impact on the value of the electrode potential. Nernst equation of electromotive force, which reflects the relationship between the potential of non-standard electrode and the potential of standard electrode. This formula shows the relationship between any state electromotive force and standard electromotive force, concentration and temperature:

因素主要有两种：一种是电极的内在因素，即电极的热力学过程（物质的本性）是影响电极电位数值的决定因素；另一种是外在因素，即溶液的浓度及pH值对电极电位的数值有影响。电势的能斯特方程，它反映了非标准电极电势和标准电极电势的关系。式（3–13）表明任意状态电势与标准电势以及浓度、温度之间的关系。

$$V_e = V^0 + RT / nF \cdot \ln\left(a_{氧化态} / a_{还原态}\right) \tag{3–13}$$

Type in the V_e、V^0——Equilibrium potential and standard potential of electrode;

R——Universal gas constant, 8.313J/(mol·K);

T——Absolute temperatureK;

n——The number of electrons participating in the electrode reaction;

F——Faraday's constant, 96500C/mol;

a——Mean activity (effective concentration).

式中 V_e、V^0——电极的平衡电位及标准电位；

R——通用气体常数，8.313J/(mol·K)；

T——绝对温度，K；

n——参加电极反应的电子数；

F——法拉第常数，等于96500C/mol；

a——平均活度（有效浓度）。

3.2.2 Balance the potential

Equilibrium electrode potential means that the charge and substance in the positive and negative process of electrode reaction have reached equilibrium, so this potential is called equilibrium electrode potential or reversible potential.

3.2.2 平衡电位

平衡电极电位是指电极反应正逆过程的电荷和物质均达到平衡，所以将这种电位称为平衡电极电位或可逆电位。

Standard electrode potential: when the substance participating in the electrode reaction is in the standard state, that is, the activity of the ion containing the metal in the solution is 1, the temperature is 298K, and the partial pressure of the gas is 1 atm, the equilibrium electrode potential of the metal is called the standard electrode potential.

标准电极电位：参加电极反应的物质处于标准状态下，即溶液中含该种金属离子的活度为1、温度为298K、气体分压为1atm时，金属的平衡电极电位称为标准电极电位。

The base of the Potential is a Standard hydrogen Electrode, or SEP, which is artificially determined to be zero at all temperatures.

电位的基准是标准氢电极（Standard Electrode Potential，SEP），该电位是人为规定的，在各温度下皆为0。

The electrode potential is usually said to be relative to the standard hydrogen electrode, and is actually equal to the electromotive force of the following cell consisting of the electrode (negative half cell) and the standard hydrogen electrode (positive half cell):

通常所说的电极电位是相对于标准氢电极而言的，实际上等于该电极（负极半电池）与标准氢电极（正极半电池）组成的下列电池的电势（金属电极）（标准氢电极）。

$$(-)M|M^{n+}(\alpha M^{n+})|H^{+}(\alpha H^{+}=1),\ H_2(p=1\ atm)|Pt(+) \qquad (3\text{-}14)$$

（金属电极）　　（标准氢电极）

According to Equation (3-14), the criterion of hydrogen evolution corrosion tendency of metal can be obtained:

根据式（3-14），可得到金属发生析氢腐蚀倾向的判据：

$\varphi_M<\varphi_H$　Metals are prone to hydrogen evolution and corrosion;

$\varphi_M>\varphi_H$　Metal does not undergo hydrogen evolution corrosion.

$\varphi_M<\varphi_H$　金属有析氢腐蚀倾向；

$\varphi_M>\varphi_H$　金属不发生析氢腐蚀。

Where φ_M is the potential of the metal. Obviously, the lower the metal potential, the greater the possibility of hydrogen evolution corrosion. However, in actual situation, the value of electrode potential of some metals is not easy to be measured. In this case, the standard electrode potential of metals at 25℃ can be used to judge the corrosion tendency of metals. Because the standard hydrogen electrode has zero

式中：φ_M为金属的电位。显然，金属的电位越低，析氢腐蚀的可能性越大。然而实际情况中，有些金属的电极电位的数值不易测得，这时可以利用25℃下金属的标准电极电位来判断金属的腐蚀倾向。因为标

potential. Therefore, the approximate criterion of hydrogen evolution corrosion is $\phi_M^\ominus$:

$\phi_M^\ominus<0$ Metals are prone to hydrogen evolution and corrosion;

$\phi_M^\ominus>0$ Metal does not undergo hydrogen evolution corrosion.

Where $\phi_M^\ominus$ is the standard electrode potential of metal.

准氢电极的电位为0。所以发生析氢腐蚀的近似判据为：

$\phi_M^\ominus<0$ 金属有析氢腐蚀倾向；

$\phi_M^\ominus>0$ 金属不发生析氢腐蚀。

式中：$\phi_M^\ominus$ 为金属的标准电极电位。

3.3 Thermodynamic criterion

3.3 热力学判据

热力学判据

The physical changes in nature are broadly divided into two categories: those that can and those that cannot. For all chemical reactions that can proceed spontaneously, they must be accompanied by a decrease in energy, or they cannot proceed spontaneously. The Gibbs free energy is usually used to judge the corrosion tendency of metals. That is, under isothermal and isoparic conditions.

自然界的物质变化大体分为两类：能够自发进行的和不能够自发进行的过程。对于所有的化学反应，凡是能自发进行的过程，一定伴随着能量的降低，否则不能自发进行。对于断金属的腐蚀倾向，通常采用吉布斯自由能进行判断，即在等温等压的条件下：

$$(\Delta G)_{T,P}<0 \text{ 自发过程} \tag{3-15}$$

$$(\Delta G)_{T,P}=0 \text{ 平衡状态} \tag{3-16}$$

$$(\Delta G)_{T,P}>0 \text{ 非自发过程} \tag{3-17}$$

Corrosion of metals includes anodic reaction and cathodic reaction.

An oxidation reaction occurs at the anode. The equilibrium electrode potential of the corresponding metal is V_{Ae}. Obviously, if the electrode potential of the metal deviates from V_{Ae} value, the dynamic equilibrium will be destroyed. If the electrode potential of metal can be kept equal to or negative to V_{Ae} value, the dissolution reaction of metal into metal ions is actually

金属腐蚀包括阳极反应和阴极反应。

阳极发生氧化反应。相应地，金属的平衡电极电位为 V_{Ae}。显然，若使金属的电极电位偏离 V_{Ae} 值，则动态平衡就遭到破坏。如果金属的电极电位能够保持在等于或负于 V_{Ae} 值，则金属成为金属离子的溶解反

in the condition from low energy to high energy, and the process cannot happen automatically. On the other hand, if the electrode potential V_A of the metal is maintained at a more positive level than V_{Ae} due to the presence of other ions or dissolved gases in the solution, then the dissolution reaction of the metal is thermodynamics possible. So the conditions for spontaneous anodic reaction are: $V_A>V_{Ae}$.

应实际上是处于从低能位到高能位的条件下，过程不可能自动发生。反之，如果由于溶液中其他离子或溶解气体的存在，使金属的电极电位V_A维持在比V_{Ae}更正的水平上，那么发生金属的溶解反应在热力学上就是可能的。所以，阳极反应自发进行的条件是：$V_A>V_{Ae}$

A reduction reaction occurs at the cathode. In the same way as the anodic reaction, when the actual electrode potential V_k deviates from the equilibrium potential V_{ke} of the cathode, the equilibrium of the cathode reaction will be destroyed. If the $V_k>V_{ke}$ reaction is in the oxidation direction (to the left) and the $V_k>V_{ke}$ reaction is in the reduction direction (to the right). Therefore, the conditions for spontaneous cathodic depolarization reaction are: $V_k<V_{ke}$.

阴极发生还原反应。与阳极反应同样的道理，当电极实际电位V_k偏离阴极的平衡电位V_{ke}时，上述阴极反应的平衡亦将遭到破坏。$V_k>V_{ke}$时反应朝氧化方向（向左）进行，而$V_k>V_{ke}$时反应朝还原方向（向右）进行。因此，阴极去极化反应自发进行的条件为：$V_k<V_{ke}$。

The metal's potential V must be positive over V_{Ae} and negative over V_{ke}: $V_{Ae}<V<V_{ke}$.

金属的电位值V必须维持在既比V_{Ae}正而又比V_{ke}负的数值上：$V_{Ae}<V<V_{ke}$。

The change in the free energy of A cell reaction, ΔG, is proportional to the electromotive force of the cell, ΔG:

一个电池反应的自由能变化ΔG与电池的电势V成正比，即：

$$\Delta G=-nF\left(V_k-V_A\right)<0 \tag{3-18}$$

It is consistent with the judgment result of Equation (3-15).

与式（3-15）判断结果相一致。

3.4 Potential-pH figure

3.4 电位-pH图

3.4.1 M. Pourbaix

3.4.1 布拜

M. Pourbaix was born in 1904. As a young student,

M.布拜生于1904年，年

Pourbaix was known for his diligence. In 1922, he finished high school with honors, won the special prize of the Belgian government, and entered the Department of Applied Science of the Free University of Brussels in the same year, graduating in 1927. Due to excellent academic performance, he has been a teaching assistant for one year. Since then, he has served in the Belgian Chemical Federation. In 1934 he returned to the Free University to teach. During this period of work, he was extensively exposed to various practical problems in the chemical industry, and served as an educator in general chemistry, electrochemistry, applied physical chemistry, analytical chemistry, metallurgy and other courses, which laid a solid foundation for his later career as a metal corrosion scientist.

In 1938, M. Pourbaix published the first potential-balance map in *Bulletin Soc. Chimaisn Belgique*. Then, in 1940, he submitted to the Free University of Brussels a doctoral dissertation entitled "Studies on the Corrosion of Iron-based Alloys", one of which was "the thermodynamics of dilute solutions-the role of pH and potential in equilibrium diagrams". Unfortunately, the paper was not appreciated by the Brussels paper Review Committee at that time, partly because of the international confusion of potential signs, which led to their misunderstanding of the paper. The Second World War delayed the propagation of potential-

布拜学术生涯

轻时的布拜是一位以勤奋著称的学生，1922年他以优异成绩完成了中学学习，获得了比利时政府的特种奖，同年进入布鲁塞尔自由大学应用科学系，1927年毕业。由于学习成绩优秀，曾留校任助教一年。此后，服务于比利时化学联合会。1934年重新回到布鲁塞尔自由大学任教。在这一段工作中，他曾广泛地接触各种化学工业中的实际问题，并担任过普通化学、电化学、应用物理化学、分析化学、冶金学等多种课程的教育工作，这些为他后来成为金属腐蚀科学家奠定了坚实的基础。

1938年，M.布拜发表了第一篇电位-pH平衡图（刊登在 *Bulletin Soc. Chimaisn Belgique* 杂志上）。1940年，他送交布鲁塞尔自由大学一篇博士论文，题目是《铁基合金的腐蚀研究》，此文包括三部分内容，其中之一是"稀溶液热力学-pH和电位在平衡图中的作用"。可惜这篇文章未被当时的布鲁塞尔论文审查委员会所理解，部分原因是当时国际上电位符号混乱而引起了他们对文章的误解。第二次世界

pH equilibrium maps.

Shortly after the war, on the recommendation of Professors F. E. C. Scheffer and W. G. Burgers, the paper above was submitted without modification to the Technical University of Delft in the Netherlands, for which he received a Doctor of Science degree, and soon a doctorate from the Free University of Brussels.

This article attracted the attention of the famous corrosion scientist U. R. Evans, he asked J.Agar to interpret the English version, and personally for this article made a long preface, published in 1949. Evans spoke highly of this creative paper, believing that this discovery contributed as much to science as the invention of Nernst,G. N. Lewis,Wagner and others[1].

大战推迟了电位-pH平衡图的传播。

战后不久，由F. E. C. Scheffer和W. G. Burgers教授推荐，上文未经任何修改提交至荷兰代尔夫特技术大学。为此，布拜获得了科学博士学位，不久又获得了布鲁塞尔自由大学的博士学位。

此文引起了著名腐蚀学家U. R. Evans的重视，他请J.Agar翻译成英文本，并亲自为此文作了长篇序言，于1949年出版。Evans高度评价这一创造性的论文，认为这一发现对科学的贡献可与Nernst，G. N. Lewis，Wagner等科学家的发明比肩[1]。

3.4.2 The principle of the potentio-pH diagram

Most of the electrochemical corrosion of metal occurs when the metal is in contact with aqueous solution. In addition to other ions in aqueous solution, and ions always exist, and the stability of metal in aqueous solution is not only related to its electrode potential, but also related to the value of aqueous solution. The value of equilibrium potential reflects the REDOX capacity of the substance and can be used to judge the possibility of electrochemical reaction. If the relationship between electrode potential and solution pH value of metal corrosion system is plotted, the possibility of corrosion reaction under given conditions can be

3.4.2 电位-pH图的原理

金属的电化学腐蚀绝大多数是金属与水溶液相接触时发生的腐蚀过程。水溶液中除其他离子外，H^+和OH^-离子始终存在，而金属在水溶液中的稳定性不但与它的电极电位有关，还与水溶液pH值有关。平衡电位的数值反映了物质的氧化还原能力，可用来判断电化学反应进行的可能性。若将金属腐蚀体系的电极电位与溶液pH值的关系绘成图，就能直接

directly judged from the diagram. This kind of graph, called potential-pH graph, was first proposed by Belgian scholar Pourbaix in 1938, also known as Pourbaix diagram. It is an electrochemical equilibrium diagram based on the principle of chemical thermodynamics, which involves temperature, pressure, composition, potential controlling electrode reaction and pH value affecting the dissolution and dissociation reaction in solution.

从图上判断在给定条件下发生腐蚀反应的可能性。该图称为电位–pH图，又称为布拜图。它是建立在化学热力学原理基础上的一种电化学平衡图，该图涉及温度、压力、成分、控制电极反应的电位，以及影响溶液中的溶解、解离反应的pH值。

The potential-pH diagram takes the solution value as the abscissa and the equilibrium electrode potential of the electrode reaction as the ordinate thermodynamic equilibrium diagram. According to the substances involved in the electrode reaction, the potential-pH map can be divided into three categories:

电位–pH图是以溶液pH值为横坐标，电极反应的平衡电极电位为纵坐标的热力学平衡图。根据参与电极反应的物质不同，可将电位–pH图分为以下三类：

(1) A class of reactions that depend only on the electrode potential and not on the pH of the solution.

（1）只与电极电位有关，与溶液pH值无关的一类反应。

(2) A class of reactions that depend on pH but not on electrode potential.

（2）与pH值有关，与电极电位无关的一类反应。

(3) A reaction that is related to both the electrode potential and the pH of the solution.

（3）既与电极电位有关，又与溶液pH值有关的反应。

3.4.3 Application of potentioal-pH maps

3.4.3 电位–pH图的应用

1. According to the “potential-pH diagram” to establish the theoretical corrosion, corrosion and passivation diagram

1.依据“电位–pH图”建立理论上的腐蚀、免蚀和钝化图

Figure 3-2 is potential pH diagram of system Fe-H_2O. The equilibrium and stable solid phases are Fe, Fe_3O_4 and Fe_2O_3. The numbers in the circles on the line represent the following reactions and equilibrium relations [2].

图3–2所示为Fe-H_2O体系电位–pH图，平衡稳定固相为Fe、Fe_3O_4和Fe_2O_3，直线上圆圈中的号码代表下列反应及平衡关系[2]。

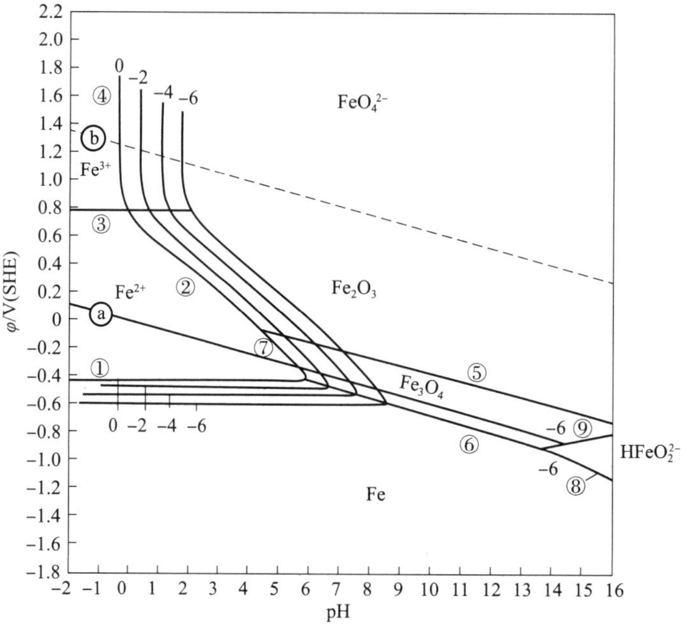

Figure 3-2 Potential pH diagram of system Fe-H_2O [2]

图 3-2 Fe-H_2O 体系电位 -pH 图 [2]

$$Fe = Fe^{2+} + 2e^- \quad (3\text{-}19)$$

① Line reaction:

That is, ϕ is pH independent, and when the activity of Fe^{2+} is given, a horizontal line is obtained. The line marked "0" represents $a_{Fe^{2+}} = 10^0$ (mol/L), the line labeled "−2" represents $a_{Fe^{2+}} = 10^{-2}$ (mol/L), and so on.

①线反应：

即 ϕ 与 pH 无关，当给定 Fe^{2+} 的活度时，得到一条水平线。标"0"的线表示 $a_{Fe^{2+}} = 10^0$（mol/L），标"−2"的线表示 $a_{Fe^{2+}} = 10^{-2}$（mol/L），依此类推。

② line reaction:

②线反应：

$$2Fe^{2+} + 3H_2O = Fe_2O_3 + 6H^+ + 2e^- \quad (3\text{-}20)$$

③ line reaction:

③线反应：

$$Fe^{2+} = Fe^{3+} + e^- \quad (3\text{-}21)$$

④ line hydrolysis reaction:

④线水解反应：

$$2Fe^{3+} + 3H_2O = Fe_2O_3 + 6H^+ \quad (3\text{-}22)$$

⑤ line reaction:

⑤线反应：

$$2Fe_3O_4 + H_2O = 3Fe_2O_3 + 2H^+ + 2e^- \quad (3\text{-}23)$$

⑥ line reaction:

⑥线反应：

$$3Fe + 4H_2O = Fe_3O_4 + 8H^+ + 8e^- \quad (3\text{-}24)$$

⑦ line reaction:

⑦线反应：

$$3Fe+4H_2O = Fe_3O_4+8H^++2e^- \quad (3\text{-}25)$$

⑧ line reaction:

⑧线反应：

$$Fe+2H_2O = HFeO_2^-+3H^++2e^- \quad (3\text{-}26)$$

⑨ line reaction:

⑨线反应：

$$3HFeO_2^-+H^+ = Fe_3O_4+2H_2O+2e^- \quad (3\text{-}27)$$

ⓐ line is the hydrogen evolution electrode reaction:

ⓐ线为析氢电极反应：

$$2H^++2e^- = H_2 \quad (3\text{-}28)$$

Below the ⓐ line, hydrogen precipitation reaction will occur.

在ⓐ线以下将发生氢气析出反应。

Line ⓑ shows the electrochemical reaction between O_2 and H_2O:

ⓑ线为O_2和H_2O之间的电化学反应：

$$2H_2O = O_2(g)+4H^++4e^- \quad (3\text{-}29)$$

This figure can be divided into three regions: corrosion zone, corrosion free zone and blunt zone.

如图3-3所示，可将此图分为3个区域：腐蚀区、免蚀区和钝态区。

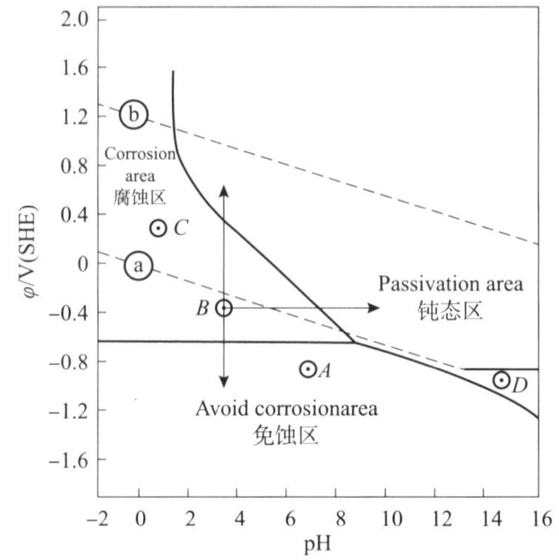

Figure 3-3 Simplified potential-diagram of system Fe-H_2O

图3-3 Fe-H_2O体系简化的电位-pH图

Corrosion zone: In the stable state, the soluble Fe^{2+}、Fe^{3+}、FeO_4^{2-} and $HFeO_2^-$ plasma metals are in the unstable state, so the metal may corrode.

腐蚀区：处于稳定状态的是可溶性的Fe^{2+}、Fe^{3+}、FeO_4^{2-}和$HFeO_2^-$等离子金属处于不稳定状态，因此金属可能发生腐蚀。

Blunt zone: The corrosion product is solid oxide, hydroxide or salt film. But whether it has a real sense of "passivation". Depends on the protection of the generated solid film.

钝态区：腐蚀产物为固态氧化物、氢氧化物或盐膜。但是否具有真正意义上的"钝化"作用，则取决于所生成的固态膜的保护性。

Corrosion-free zone: The metal in this zone is in a thermodynamically stable state and will not corrode.

免蚀区：在此区域内的金属处于热力学稳定状态，不会发生腐蚀。

As shown in Figure 3-3, point *A* is located in the no-corrosion zone and no corrosion will occur.

如图3–3所示：*A*点位于免蚀区，不会发生腐蚀。

Point *B* is located in the corrosion zone and below line ⓐ, which is the stable zone of Fe^{2+} and H_2. Therefore, the chemical reaction is as follows:

*B*点位于腐蚀区，且位于ⓐ线以下，是Fe^{2+}和H_2的稳定区，化学反应如下：

$$\text{阳极}: Fe \longrightarrow Fe^{2+} + 2e^- \tag{3–30}$$

$$\text{阴极}: 2H^+ + 2e^- \longrightarrow H_2 \tag{3–31}$$

$$\text{电池反应}: Fe + 2H^+ \longrightarrow F^{2+} + H_2 \tag{3–32}$$

If iron is at point *C*, not only in the corrosion zone but also on line ⓐ and line ⓑ, it is stable for $Fe^{2+}$and H_2O, so the chemical reaction is as follows:

若Fe处于点*C*，既在腐蚀区又在ⓐ线上ⓑ线下，则对于Fe^{2+}和H_2O是稳定的，化学反应如下：

$$\text{阳极}: Fe \longrightarrow Fe^{2+} +2e^- \tag{3–33}$$

$$\text{阴极}: 2H^+ + \frac{1}{2}O_2 +2e^- \longrightarrow H_2O \tag{3–34}$$

Point *D* corresponds to the region where Fe is corroded to form $HFeO_2^-$.

点*D*对应的是Fe被腐蚀，生成$HFeO_2^-$的区域。

In order to protect iron from corrosion, it can be moved out of the corrosion zone. Three possible ways for point *D* to move out of the corrosion zone are:

为使Fe免于腐蚀，可设法使其移出腐蚀区。点*D*移出腐蚀区的三种可能途径为：

① Cathodic protection is applied to iron by lowering its electrode potential to the no-corrosion zone. Sacrificial anode method is adopted, that is, iron is connected with zinc or aluminum alloy with low electrode potential to form corrosion galvanic couple.

①把Fe的电极电位降低至免蚀区，即对Fe施行阴极保护。采用牺牲阳极法，即将Fe与电极电位较低的锌或铝合金相连，构成腐蚀电偶；或将外

Or the negative end of the external DC power supply is connected to the iron and the positive end is connected to the auxiliary anode, forming a loop, both methods can protect the iron from corrosion.

加直流电源的负端与铁相连而正端与辅助阳极相连，构成回路，两种方法都可保护铁免遭腐蚀。

②Raise the potential of the iron into the passivation zone. This can be achieved by anodic protection or by adding anodic corrosion inhibitors or passivators to the solution.

②把铁的电位升高，使之进入钝化区。可通过阳极保护法或在溶液中添加阳极型缓蚀剂或钝化剂来实现。

③ Adjusting the pH of the solution to between 9 to 13 also allows iron to enter the passivation zone.

③将溶液pH值调整至9~13，也可使铁进入钝化区。

2. For the study of local corrosion principle

2.用于局部腐蚀原理的研究方面

The EQUILIBRIUM diagram can be used to estimate the properties of the medium, the effect of alloying elements, and the critical potential for local corrosion in the localized region of the occluded cell due to crevices, pitting and cracks. Figure 3-4 shows the theoretical corrosion protection, corrosion and passivation conditions of copper in water at 25℃. The equilibrium condition inside the copper pit is E=+270mV, pH=3.5. In water pH 8, the potential value of copper can vary from 0 to +600mV, depending on the oxygen content of the water and the surface conditions of the electrode. The protective potential against copper pitting is about +50 to +420mV, which is within the above potential value of copper (0 to +600mV).

应用平衡图可以估计在由于缝隙、点蚀和裂纹而形成的闭塞电池中的局部区域内介质的性质、合金元素的效应、发生局部腐蚀的临界电位等。图3-4所示为铜在水中于25℃时的理论免蚀、腐蚀与钝化条件。在铜的蚀坑内部的平衡条件为E=+270mV，pH=3.5。在pH为8的水中，铜的电位值可在0~+600mV变化，这依赖于水中的氧含量和电极的表面条件。而防止铜点蚀的保护电位为+50~+420mV，这一电位处于上述铜的电位值（0~+600mV）之间。

3. Application in metallurgical industry

3.在冶金行业中的应用

Hydrometallurgy is one of the most important methods in the process of non-ferrous metal metallurgy. In recent years, due to its unique advantages, such as no flue gas and low energy consumption, it has been

湿法冶金是有色金属冶金过程中非常重要的方法之一，近年来，由于其特有的优点，如不产生烟气、能耗较低等，

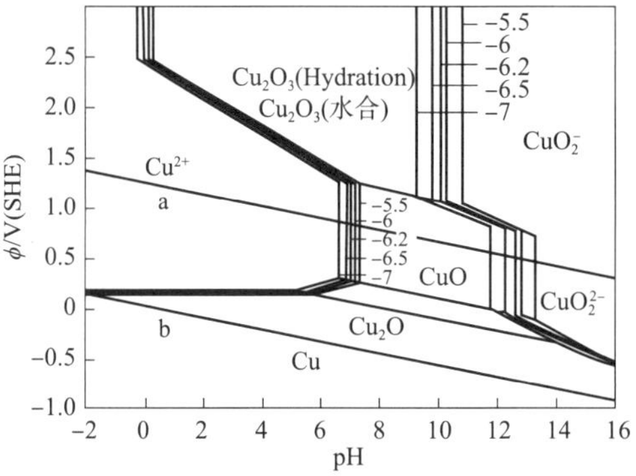

Figure 3-4 Simplified potentio-pH diagram of Cu-water system with Cu、CuO、Cu_2O and Cu_2O_3 as the equilibrium solid phase

图 3-4 以 Cu、CuO、Cu_2O 和 Cu_2O_3 为平衡固相的铜水体系的简化电位 -pH 图

gradually applied to the treatment of non-ferrous metal sulfide ore. In order to improve the reaction speed, leaching rate and reduce environmental pollution, high temperature pressure leaching method came into being, especially the treatment of metal sulfide ore by pressure acid leaching. For the oxidation leaching of zinc sulfide ore and copper sulfide ore with relatively single composition at high temperature and pressure, many scholars have carried out technological research and achieved high zinc and copper leaching rates, and some of the results have been applied to industrial practice. On the thermodynamics of various single sulfide, many scholars have done some research. For the study of the pressure leaching of complex polymetallic sulfide ore, the potentio-pH diagram of several typical sulfide at 25℃ and 150 ℃ was drawn by thermodynamic calculation method. It provides a certain thermodynamic basis for direct high-temperature pressure oxidation leaching of complex polymetallic sulfide concentrates containing copper, lead, zinc and iron. Figure 3-5 shows the ZnS-H_2O system φ-pH at 25℃ and 150℃.

逐渐被应用于有色金属硫化矿的处理。为提高反应速度、浸出率及减少环境污染，高温加压浸出法应运而生，尤其是加压酸浸对金属硫化矿的处理。对于成分相对单一的硫化锌矿、硫化铜矿等的高温加压氧化浸出，许多学者已进行了工艺方面的研究，并取得了较高的锌、铜浸出率，部分成果已应用到工业实践中。关于各种单一硫化物的热力学，许多学者进行了研究。对于复杂多金属硫化矿加压浸出的研究也通过热力学计算的方法绘制了在25℃和150℃时几种典型硫化物的电位-pH图，为含铜、铅、锌、铁的复杂多金属硫化物精矿直接高温加压氧化浸出提供一定的热力学依据。图3-5所示为25℃和150℃时ZnS-H_2O系φ-pH。

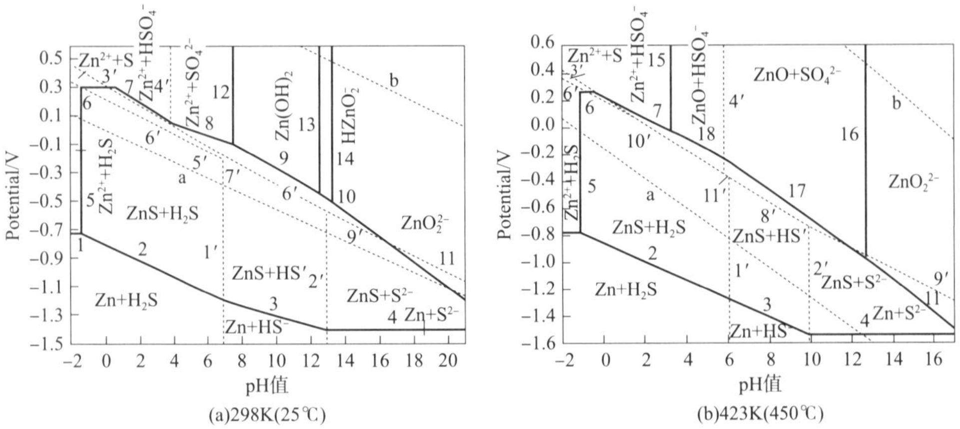

Figure 3-5 φ-pH diagram of ZnS-H_2O system at different temperatures

图 3-5 不同温度时 ZnS-H_2O 系 φ-pH 图

According to Figure 3-5, zinc mainly exists in the form of ZnS, Zn^{2+} and $HZnO_2^-$ in the stable region of water. During the leaching of sphalerite, the acidity of the aqueous solution system is very high, and the pH value is about −1. Therefore, in order to improve the possibility of the leaching process, the leaching of sphalerite can be realized by raising the potential (adding oxygen pressure) and increasing the temperature. According to Figure 3-5(b), when the oxygen pressure is high and the temperature is 150℃, the temperature of pH&lt is high. Under the condition of 3, sphalerite leaching can be realized. From the perspective of thermodynamics, sphalerite can be oxidized by oxygen almost in the whole range of pH value, and the trend of oxidation is determined by the potential difference between oxygen electrode and sulfide electrode.

由图 3－5 可知：在水的稳定区域内，锌主要以ZnS、Zn^{2+}、$HZnO_2^-$等形式存在。闪锌矿浸出时对水溶液体系的酸度要求很高，pH值约为−1。因此，为提高浸出过程发生的可能性，可采用提高电位（加氧压）和升温的方式实现闪锌矿浸出。由图3－5（b）可知：当氧压较高，温度为150℃时，在pH<3的条件下，就能实现闪锌矿浸出。对于热力学来说，闪锌矿几乎在整个pH值的范围内都能被氧化，被氧化的趋势决定于氧电极与硫化物电极之间的电位差。

4. Application of electrochemical protection to corrosion

4.对腐蚀进行电化学保护方面的应用

The equilibrium diagram can be used to evaluate the possibility of applying cathodic protection to a metal or alloy and the selection of protection conditions. The

平衡图可用来对金属或合金施加阴极保护的可能性和保护条件的选择方面进行评价，

metal can be protected by anodic polarization, cathodic polarization or alkalization in different ways.

5. Other

In addition to the above aspects, the equilibrium diagram can also be used for practical corrosive media (multi-system equilibrium diagram, concentrated solution equilibrium diagram, etc.), atmospheric corrosion, corrosion caused by microorganisms. The selection of acid pickling and chemical etching conditions of metals, the study of corrosion and corrosion inhibitor pigments under organic film.

布拜图在生物方面的应用

In conclusion, more and more research work has shown that the "potential-pH" diagram is an extremely useful tool. However, since it is based on the basic principles and data of thermodynamics, several problems must be paid attention to when drawing, studying and applying these diagrams.

First of all, it is only in a dilute solution where soluble ions dominate that activity can be assumed to be actually equal to the volumetric molar concentration. Therefore, when these figures are used to study concentrated solutions, they must be corrected by activity instead of concentration. At the same time, the electrochemical properties of the metal phase immersed in the electrolyte solution mainly depend on the composition of the liquid layer directly in contact with the phase, which is often greatly different from the composition or initial composition of the solution due to the difference of the electrode process and the speed of the material migration process. The same is true for phases formed by metallic compounds. Therefore, in the

可借不同方法阳极极化、阴极极化或碱化使金属达到保护。

5.其他

除上述几方面外，平衡图还可用于实际腐蚀介质（多元系平衡图和浓溶液平衡图等）、大气腐蚀，由微生物引起的腐蚀，以及金属酸洗和化学浸蚀条件的选择、有机膜层下腐蚀和缓蚀性颜料等方面的研究。

总之，越来越多的研究表明，“电位-pH”图是一个极为有用的工具。但由于它是以热力学基本原理和数据为基础而建立的，因此在绘制、研究及应用电位-pH图时，必须注意以下几个方面的问题。

首先，只有在一种可溶性离子处于支配地位的稀溶液中，才可以假设活度实际上等于物质的量浓度。因此，在使用电位-pH图研究浓溶液时，必须用活度代替浓度加以校正。同时，浸在电解质溶液中金属相的电化学性质主要取决于直接与该相接触液层的组成，这一组成却因电极过程和物质迁移过程速度的不同，而与溶液本体的组成或初始组成有极大差异。对于由金属化合物形成的相也是如此。所以，

application of "potential -pH" diagram, must be taken into account.

在应用"电位-pH"图时，必须加以考虑。

Secondly, if the metal and other substances such as cyanide, ammonia, etc. formed a highly soluble and stable complex, we should take into account the equilibrium conditions of these complexes, the metal and water binary system of equilibrium diagram changed to a ternary or multivariate system of equilibrium diagram, to be modified.

其次，若金属其他物质如氰化物、氨水等形成了易溶的高度稳定的络合物时，我们就应该考虑这些络合物的平衡条件，将金属与水的二元系平衡图改绘为三元系平衡图或多元系平衡图，加以修正。

In fact, the formation of the surface layer is the most important factor affecting the electrochemical reaction. The solid compounds oxides, hydrides and basic salts formed by different metals and water may be the most common and important passivation layers formed by metals and water. According to the thermodynamic stability conditions of these compounds, it is possible to speculate the range of passivation zone.

实际上，在进行电化学反应时，金属相表面层的形成是影响反应最重要的因素。而不同金属与水形成的固态化合物氧化物、氢化物、碱式盐可能是金属与水形成的最常见的和最重要的钝化层，根据这些化合物的热力学稳定条件，我们有可能推测出钝化区的范围。

However, we must realize that the theoretical and applied research of electrochemistry can be perfected only by considering the three factors of thermodynamics, kinetics and structure simultaneously. In kinetic studies, special attention should be paid to the influence of anions and those metal ions with slow exchange rate on the reaction kinetics. It should be taken into account that in corrosion systems, often several reactions occur simultaneously, including extremely complex and special kinetic phenomena. We must complement the information obtained from the equilibrium diagram with knowledge of these dynamics. As for the structure of the research such as application of light diffraction, electron diffraction, energy spectrum analysis and other methods to study can help us to find theoretical and actual of

然而，我们必须认识到，只有同时考虑热力学、动力学和结构上的这三方面因素，才能使电化学的理论研究和应用研究得以完善。在动力学研究中，要特别注意阴离子以及那些交换速度缓慢的金属离子对反应动力学的影响。要考虑在腐蚀体系中，往往是几个反应同时发生，包括了极其复杂和特殊的动力学现象。我们必须用这些动力学方面的知识对从平衡图得到的资料加以补充。至于结构上的研究如应用光衍射、电子衍射、能谱分析等方

facies of soil, the deviation between the analysis, can be anodized or the nature of the passive film, understanding of the selective corrosion process, access to the study of electrochemical process of a parameter, for metal protection measures should be taken.

法进行的研究则可以帮助我们发现理论上的相与实际中的相之间的偏差，分析、预测电镀膜或钝化膜的性质，认识腐蚀过程的选择性，获取所研究的电化学过程的个别参数，得到对金属应采取的保护措施。

Finally, it must be noted that "potential-pH" maps drawn from pure metals and non-metals do not provide us with direct information about the properties of alloys. On the one hand, the corrosion behavior of an alloy is an extremely complex synthesis of its component metal properties. On the other hand, the thermodynamic properties of the corrosion products of the metal forming the alloy provide us with the possibility to test the passivation conditions of the alloy. All these need to be considered comprehensively.

最后还必须指出，由纯金属和非金属绘制出的“电位-pH”图，不能直接提供关于合金性能的资料。一方面，因为合金的腐蚀性能是其组分金属性质极其复杂的综合表现；另一方面，构成合金的金属腐蚀产物的热力学性质，又提供了预测合金钝化条件的可能性。这些都需要全面予以考虑。

3.5 References

[1] YANG X Z, YANG W. Distinguished corrosologist M. Bubye and the Belgian Centre for Corrosion Research (CEBELCOR) [J]. Corrosion and Protection, 1980: 48-53.

[2] SUN Y, HU J. Metal Corrosion and Control [M]. Harbin: Harbin Institute of Techonlogy press, 2003.

3.5 参考文献

[1] 杨熙珍，杨武．杰出的腐蚀学家 M. 布拜和比利时腐蚀研究中心（CEBELCOR）[J]．腐蚀与防护，1980：48-53.

[2] 孙跃，胡津．金属腐蚀与控制 [M]．哈尔滨：哈尔滨工业大学出版社，2003.

4 Electrochemical Corrosion

电化学腐蚀

4.1 Corroded battery

The corrosion phenomenon of metals in electrolyte solution is an electrochemical corrosion process.

The three types of corrosion in natural environments-the electrochemical corrosion process of the atmosphere, soil and seawater are inseparable from the anodic dissolution of metal at the metal/electrolyte interface, accompanied by the reduction of certain substances in the solution on the metal surface.Therefore, the cause of electrochemical corrosion of metals is the result of a spontaneous, short-circuited corrosion galvanic action [1,2].

4.1.1 Composition of corrosion cells

After immersing the zinc sheet and the copper sheet in the dilute sulfuric acid solution, connect them with wires (Figure. 4-1). Due to the lower potential of zinc

4.1 腐蚀电池

金属在电解质溶液中的腐蚀现象是一个电化学腐蚀过程。

三类自然环境下的腐蚀——大气、土壤、海水的电化学腐蚀过程都离不开金属/电解质界面上金属的阳极溶解，同时伴随着溶液中某些物质在金属表面的还原。因此，金属发生电化学腐蚀的原因是一种自发的、短路的腐蚀原电池作用的结果[1,2]。

4.1.1 腐蚀电池的构成

将锌片和铜片浸入稀硫酸溶液中后，再用导线把它们连接起来（图4-1），由于锌的电

and the higher potential of copper, they are established on the electrode/solution interface. The electrode equilibration process is disrupted, and the following electrode reactions are carried out on the two electrodes respectively

势较低，铜的电势较高，它们各自在电极/溶液界面上建立的电极平衡过程遭到破坏，并在两个电极上分别进行以下的电极反应。

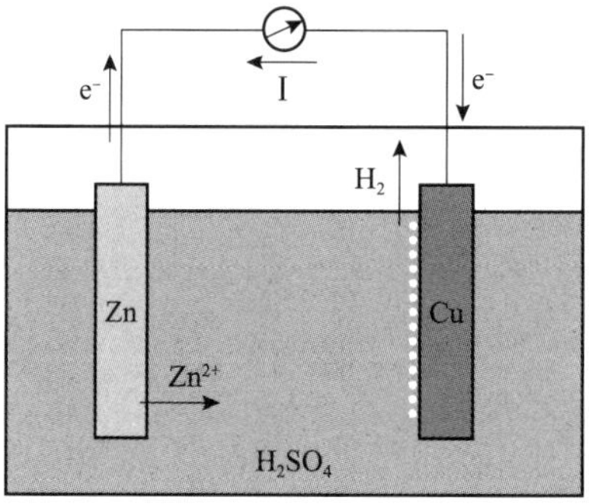

Figure 4-1 Copper-zinc battery
图 4-1 铜锌电池

On the zinc electrode, the metal loses electrons and is oxidized:

锌电极上，金属失去电子被氧化：

$$Zn \longrightarrow Zn^{2+}+2e^- \quad (4-1)$$

On the copper electrode, the hydrogen ions in the dilute sulfuric acid solution get electrons and are reduced:

铜电极上，稀硫酸溶液中的氢离子得到电子被还原：

$$2H^++2e^- \longrightarrow H_2\uparrow \quad (4-2)$$

The whole electrode reaction is

整个电极反应为：

$$Zn+2H^+ \longrightarrow Zn^{2+}+H_2\uparrow \quad (4-3)$$

It can be seen that after the copper-zinc battery is turned on, due to the dissolution of the zinc sheet, the electrons flow along the wire to the copper sheet, that is, the current flows from the copper sheet to the zinc sheet.

可见，铜锌电池接通后，由于锌片溶解，电子沿导线流向铜片，即电流由铜片流向锌片。

Corrosion battery is an irreversible battery, and it can not convert chemical energy into electrical energy, and the chemical energy released by the oxidation reaction is completely dissipated as thermal energy and canot be converted into electrical work.

腐蚀电池是一种不可逆电池，也不能将化学能转变为电能，氧化还原反应释放出的化学能全部以热能方式散发，而不能转化为电功。

As shown in Figure 4-2, The zinc sheet and the copper sheet are the cathode and anode of the battery.

如图4-2所示，锌片与铜片为电池的阴、阳两极。锌片

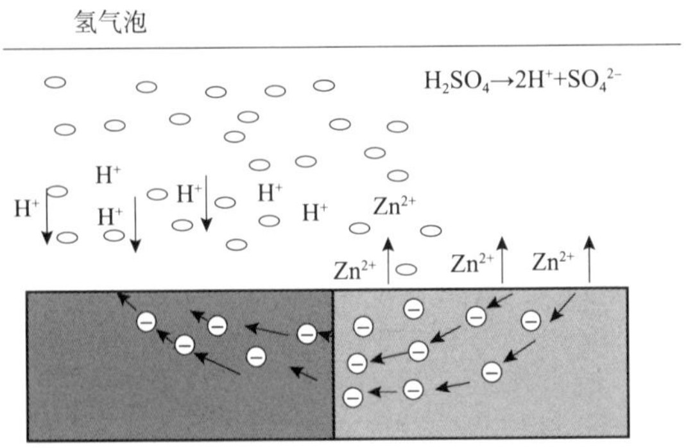

Figure 4-2 Dissolution of copper-zinc in sulfuric acid

图 4-2 铜 - 锌在硫酸中溶解

The remaining electrons provided by the dissolution of the zinc flakes flow through the copper in direct contact with it, and are accepted on the copper surface by hydrogen ions in solution, where the hydrogen gas forms and escapes. Zinc will continue to dissolve as long as the hydrogen ions are continuously reduced. Batteries like this are called corrosion galvanic cells or corrosion cells.

溶解所提供的剩余电子流过与它直接接触的铜，并在铜表面被溶液中的氢离子所接受，氢气便在铜表面形成并逸出。只要氢离子能够不断被还原，锌就会继续溶解，类似这样的电池称为腐蚀原电池或腐蚀电池。

For the macroscopic voltaic corrosion cell, two metals with different electrode potentials are in contact with each other and put into the electrolyte solution at the same time, forming a corrosion cell. A corrosion battery must include four parts: anode, cathode, electrolyte solution and circuit, which constitute the three links of corrosion battery work. Figure 4-3 shows the schematic diagram of the electrochemical corrosion process.

对于宏观的伏特腐蚀原电池而言，两种电极电位不同的金属互相接触并同时放入电解质溶液中，就组成一个腐蚀电池。一个腐蚀电池必须包括阳极、阴极、电解质溶液和电路四个部分，由此构成腐蚀电池工作的三个环节。电化学腐蚀过程示意如图 4-3 所示。

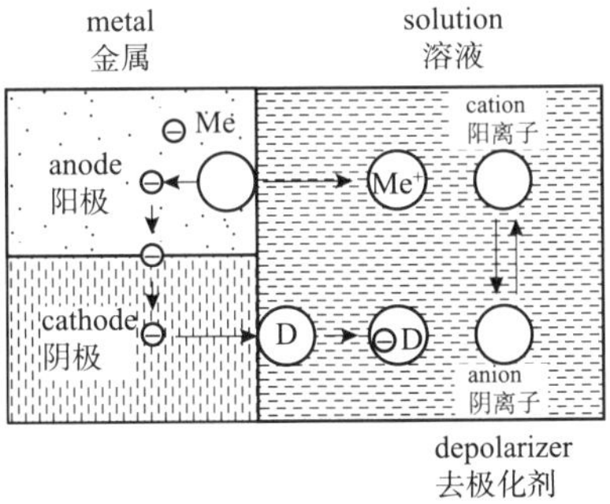

Figure 4-3 Schematic diagram of electrochemical corrosion process

图 4-3 电化学腐蚀过程示意

1. Anode process

After the metal is dissolved, it is transferred into the solution in the form of metal ions or hydrated ions, while leaving the same amount of electrons on the substrate.

1.阳极过程

金属溶解后，以金属离子或水化离子形式转入溶液，同时将等量电子留在基体上。

$$M \longrightarrow M^{n+} + ne^- \quad (4\text{-}4)$$

2. Cathodic process

Electrons transferred from the anode are absorbed by electron-accepting substances (oxidizing substances) in the electrolyte solution. Electron acceptors in metal corrosion solutions are also. known as depolarizers.

2.阴极过程

从阳极迁移来的电子被电解质溶液中能够接受电子的物质（氧化性物质）所吸收。金属腐蚀溶液中的电子受体也称去极化剂。

$$D + ne^- \longrightarrow \left[D \cdot ne^-\right] \quad (4\text{-}5)$$

3. Flow of electrons

The flow of current in metals depends on the flow of electrons from the anode to the cathode, and in the solution, it depends on the migration of ions, that is, the movement of cations from the anode area to the cathode area, and the migration of anions from the cathode area to the anode area, so that the current in the electrode system constitutes a path.

3.电子的流动

电子的流动在金属中依靠电子从阳极流向阴极，在溶液中则依靠离子迁移，即阳离子从阳极区向阴极区移动，阴离子从阴极区向阳极区迁移，使电极系统中的电流构成通路。

The three links of the corrosion battery are not only independent of each other, but also closely related and interdependent. If one of the links is blocked and stops working, the entire corrosion process stops. It should be pointed out that the spatial separation of cathodic and anodic reactions caused by the electrochemical inhomogeneity of the corrosion system is not the only cause of electrochemical corrosion. In fact, in many corrosion cases, cathodic and anodic processes can also alternate with time on the same metal surface. The only cause of electrochemical corrosion of metals is the presence of oxidants (ie depolarizers) in solution that can

腐蚀电池的三个环节既相互独立又彼此紧密联系和相互依存。其中一个环节受到阻碍停止工作，则整个腐蚀过程就停止。应该指出，腐蚀体系的电化学不均匀性所造成的阴阳极反应的空间分隔并不是发生电化学腐蚀的唯一原因。实际上，在许多腐蚀情况下，阴、阳极过程也可在同一金属表面随着时间变化相互交替地进行。使金属发生电化学腐蚀的唯一

accept excess electrons on the metal surface. Without these substances, even if there are different cathode and anode regions on the metal surface, it will not occur Electrochemical corrosion.

原因是溶液中存在能够接受金属表面过剩电子的氧化剂（去极化剂），如果没有这些物质，即使金属表面有不同的阴、阳极区，也不会发生电化学腐蚀。

4. Types of corrosion batteries

4.腐蚀电池的类型

According to the electrode size of the corrosion cell and the stability of the distribution of the anode and cathode regions over time, as well as the influencing factors and the characteristics of corrosion damage that promote the formation of the corrosion cell, the corrosion cell is divided into three categories: macroscopic corrosion cells, microscopic corrosion cells and submicroscopic corrosion cells.

根据腐蚀电池的电极尺寸大小与阳极区和阴极区分布随着时间的稳定性，以及促使形成腐蚀电池的影响因素和腐蚀破坏的特征，即将腐蚀电池分为三类：宏观腐蚀电池、微观腐蚀电池和亚微观腐蚀电池。

4.1.2 Macroscopic corrosion cell

4.1.2 宏观腐蚀电池

1. Galvanic corrosion

1.接触腐蚀

contact corrosion also known as galvanic corrosion. Two or more metals are in contact and form a corrosion cell in an electrolyte solution. In the actual metal structure, there are often different metals in contact, and it can be observed that the metal with a lower electrode potential (anode) corrodes faster, while the metal with a higher electrode potential (cathode) corrodes slowly and even gets protected. The greater the potential difference between the two metal electrodes constituting the dissimilar metal battery, the more serious the local corrosion. This corrosion damage is called galvanic corrosion. Figure 4-4 Daniel battery: $Zn|ZnSO_4||CuSO_4|Cu$

接触腐蚀又称电偶腐蚀。两种或两种以上的金属相互接触，在电解质溶液中构成的腐蚀电池。在实际金属结构中常常有不同金属相互接触，可观察到电极电势较低的金属（阳极）腐蚀加快，而电极电势较高的金属（阴极）腐蚀减慢，甚至得到保护。构成异金属电池的两种金属电极电势相差越大，局部腐蚀越严重。这种腐蚀破坏称为接触腐蚀。图4-4所示为丹聂尔电池：$Zn|ZnSO_4||CuSO_4|Cu$。

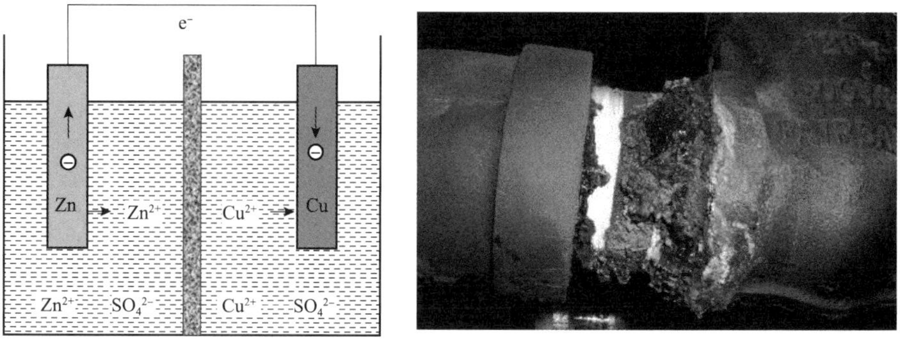

Figure 4-4 Galvanic corrosion
图 4-4 接触腐蚀

The electrode potential of Zn is lower than that of Cu. In this corrosion cell, Zn is the anode and Cu is the cathode. Electrons are transferred from the zinc plate to the copper plate. The following reactions take place on the two electrodes respectively.

Zn的电极电势比Cu的电极电势低，在这个腐蚀电池中Zn是阳极，Cu是阴极，电子从锌板转移到铜板上。在两个电极上分别发生以下反应。

Anode: 阳极：

$$Zn = Zn^{2+} + 2e^- \quad (4\text{-}6)$$

Cathode: 阴极：

$$Cu^{2+} + 2e^- = Cu \quad (4\text{-}7)$$

electrode reaction: 电极反应：

$$Cu^{2+} + Zn = Cu + Zn^{2+} \quad (4\text{-}8)$$

2. Concentration battery

2.浓差电池

Corrosion batteries composed of the same metal in contact with electrolyte solutions of different compositions can be further divided into differential gas-filled batteries (oxygen concentration batteries) and salt concentration batteries[3].

同一金属与不同成分的电解质溶液相接触构成的腐蚀电池，还可进一步分为差异充气电池（氧浓差电池）和盐浓差电池[3]。

Differentially gas-charged batteries are a ubiquitous and harmful destructive factor that causes corrosion. In general, this kind of battery is formed after the metal is in contact with the medium with different oxygen content, also known as the oxygen concentration battery. For example, two iron pieces are immersed in the same concentration of dilute salt water, one of the iron pieces is in contact with the air, and the other is in contact

差异充气电池是一种普遍存在、危害性很大的引起腐蚀的破坏因素。一般情况下，这种电池是由于金属与含氧量不同的介质接触后形成的，也称为氧浓差电池。例如，两块铁片浸在同一浓度的稀盐水中，其中一块铁片与通入的空气接

with the nitrogen gas (Figure 4-5). The negative charge density on the two iron pieces It should be the same, but due to the different oxygen levels in the solution, the following reactions occur to different degrees.

触，另一块铁片与通入的氮气接触（图 4－5）。在这两块铁片上负电荷密度本来应该是相同的，但由于溶液中含氧量不同，在发生下列反应时所进行的程度也不同。

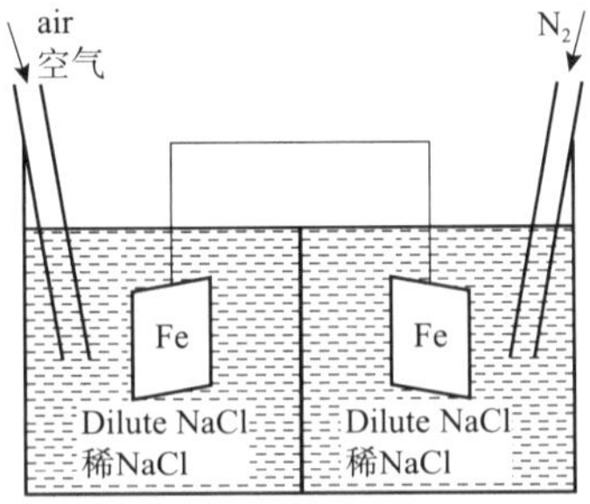

Figure 4−5 Oxygen concentration battery

图 4−5 氧浓差电池

On the iron electrode fed with air, the reduction reaction rate of oxygen is faster, so the surface electrons are neutralized, and the negative charge density is greatly reduced. On the iron electrode where nitrogen gas was introduced to reduce oxygen gas, the oxygen reduction reaction was difficult to proceed, so that the negative charge density did not change much. In this way, a potential difference is created due to the difference in oxygen concentration in the solution, resulting in a corrosion current. In the above battery, the iron pieces in contact with nitrogen gas were dissolved, and the iron pieces in contact with air were not corroded.

在通入空气的铁电极上，氧的还原反应速度较快，因而表面电子被中和，负电荷密度降低较多。在通入氮气而使氧气较少的铁电极上，氧还原反应难于进行，因此负电荷密度没有什么变化。这样，由于溶液中氧浓度的不同造成了电势差，产生了腐蚀电流。在上述电池中与氮气接触的铁片发生溶解，与空气接触的铁片没有腐蚀。

Common oxygen concentration corrosion cells are caused by the contact of various parts of the metal with solutions with different oxygen contents. For example, the most easily corroded part of an iron bucket containing water is the area below the waterline. Above the waterline, the metal iron is in direct contact with the air,

常见的氧浓差腐蚀电池是由于金属各部分与不同含氧量的溶液相接触而引起的。例如，盛水的铁桶最易腐蚀的部位是水线下面的区域，在水线上面金属铁直接与空气接触，

and the oxygen content in the water film on the surface is high. The solubility of oxygen in the solution below the waterline is limited, and the diffusivity of oxygen in the aqueous solution is limited, and the oxygen content on the iron surface is much lower than the oxygen content on the waterline, which results in an oxygen concentration battery. The metal iron becomes the anode and is corroded. Due to the influence of the solution resistance, the corrosion near the waterline is more serious, so it is called waterline corrosion. Metal pipes or equipment buried in soils of different densities or depths can also cause corrosion of oxygen concentration cells due to uneven aeration of oxygen in the soil.

其表面水膜中的含氧量较高。水线以下氧在溶液中的溶解度有限，且氧在水溶液中的扩散能力有限，铁表面的氧含量比水线上的氧含量要低得多。这样就产生了氧浓差电池，水线下的金属铁成为阳极而遭到腐蚀。由于溶液电阻的影响，水线附近的腐蚀较为严重，故称水线腐蚀。埋在不同密度或深度的土壤中的金属管道或设备，也会因土壤中氧的充气不均匀而造成氧浓差电池的腐蚀。

In engineering parts, riveting, welding, screwing and other methods are commonly used to connect, and the deep gap formed at the joint will cause serious local damage due to the difficulty of oxygen supply. Therefore, crevice corrosion is also caused by oxygen concentration cells.

在工程部件中常用铆、焊、螺钉等方法连接，在其连接处形成的缝隙深处因供氧困难导致严重的局部破坏。所以，缝隙腐蚀也是氧浓差电池造成的。

Salt concentration battery is another form of concentration battery. For example, both ends of a long copper rod are in contact with a dilute copper sulfate solution and a concentrated copper sulfate solution. According to the Nernst formula, the electrode potential of the part in contact with the dilute copper sulfate is lower, which is the anode, and the electrode potential of the other end in the concentrated copper sulfate solution is higher, which is the cathode, which constitutes a copper salt concentration battery. H^+ ions are precipitated at one end of the copper rod cathode, and the other end of the copper rod will dissolve. When the copper sulfate concentration in the solution at both ends of the copper rod is equal, the electrode potentials on the two copper

盐浓差电池是浓差电池的另一种形式，如一长铜棒两端分别接触稀硫酸铜溶液和浓硫酸铜溶液。根据能斯特公式可知：与稀硫酸铜相接触的部位其电极电势较低，为阳极，浓硫酸铜溶液中另一端的电极电势较高，为阴极，构成了铜盐浓差电池。H^+离子在铜棒阴极一端析出，铜棒另一端将发生溶解。铜棒两端所处溶液中的硫酸铜浓度相等时，两个铜电极上的电极电势相等，电池电势等于0，不再有腐蚀电流流过。

electrodes are equal, the battery electromotive force is equal to zero, and no corrosion current flows.

3. Thermoelectric battery

The various parts of the metal immersed in the electrolyte solution may form a thermoelectric battery due to different temperatures. For example, when inspecting the heat exchanger made of carbon steel, it was found that the corrosion at the high temperature end is more serious than that at the low temperature end, because the electrode potential of the carbon steel at the high temperature part is lower than that at the low temperature part, which becomes the anode of the corrosion battery. However, the electrode behavior of copper, aluminum, etc. at different temperatures in the relevant solution is opposite to that of carbon steel. For example, in a copper sulfate solution, the low temperature end of the copper electrode is the anode, and the high temperature end is the cathode.

3.温差电池

浸入电解质溶液中的金属各部分由于温度不同可能形成温差电池。例如，在检查碳钢制成的换热器时，发现其高温端腐蚀比低温端严重，这是因为高温部位碳钢的电极电势比低温部位低而成为腐蚀电池的阳极。可是，铜、铝等在有关溶液中不同温度下的电极行为与碳钢却相反。如在硫酸铜溶液中，铜电极低温端是阳极，高温端是阴极。

4.1.3　Microscopic corrosion cells

The size of the electrode in this type of battery can be similar to the grain size (0.1~1mm), and the polarity of the electrode can be unchanged for a period of time during the corrosion process. This can of course cause localized corrosion on the microscopic scale (e.g. pitting, intergranular corrosion), but macroscopically the corrosion damage can be considered uniform. There are many reasons for the electrochemical inhomogeneity of metal surfaces, which can generally be summarized as the following situations.

4.1.3　微观腐蚀电池

微观腐蚀电池中电极尺寸可与晶粒尺度相近（0.1~1mm），电极的极性在腐蚀过程中能够在一段时间内不变。这当然可引起微观范围内的局部腐蚀（如点蚀、晶间腐蚀），但宏观上可认为腐蚀破坏是均匀的。引起金属表面电化学不均匀性的原因很多，一般可归纳为以下几种情况。

Inhomogeneity of metal chemical composition: Generally, industrial pure metals often contain various

金属化学成分的不均匀性：一般工业纯金属中常常含

impurities. When they are in contact with the electrolyte solution, the impurities in the metal will form many short-circuit microbatteries with the base metal in the form of microelectrodes. If the impurities have cathodic properties, they will accelerate the corrosion of the base metal.

有各种杂质，当它们与电解质溶液接触时，金属中的杂质将以微电极的形式与基体金属构成许多短路微电池。若杂质具有阴极性质时，将加速基体金属的腐蚀。

Inhomogeneity of metal structure: Most of the metal materials currently used are polycrystalline, so there are a large number of grain boundaries on the surface of the material. The metal grain boundary is an area where the atomic arrangement is relatively loose and disordered, and it is easy to enrich impurity atoms and produce grain boundary precipitation, and the density of crystal defects here is high, such as dislocations, holes and distortion. Therefore, the electrode potential of the grain boundary is usually lower than that of the inside of the grain, and it becomes the anodic region preferentially dissolved in the corrosive medium.

金属组织的不均匀性：目前所使用的金属材料大多数是多晶组织，因此材料表面存在大量的晶界。金属晶界是原子排列较为疏松而紊乱的区域，容易富集杂质原子和产生晶界沉淀，而且此处晶体缺陷密度大，如位错、空穴和畸变。因而，晶界的电极电势通常比晶粒内部的电极电势低，在腐蚀介质中成为优先溶解的阳极区。

Inhomogeneity of the physical state of the metal: In the process of metal machining or assembly, uneven deformation of various parts of the metal and uneven internal stress are often caused. In general, the part with larger deformation is called the anode.

金属物理状态的不均匀性：金属机械加工或装配过程中常常会造成金属各部分变形不均匀及内应力的不均匀。一般情况下，变形较大的部分称为阳极。

Incompleteness of the metal surface film (or surface coating): If the metal surface film and the surface coating have pores or damage, a potential difference will be generated between the surface film (or surface coating) and the metal substrate under the pores and damage, Constitutes microscopic corrosion cells. In most cases, the potential of the metal phase at the pores or breakage is lower than that of the surface film and becomes the anode of the microbattery. Stainless steel in the medium containing Cr, due to the strong destructive effect of Cr

金属表面膜（或表面涂层）的不完整性：金属表面膜和表面涂层如果存在孔隙或破损，则在表面膜（或表面涂层）与孔隙、破损下的金属基体之间产生电势差，构成微观腐蚀电池。多数情况下，孔隙或破损处金属相的电势比表面膜低，成为微电池阳极。不锈钢在含有Cr的介质中，由于Cr对

on the passivation film, accelerates the destruction of the film at the weak point of the film and causes pitting corrosion.

钝化膜的强烈破坏作用，在膜的薄弱处加速膜的破坏而造成点蚀。

4.1.4 Submicroscopic corrosion cells

4.1.4 亚微观腐蚀电池

Submicroscopic corrosion cells usually refer to corrosion cells in which it is difficult to distinguish the cathode and anode with the naked eye and ordinary microscope. The surface area of each electrode is very small (100~1000nm) and spreads over the entire metal surface. The anodic and cathodic regions are constantly changing over time. The destruction of submicroscopic cells results in a uniform corrosion. Therefore, the entire surface of the metal that is undergoing corrosion can be regarded as an electrode undergoing anodic and cathodic reactions at the same time, that is, the entire surface is both an anode and a cathode. This can be explained by various differences in the surface crystal structure. Stress distribution is also an important reason for the formation of submicroscopic cells. Macroscopic corrosion cells and microscopic corrosion often coexist in practical metal corrosion.

亚微观腐蚀电池通常是指用肉眼和普通显微镜都难以分辨出阴、阳极的腐蚀电池。每个电极表面积十分微小（100~1000nm），遍及整个金属表面。阳极区和阴极区随着时间不断变化。亚微观腐蚀电池的破坏结果是一种均匀腐蚀。因此可以把这种正在进行腐蚀的金属整个表面看作同时进行阳极反应和阴极反应的一个电极，即整个表面既是阳极，又是阴极。可从表面晶体结构的各类差异作出解释。应力分布也是形成亚微观腐蚀电池的一个重要原因。在实际金属腐蚀中宏观腐蚀电池和微观腐蚀常常同时存在。

The difference and connection between primary battery and corrosion battery:

原电池与腐蚀电池的区别与联系：

The redox reaction of the primary battery is carried out separately, the positive electrode of the primary battery corresponds to the cathode of the corrosion battery, and the negative electrode of the primary battery corresponds to the anode of the corrosion battery, that is, anodic oxidation and cathodic reduction. Corrosion cells, on the other hand, react simultaneously at the same location where corrosion occurs. As shown in Figure

原电池的氧化还原反应是分开进行的，原电池正极对应腐蚀电池的阴极，原电池负极对应腐蚀电池的阳极，即阳极氧化，阴极还原。而腐蚀电池则是在发生腐蚀的位置同时同地进行反应。如图4–6所示，原电池是把化学能转化为电能

4-6, A primary battery is a device that converts chemical energy into electrical energy and can do external work. Corrosive batteries can only cause metal corrosion and damage, and cannot be used as useful short-circuit primary batteries. The released electrical energy is not utilized and becomes heat energy lost.

并可以对外做功的装置。腐蚀电池只能导致金属腐蚀破坏，不能对外做有用的短路原电池，其释放电能不被利用，而变成热能散失。

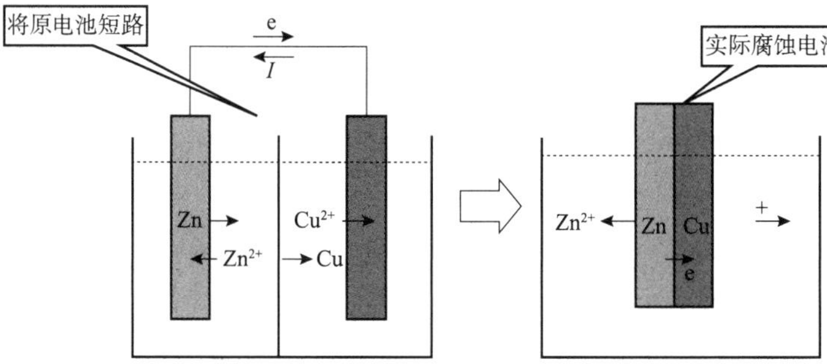

Figure 4-6　Galvanic and corrosion cell models

图 4-6　原电池与腐蚀电池模型

4.2　Polarization and depolarization

4.2　极化与去极化

4.2.1　Polarization of corrosion cells

4.2.1　腐蚀电池的极化现象

As shown in Figure 4-7, after the battery is turned on, the cathode potential changes in the negative direction and the anode potential changes in the positive direction. Both changes result in a smaller potential difference in the corrosion cell, resulting in a sharp reduction in corrosion current. As shown in Figure 4-8, this phenomenon is called the polarization of the battery.

As shown in Figure 4-9, its essence is the result of polarization of the two electrodes of the battery, respectively. When there is a net current passing through the electrode, the electrode potential deviates significantly from the open circuit potential (equilibrium potential or non-equilibrium steady-state potential)

如图4-7所示，电池接通后，阴极电势向负方向变化，阳极电势向正方向变化。两种变化的结果均使腐蚀电池的电势差变小，导致腐蚀电流急剧降低，如图4-8所示。这种现象称为电池的极化作用。

如图4-9所示，其实质是电池的两个电极分别发生极化的结果。当电极上有净电流通过时，电极电势显著偏离了未通电时的开路电势（平衡电势或非平衡的稳态电势），这种现

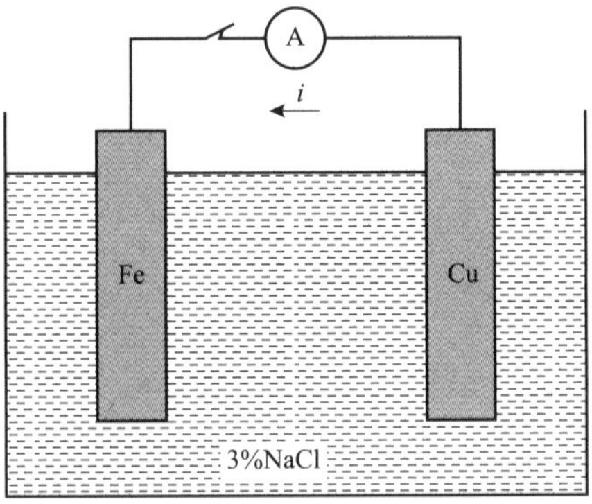

Figure 4-7 Polarization phenomenon observation device

图 4-7 极化现象观测装置

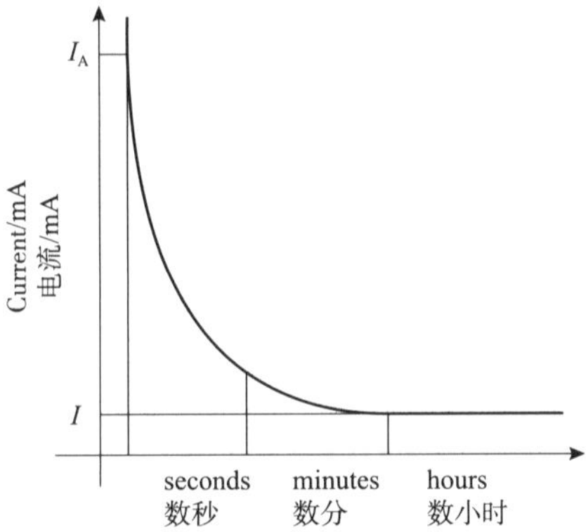

Figure 4-8 Current change due to polarization

图 4-8 极化引起的电流变化

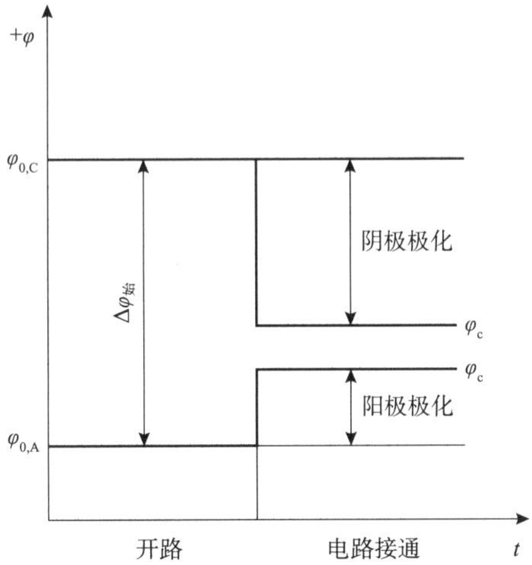

Figure 4-9 Schematic diagram of cathode and anode potential changes before and after the corrosion cell is connected

图 4-9 腐蚀电池接通前后阴、阳极电势变化示意

when it is not energized, a phenomenon called electrode polarization.

象叫作电极的极化。

4.2.2 Polarization curve

4.2.2 极化曲线

As shown in Figure 4-10, the curve representing the relationship between polarization potential and current or polarization current density is called polarization curve.

如图4-10所示，表示极化电位与电流或极化电流密度之间关系的曲线，称为极化曲线。

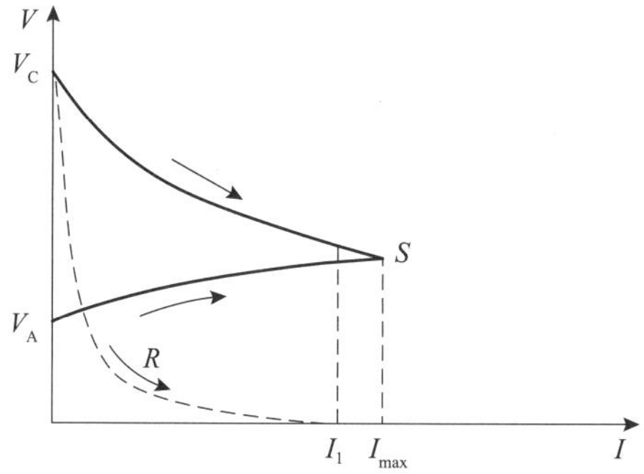

Figure 4-10 Corrosion battery current versus ohmic resistance and cathode and anode potentials

图4-10 腐蚀电池电流与欧姆电阻及阴、阳极电位的关系

The polarization curves are divided into theoretical polarization curves and measured polarization curves. The theoretical polarization curve is derived from an ideal electrode. The so-called ideal electrode means that only one electrode reaction occurs when the electrode is in an equilibrium state or a polarization state, that is, only an oxidation reaction or a reduction reaction occurs. However, in fact, due to the electrochemical inhomogeneity of metals, there are always cathode and anode regions at the same time. There are often two or more conjugated electrodes on the electrode surface, and the local cathode region and the anode region are not separated at all. not open, so the theoretical polarization curve is often not directly obtained.

极化曲线分为理论极化曲线和实测极化曲线。理论极化曲线是以理想电极得出的。理想电极是指该电极无论处于平衡状态或极化状态时只发生一种电极反应，即只发生氧化反应或者还原反应。但是，实际上金属由于电化学不均匀性，总是同时存在阴极区和阳极区，在电极表面常常有两个或两个以上相互共轭的电极反应，而局部的阴极区和阳极区根本就分不开，所以理论极化曲线无法直接得到。

Initial electrode potential: It is the electrode potential when the current density is equal to zero. For reversible electrodes, it is the equilibrium electrode potential; for irreversible electrodes, it is the stable potential.

起始电极电位：即电流密度等于0时的电极电位。对于可逆电极就是平衡电极电位；对于不可逆电极就是稳定电位。

Overpotential (overvoltage): The potential change value caused by a certain current density is the overpotential, which means that the specified overpotential takes a positive value.

过电位（超电压）：一定电流密度所引起的电位变化值即过电位，用来表示，规定超电位取正值。

Cathodic polarization:

阴极极化值：

$$\eta_e = V_{ec} - V_c \quad (4\text{–}9)$$

Anodic polarization value:

阳极极化值：

$$\eta_a = V_c - V_{ea} \quad (4\text{–}10)$$

Polarizability: also known as polarization degree and polarization resistance, it refers to its slope on the polarization curve, which reflects the trend of potential change under any current density.

极化率：又称极化度、极化电阻，在极化曲线上是指其斜率，反映出任一电流密度下电位变化的趋势。

4.2.3 Anodic polarization

4.2.3 阳极极化

When there is current passing through the anode, its potential moves to the positive direction, which is anodic polarization. The reason for anodic polarization is:

阳极上有电流通过时，其电势向正方向移动，即为阳极极化。产生阳极极化的原因如下：

(1) Activation polarization

（1）活化极化

When the reaction speed of metal ions into the solution is lower than the speed of electrons flowing from the anode to the cathode through the wire, the anode will have excessive positive charge accumulation, thereby changing the electric double layer charge distribution and the potential difference between the electric double layers, making the anode potential Move forward. This polarization due to the slow progress of the anodic process is called activation polarization.

当金属离子进入溶液的反应速度小于电子由阳极通过导线流向阴极的速度，则阳极就会有过多的正电荷积累，从而改变双电层电荷分布及双电层间的电势差，使阳极电势向正向移动。这种由于阳极过程进行缓慢而引起的极化，称为活化极化。

(2) Concentration polarization

（2）浓差极化

The metal ions generated by the dissolution of the anode first enter the liquid layer near the surface of the anode to create a concentration difference between them and the depth of the solution. Due to the restriction of the diffusion rate of metal ions on the surface of the anode, the concentration of metal ions near the anode

阳极溶解产生的金属离子，首先进入阳极表面附近的液层中，使之与溶液深处产生浓差。由于阳极表面金属离子受扩散速度制约，阳极附近金属离子浓度逐渐升高，相当于

gradually increases, which is equivalent to the insertion of the electrode into the solution of high concentration of metal ions, resulting in a positive potential and anodic polarization.

(3) Resistive polarization

When there is an oxide film on the metal surface, or when a film is formed during the corrosion process, the resistivity of the film is much higher than that of the base metal, and when the anode current passes through the film, a voltage drop will occur, making the potential significantly positive, and the resulting extreme, called resistance polarization. It can be seen that the stronger the anodic polarization, the more serious the reaction of the anodic process is blocked, which is beneficial to preventing metal corrosion; on the contrary, removing the anodic polarization will accelerate the progress of the corrosion process. This process of eliminating or weakening anodic polarization is called anodic depolarization. For example, stirring the solution to accelerate the diffusion of metal ions; making the anodic product form precipitates or complex ions; adding activated ions (such as Cr) to eliminate anodic passivation, etc., can promote anode depolarization, thereby accelerating corrosion.

The degree of polarization is usually judged by the polarization curve. The polarization curve is a curve showing the relationship between the electrode potential and the current density through it. Figure 4-11 shows the polarization curve of 1008 steel in 0.5mol/L Na_2SO_4 solution. The slope of the curve indicates the degree of polarization, which is called the degree of polarization. Obviously, the steeper the curve, the greater the degree of polarization, which means that the polarization

电极插入高浓度金属离子的溶液中，导致电势变正，产生阳极极化。

（3）电阻极化

当金属表面有氧化膜，或在腐蚀过程中形成膜时，膜的电阻率远高于基体金属，则阳极电流通过此膜时，将产生压降，使电势显著变正，由此引起的极化，称为电阻极化。可见，阳极极化越强烈，说明阳极过程的反应受阻滞越严重，这对防止金属腐蚀是有利的；相反，去除阳极极化，会加速腐蚀过程的进行。这种消除或减弱阳极极化的过程，称为阳极去极化。例如，搅拌溶液，加速金属离子扩散；使阳极产物形成沉淀或配离子；加入活化离子（如Cr），消除阳极钝化等，都可促进阳极去极化，从而加速腐蚀。

通常由极化曲线来判断极化程度的大小。极化曲线是表示电极电势与通过的电流密度间的关系曲线。图4-11所示为1008钢在0.5mol/L Na_2SO_4溶液中的极化曲线。曲线的倾斜程度表示极化程度，叫作极化度。很显然曲线越陡，极化度就越大，表示电极极化过程受

process of the electrode is hindered to a greater extent, and the corrosion is more difficult to proceed. Generally, in the activated state, the degree of anodic polarization of metals is not large, and the anodic polarization curve is relatively flat at this time. If the metal is passivated, the anodic polarization curve is steep and the degree of polarization is large.

阻滞程度越大，腐蚀进行越困难。一般来说，金属在活化状态下，阳极极化程度不大，这时阳极极化曲线较平坦。如果金属发生钝化，则阳极极化曲线很陡，极化度很大。

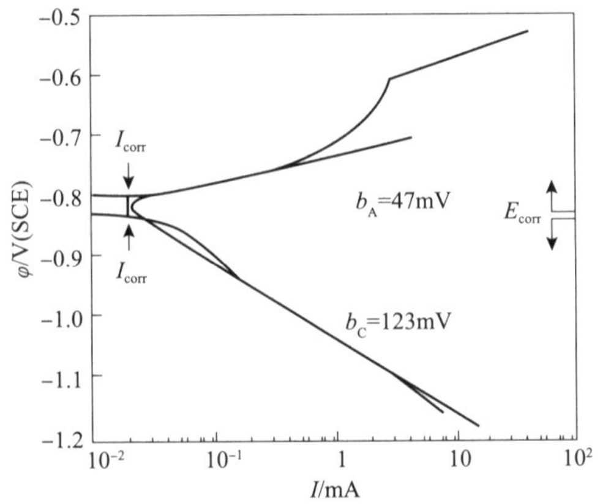

Figure 4-11 Polarization curve of 1008 steel in 0.5mol/L Na_2SO_4 solution (pH =6.3, 30℃, scanning speed 20mV/min, electrode area 4.79cm^2)

图 4-11 1008 钢在 0.5mol/L Na_2SO_4 溶液中的极化曲线（pH=6.3，30℃，扫描速度 20mV/min，电极面积 4.79cm^2）

4.2.4 Cathodic polarization

When a current passes through the cathode, its potential moves to the negative direction, which is cathodic polarization. The causes of cathodic polarization are:

(1) Activation polarization (or electrochemical polarization)

Since the cathode reduction reaction needs to reach a certain activation energy, when the cathode reduction reaction speed is less than the speed of electrons entering the cathode, electrons will accumulate at the cathode, and the electron density will increase, resulting in the cathode potential moving in the negative direction, resulting in a cathode electrode change.

4.2.4 阴极极化

阴极上有电流通过时，其电势向负方向移动，即为阴极极化。阴极极化的原因如下：

（1）活化极化（或电化学极化）

由于阴极还原反应需达到一定的活化能才能进行，当阴极还原反应速度小于电子进入阴极的速度时，使得电子在阴极堆积，电子密度增高，结果使阴极电势向负方向移动，产生了阴极极化。

(2) Concentration polarization

The slow diffusion rate of reactants or reaction products near the cathode can cause cathode concentration polarization, making the electrode potential negative.

Cathodic polarization means that polarization is blocked, preventing metal corrosion from progressing. The process of eliminating cathodic polarization is called cathodic depolarization, which can make the anodic polarization process proceed smoothly and maintain or accelerate the corrosion process. The cathodic depolarizers of most electrochemical ions are hydrogen ions and oxygen molecules in solution.

（2）浓差极化

阴极附近反应物或反应产物扩散速度缓慢可引起阴极浓差极化，使电极电势变负。

阴极极化表示极化受到阻滞，阻碍金属腐蚀的进行。消除阴极极化的过程叫作阴极去极化，可使阳极极化过程顺利进行，维持或加速腐蚀过程。绝大多数电化学离子的阴极去极化剂为溶液中的氢离子和氧分子。

4.2.5 Depolarization

The effect of reducing or eliminating polarization is called depolarization, or depolarization for short. Polarization is caused by many factors. When these factors change in opposite directions, the polarization effect can be reduced or eliminated, thus resulting in depolarization. In a word, the cause of depolarization may be factors such as temperature, concentration, stirring, pH value of solution, etc., or some substances. People call substances that can cause depolarization as depolarizers.

Raising the temperature produces a significant depolarization effect, because on the one hand, it will increase the electrochemical reaction speed and reduce the activation polarization, on the other hand, it will promote the diffusion of related substances, accelerate the mass transfer process, and weaken the concentration polarization effect. Stirring or aeration can effectively

4.2.5 去极化作用

减小或消除极化的作用称为去极化作用，简称去极化。极化是多方面因素造成的，当这些因素向相反方向变化，极化效应得以减小或消除，于是产生去极化作用。总之，造成去极化的原因可能是如温度、浓度、搅拌、溶液pH值等因素，也可能是一些物质，人们把能引起去极化作用的物质称为去极化剂。

升高温度产生明显的去极化效应，这是因为一方面会增大电化学反应速度，减小活化极化；另一方面促进有关物质的扩散，加速传质过程，削弱浓差极化效应。搅拌或充气能有效减小电极表面与溶液本体

reduce the concentration difference between the electrode surface and the solution body, and can reduce or even eliminate concentration polarization. The pH value of the solution directly affects the formation or dissolution of insoluble products on the electrode surface, thereby affecting the ohmic polarization.

的浓度差，可以减小甚至消除浓差极化。溶液pH值直接影响电极表面难溶产物的形成或溶解，从而对欧姆极化产生影响。

There are many types of depolarizers, and precipitants or complexing agents precipitate or complex anode reaction products, resulting in anode depolarization. The cathodic depolarizers can be various oxidants such as cations, anions, neutral molecules, and the like.

去极化剂的种类很多，沉淀剂或络合剂使阳极反应产物沉淀或络合，造成阳极去极化。而阴极去极化剂可以是如阳离子、阴离子、中性分子等各种氧化剂。

Polarization and depolarization are interdependent and contradictory to each other. In scientific research and production, people achieve a certain degree of polarization and polarization characteristics through various ways to achieve a given purpose. For example, the degree of polarization and polarization characteristics in electroplating production have a significant impact on the quality of the coating, while in corrosion science, polarization and depolarization directly determine the size of the corrosion rate. It can be said that all factors that affect polarization and depolarization The factors are all factors that affect the corrosion rate.

极化和去极化是互为依存又互为矛盾，在科研和生产中，人们通过各种途径实现一定的极化程度和极化特征，以达到既定的目的。如电镀生产中极化程度和极化特征对镀层质量有重大影响，而在腐蚀科学中，极化与去极化直接决定了腐蚀速度的大小，因此，一切影响极化和去极化的因素都是影响腐蚀速度的因素。

4.3 Evans polarization diagram

4.3 Evans 极化图

4.3.1 Corrosion polarization diagram

4.3.1 腐蚀极化图

Figure 4-12 shows a corrosion cell. When the circuit is open, the potentials of the cathode and anode are measured as $\varphi_{0,C}$ and $\varphi_{0,A}$, respectively. Connect the resistance *R* and make the value of the resistance *R*

图4–12所示为一腐蚀电池。开路时测得阴、阳极的电势分别为$\varphi_{0,C}$和$\varphi_{0,A}$。接入电阻*R*并使电阻*R*值由大变小，电流

change from large to small, and the current gradually increases from small to large. Correspondingly, measure the electrode potential under each current intensity, and draw the relationship between the cathode and anode potentials and the current intensity, as shown in Figure 4-13, It is the corrosion polarization map.

则由小逐渐变大，相应地测出各电流强度下的电极电势，绘出阴、阳极电势与电流强度的关系图，图4-13所示为腐蚀极化图。

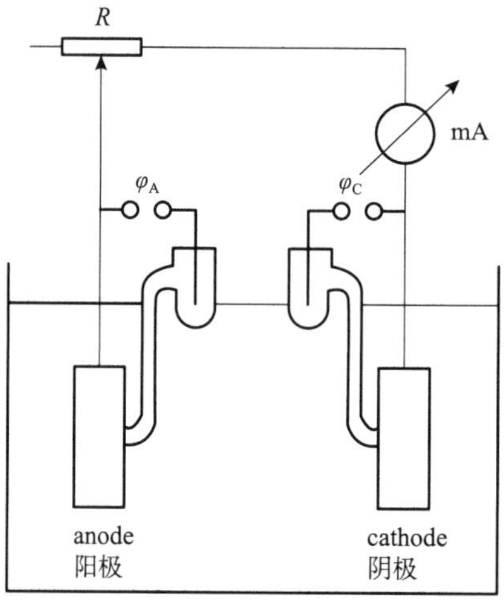

Figure 4-12 Polarization behavior of corrosion cells schematic diagram of the measuring device

图 4-12 腐蚀电池极化行为测量装置示意

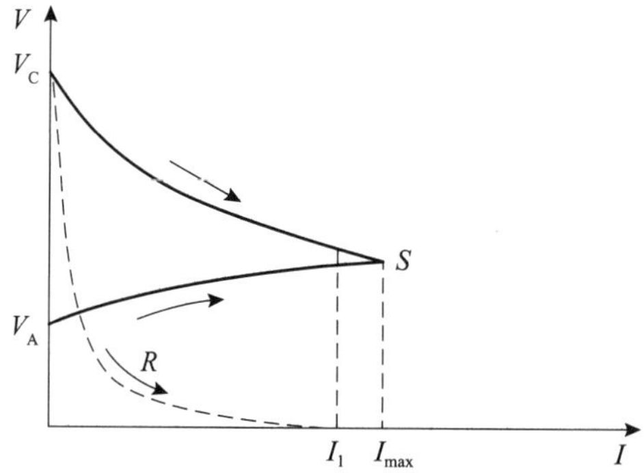

Figure 4-13 Variation of cathode and anode potentials with current intensity in corrosion cells

图 4-13 腐蚀电池的阴、阳极电势随电流强度的变化

Corrosion polarization diagram is a potential-current diagram, which is formed by drawing the cathodic and anodic polarization curves that characterize the corrosion cell on the same diagram. If the two polarization curves are simplified into a straight line, the

腐蚀极化图是一种电势-电流图，它是把表征腐蚀电池特征的阴、阳极极化曲线画在同一张图上构成的。若将两个极化曲线简化成直线，可得到

simplified corrosion polarization diagram as shown in Figure 4-14 can be obtained. This polarization diagram was first proposed and applied by the British corrosion scientist U.R. Evans and his students in 1929, so Called the Evans diagram. In the figure, the initial potentials of the cathode and anode are the equilibrium potentials of the cathode reaction and the anode reaction, which are represented by $\varphi_{0,C}$ and $\varphi_{0,A}$ respectively. The simplified polarization curves intersect at a point *S*. The potential corresponding to the intersection is called the mixed potential. Since this cathodic and anodic reaction constitutes the corrosion process, the mixed potential is the self-corrosion potential, referred to as the corrosion potential for short, and is represented by φ_{corr}. Corrosion potential is an irreversible non-equilibrium potential that can be measured experimentally.

如图4–14所示的简化的腐蚀极化图，此极化图是英国腐蚀科学家U.R.Evans及其学生于1929年首先提出和应用的，故称为Evans图。图中阴、阳极的起始电势为阴极反应和阳极反应的平衡电势，分别以$\varphi_{0,C}$和$\varphi_{0,A}$表示。简化的极化曲线可交于一点*S*。交点对应的电势，称为混合电势。由于此阴、阳极反应构成了腐蚀过程，所以混合电势就是自腐蚀电势，简称为腐蚀电势，用φ_{corr}表示。腐蚀电势是一种不可逆的非平衡电势，可通过实验测得。

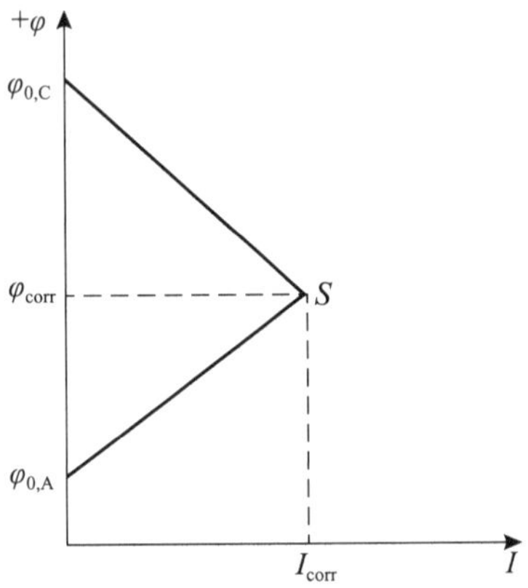

Figure 4–14 Simplified corrosion polarization diagram

图4–14 简化的腐蚀极化图

Figure 4-14 shows the current corresponding to the corrosion potential in the figure is called corrosion current. The cathode and anode areas in metal corrosion cells are not equal, but the current intensity flowing in the steady state is equal, so it is more convenient to use the

图4–14中与腐蚀电势相对应的电流，称为腐蚀电流，金属腐蚀电池中阴极和阳极面积是不相等的，但稳态下流过的电流强度相等，因此用φ–I极

φ-I polarization diagram., for both uniform corrosion and localized corrosion. Under uniform corrosion conditions, the entire metal surface acts as both cathode and anode, and a potential-current density polarization diagram can be used. When both the cathodic and anodic reactions are controlled by electrochemical polarization, it is more accurate to use the semi-logarithmic coordinate φ-lg$i$ polarization diagram. Because under electrochemical polarization control, the logarithm of the current density has a linear relationship with the change in potential.

化图较方便，对于均匀腐蚀和局部腐蚀都适用。在均匀腐蚀条件下，整个金属表面同时起阴极和阳极的作用，可采用电势－电流密度极化图。当阴、阳极反应均由电化学极化控制时，则用半对数坐标φ-lg$i$极化图更为确切。因为在电化学极化控制下，电流密度的对数与电位的变化呈直线关系。

4.3.2 Application of corrosion polarization map

4.3.2 腐蚀极化图的应用

Corrosion polarization map is an important tool for studying electrochemical corrosion and has a wide range of uses. For example, the polarization diagram can be used to determine the main controlling factors of corrosion, explain the corrosion phenomenon, analyze the nature and influencing factors of the corrosion process, judge the action mechanism of additives, and use the graphical method to calculate the corrosion rate of the multi-electrode system. Arguably, polarization diagrams are one of the most important theoretical tools in corrosion science.

腐蚀极化图是研究电化学腐蚀的重要工具，用途很广。例如，利用极化图可以确定腐蚀的主要控制因素、解释腐蚀现象、分析腐蚀过程的性质和影响因素、判断添加剂的作用机理，以及用图解法计算多电极体系的腐蚀速度等。因此，极化图是腐蚀科学最重要的理论工具之一。

As shown in Figure 4-15, when a current is passed through the corrosion cell, the two electrodes are polarized. If the movement of the electrode potential is not large when the current increases, it indicates that the electrode process is less hindered, that is, the polarizability of the electrode is small, or the polarization performance is poor. The polarizability of the electrode is determined by the slope of the polarization curve.

如图4－15所示，当腐蚀电池通过电流时，两电极发生极化。如果电流增加时，电极电势的移动不大，则表明电极过程受到的阻碍较小，即电极的极化率较小，或者是极化性能较差。电极的极化率由极化曲线的斜率决定。

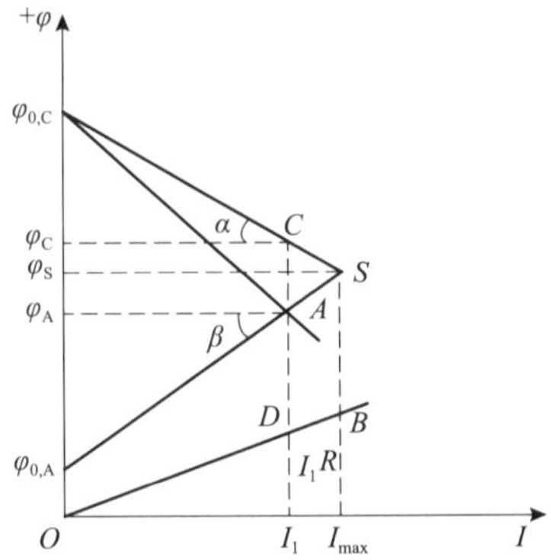

Figure 4-15 Corrosion polarization diagram

图 4-15 腐蚀极化图

The cathodic polarizability P_c is defined as:

阴极极化率P_c定义为：

$$P_c = \frac{\varphi_C - \varphi_{0,C}}{I_1} = \frac{\Delta\varphi_C}{I_1} = \tan\alpha \tag{4-11}$$

Polarizability P_a is defined as:

阳极极化率P_a定义为：

$$P_a = \frac{\varphi_A - \varphi_{0,A}}{I_1} = \frac{\Delta\varphi_A}{I_1} = \tan\beta \tag{4-12}$$

where: φ_C and φ_A are the cathodic and anodic polarization potentials at the current I_1;

$\Delta\varphi_C$ and $\Delta\varphi_A$ are the cathode and anode polarization values at the current I_1;

α and β are the inclination angles of the cathodic and anodic polarization curves and the horizontal line.

Since the corrosion system has an ohmic resistance R, the influence of the ohmic potential drop IR on the corrosion current should be considered. OB in Figure 4-15 represents the change of ohmic potential drop with current. When drawing the polarization diagram, the ohmic potential drop can be summed up with the cathode (or anode) polarization curve, as shown the $\varphi_{0,C}A$ line in the Figure 4-15, and then intersect with the anode polarization curve $\varphi_{0,A}S$ at point A, then the point The current I_1 corresponding to A is the corrosion current in this case, because

式中 φ_C，φ_A——电流I_1时的阴、阳极极化电势；

$\Delta\varphi_C$，$\Delta\varphi_A$——电流I_1时的阴、阳极极化值；

α，β——阴、阳极极化曲线与水平线的倾角。

由于腐蚀体系有欧姆电阻R，因此要考虑欧姆电势降IR对腐蚀电流的影响。图4-15中OB表示欧姆电势降随电流的变化。在绘制腐蚀极化图时，可将欧姆电势降与阴极（或阳极）极化曲线加和，如图4-15中的$\varphi_{0,C}A$线，然后与阳极极化曲线$\varphi_{0,A}S$相交于点A，则点A对应的电流I_1就是这种情况下的腐蚀电流，因为

$$\varphi_{0,C}-\varphi_{0,A}=\Delta\varphi_C+\Delta\varphi_A+I_1R=I_1P_c+I_1P_a+I_1R \quad (4\text{-}13)$$

$$I_{corr}=\frac{\varphi_{0,C}-\varphi_{0,A}}{P_c+P_a+R} \quad (4\text{-}14)$$

This formula shows that the smaller the initial potential difference $\varphi_{0,C}-\varphi_{0,A}$ of the corrosion cell, the larger the cathode and anode polarizability P_c and P_a, and the larger the ohmic resistance of the system, the smaller the corrosion current, that is, the smaller the electrochemical corrosion rate. For many corrosion systems, there is a short circuit between the cathode and the anode. If the solution resistance is not large and R is negligible, the corrosion current mainly depends on the polarization properties of the cathode and anode reactions.

式(4–14)表明，腐蚀电池的初始电势差$\varphi_{0,C}-\varphi_{0,A}$越小，阴、阳极极化率$P_c$和$P_a$越大，系统的欧姆电阻越大，则腐蚀电流越小，即电化学腐蚀速度越小。对于许多腐蚀体系，阴、阳极间为短路，如果溶液电阻不大，R可忽略不计，则腐蚀电流主要取决于阴、阳极反应的极化性能。

$$I_{corr}=\frac{\varphi_{0,C}-\varphi_{0,A}}{P_c+P_a} \quad (4\text{-}15)$$

Since the ohmic potential drop between the cathode and anode is ignored, the corrosion current is equivalent to the current corresponding to the intersection S of the cathode and anode polarization curves in Figure 4-11, and the potential is the corrosion potential I_{corr}.

由于忽略了阴、阳极间的欧姆电势降，腐蚀电流就相当于图4–11中阴、阳极极化曲线交点S对应的电流，而电势就是腐蚀电势I_{corr}。

It is easy to analyze the controlling factors of the corrosion process and the influence of various factors on the corrosion rate by using the corrosion polarization diagram. For simplicity, the ohmic resistance of the solution is usually ignored, and the polarization curves of the cathode and anode are represented by straight lines. In the following, the influence of various factors on corrosion will be analyzed using the polarization diagram.

利用腐蚀极化图很容易分析腐蚀过程的控制因素以及各种因素对腐蚀速度的影响。为简便起见，通常忽略溶液的欧姆电阻，并以直线表示阴、阳极的极化曲线。下面就利用腐蚀极化图分析各种因素对腐蚀的影响。

(1) Corrosion Rate vs. Initial Potential Difference

（1）腐蚀速度与初始电势差的关系

The initial potential difference $\varphi_{0,C}-\varphi_{0,A}$ is the

初始电势差$\varphi_{0,C}-\varphi_{0,A}$是腐

driving force for corrosion. With all other conditions being equal, a larger initial potential difference leads to a higher corrosion current. The curve in Figure 4-16 represents the anodic polarization curve of Fe corrosion in H_2SO_4, $\varphi_{0,Fe}S_1$ is the cathodic polarization curve of the hydrogen evolution reaction, and the corrosion rate is I_1. If HNO_3 is used instead of H_2SO_4, the cathodic polarization curve of HNO_3 reduction is $\varphi_{0,HNO_3}S_2$, and the equilibrium potential φ_{0,HNO_3} is much higher than $\varphi_{0,OH}$. It will greatly increase the corrosion current, that is, $I_2 \gg I_1$. However, if Fe is passivated in HNO_3, its anodic polarization curve becomes $\varphi'_{0,Fe}S_1$, which is much more positive than $\varphi'_{0,Fe}S_2$, the corrosion rate will be greatly reduced, that is, $I_3 \ll I_2$.

蚀的驱动力。在其他条件相同的情况下，初始电势差越大，腐蚀电流越大。图4-16中的曲线表示Fe在H_2SO_4中腐蚀的阳极极化曲线，$\varphi_{0,Fe}S_1$为析氢反应的阴极极化曲线，腐蚀速度为I_1。若用HNO_3代替H_2SO_4，则HNO_3还原的阴极极化曲线为$\varphi_{0,HNO_3}S_2$，平衡电势φ_{0,HNO_3}比$\varphi_{0,OH}$高得多，即使阴极极化率近似，也会使腐蚀电流大大增加，即$I_2 \gg I_1$。然而，若Fe在HNO_3中发生钝化，其阳极极化曲线变为$\varphi'_{0,Fe}S_1$，比$\varphi'_{0,Fe}S_2$正得多，则腐蚀速度将大大降低，即$I_3 \ll I_2$。

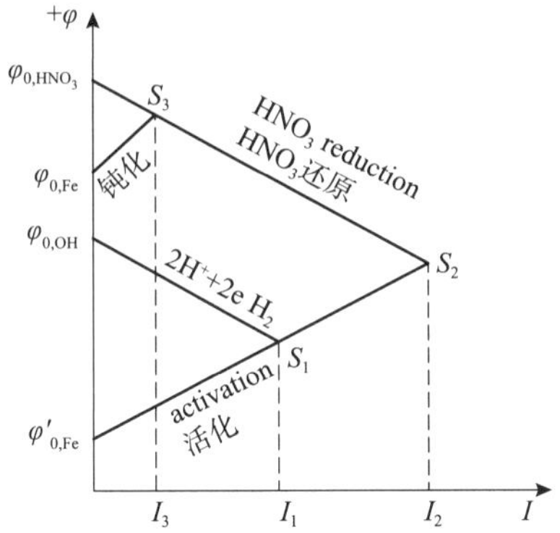

Figure 4-16 Corrosion diagram of iron by oxidizing acid

图4-16 氧化性酸对铁的腐蚀

Due to the different equilibrium potentials of different metals, their anodic polarization is small. When the cathodic reaction and its polarization curve $\varphi_{0,O_2}S$ are the same, the lower the equilibrium potential of the metal, the greater the corrosion current, namely:

如图4-17所示，由于不同金属的平衡电势不同，它们的阳极极化较小，当阴极反应及其极化曲线$\varphi_{0,O_2}S$相同时，金属的平衡电势越低，其腐蚀电流越大，即：

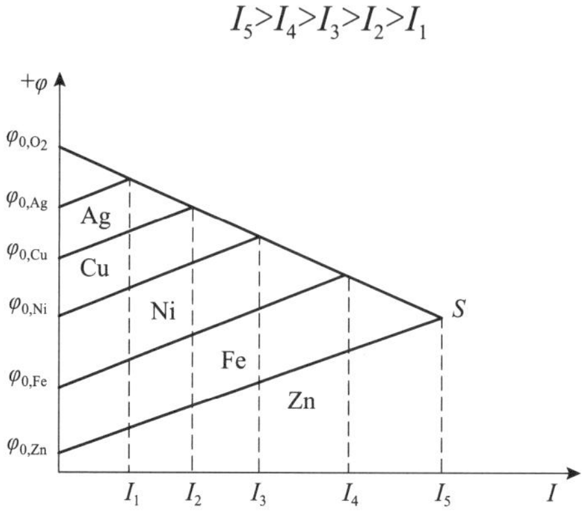

Figure 4-17 Effect of metal equilibrium potential on corrosion current

图 4-17 金属平衡电位对腐蚀电流的影响

(2) The effects of polarization and depolarization

As shown in Figure 4-18, regardless of whether it is a cathode or an anode, as long as polarization occurs, the slope of the corresponding polarization curve increases, and the corrosion rate decreases; as long as depolarization occurs, the slope of the corresponding polarization curve decreases, and the corrosion rate increases. When the cathode and anode are polarized or depolarized at the same time, the corrosion rate changes more significantly.

（2）极化、去极化的影响

如图 4-18 所示，不论是阴极或阳极，只要发生极化，相应的极化曲线斜率增大，腐蚀速度即减小；只要发生去极化，相应的极化曲线斜率减小，腐蚀速度就增大。而阴、阳极同时发生极化或去极化时，腐蚀速度变化更为显著。

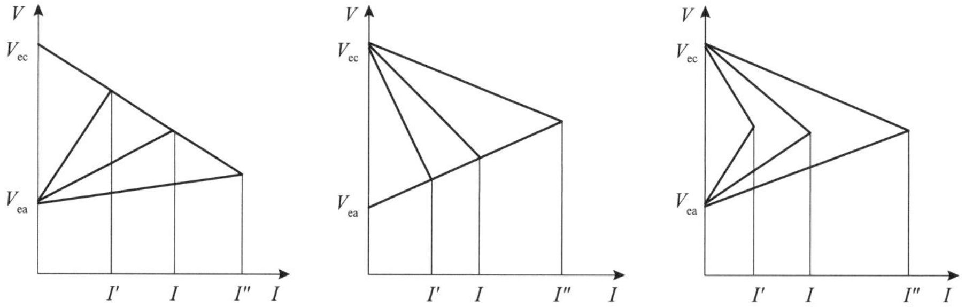

(a)Influence of anodic polarization and depolarization
阳极极化去极化的影响

(b)Influence of cathodic polarization and depolarization
阴极极化去极化的影响

(c)Influence of co-polarization and depolarization of cathode and anode
阴、阳极共同极化、去极化的影响

Figure 4-18 Effect of polarization and depolarization on corrosion rate

图 4-18 极化、去极化对腐蚀速度的影响

(3) Influence of internal resistance

As shown in Figure 4-19, when the internal resistance is large, the corrosion rate is small; when the internal resistance is small, the corrosion rate is large.

（3）内电阻$R_{内}$的影响

如图4-19所示，内电阻较大时，腐蚀速度较小；内电阻较小时，腐蚀速度较大。

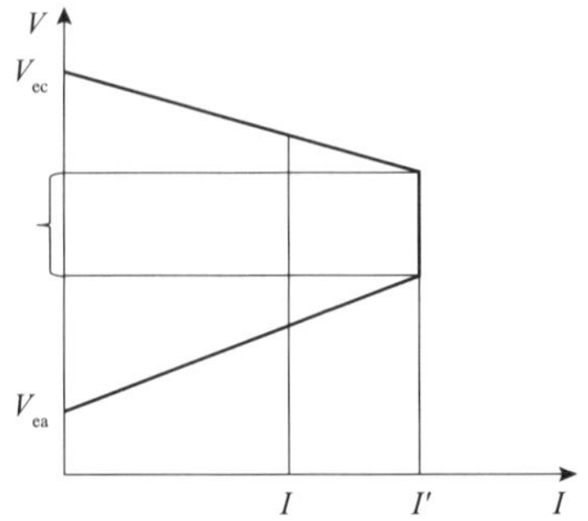

Figure 4-19 Effect of resistance on corrosion rate

图 4-19 电阻对腐蚀速度的影响

4.3.3 Corrosion control factors

The corrosion rate depends not only on the driving force of corrosion-the initial potential difference $\varphi_{0,C}-\varphi_{0,A}$ of the corrosion cell, but also on the resistance to corrosion, the polarizability P_c and P_a of the cathode and anode, and the ohmic resistance R. Any one of these three resistances may significantly exceed the other two, and control the speed during the corrosion process, which is called the controlling factor. As shown in Figure 4-20, Using the polarization diagram, the controlling factors of corrosion can be judged very intuitively. When $P_c \gg P_a$, it is mainly determined by the cathode polarization, which is called cathode control; if $P_a \gg P_c$, it mainly depends on the size of P_a, which is called anode control. If P_a and P_c are close and determine the size of the corrosion rate at the same time, it is called mixed control; if the ohmic resistance of the corrosion system is large, the corrosion current is mainly determined by the resistance, which is called ohmic control.

The corrosion polarization diagram can be used to judge not only various control factors, but also the degree of control of corrosion by each factor. The ratio

4.3.3 腐蚀控制因素

腐蚀速度既取决于腐蚀的驱动力-腐蚀电池的起始电势差 $\varphi_{0,C}-\varphi_{0,A}$，也决定腐蚀阻力阴、阳极的极化率$P_c$和$P_a$，以及欧姆电阻$R$。这三项阻力中任意一项都可能明显地超过另两项，在腐蚀过程中对速度起控制作用，称为控制因素。如图4-20所示，利用极化图可以非常直观地判断腐蚀的控制因素。当$P_c \gg P_a$时，主要由阴极极化决定，称为阴极控制；若$P_a \gg P_c$，主要取决P_a的大小，称为阳极控制。如果P_a和P_c接近，同时决定腐蚀速度的大小，则称为混合控制；如果腐蚀系统的欧姆电阻很大，则腐蚀电流主要由电阻决定，称为欧姆控制。

利用腐蚀极化图不仅可判断各种控制因素，还可判断各因素对腐蚀控制的程度，各个

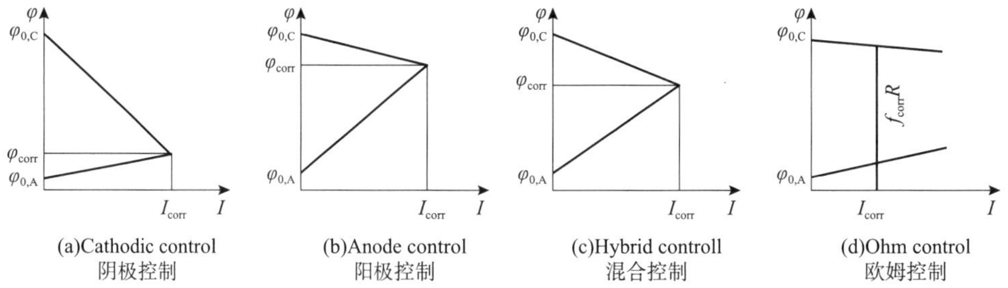

Figure 4-20 Corrosion polarization diagram for different control factors

图 4-20 不同控制因素的腐蚀极化图

of each resistance to the total resistance of the process is called the degree to which the total process is controlled by each resistance, which is commonly expressed as a percentage.

阻力对过程总阻力之比称为总过程受各阻力控制的程度，常用百分数表示。

Degree of anode control:

阳极控制程度：

$$C_a = \frac{R_{pa}}{R_{内} + R_{pa} + R_{pc}} = \frac{I'R_{pa}}{V_{ec} - V_{ea}} \times 100\% \tag{4-16}$$

Degree of cathode control:

阴极控制程度：

$$C_c = \frac{R_{pc}}{R_{内} + R_{pa} + R_{pc}} = \frac{I'R_{pc}}{V_{ec} - V_{ea}} \times 100\% \tag{4-17}$$

Egree of ohmic control:

欧姆控制程度：

$$C_R = \frac{R_{内}}{R_{内} + R_{pa} + R_{pc}} = \frac{I'R_{内}}{V_{ec} - V_{ea}} \times 100\% \tag{4-18}$$

4.3.4 Conjugated system and corrosion potential

4.3.4 共轭体系与腐蚀电位

As shown in Figure 4-21, the phenomenon that an anodic reaction and a cathodic reaction are simultaneously carried out on an isolated metal electrode at the same speed is called the coupling of electrode reactions, and the mutually coupled reactions are called conjugated reactions, and the corresponding corrosion system is called a conjugated system.

如图 4-21 所示，在一个孤立金属电极上同时以相等的速度进行一个阳极反应和阴极反应的现象称为电极反应的耦合，而互相耦合的反应称为共轭反应，相应的腐蚀体系称为共轭体系。

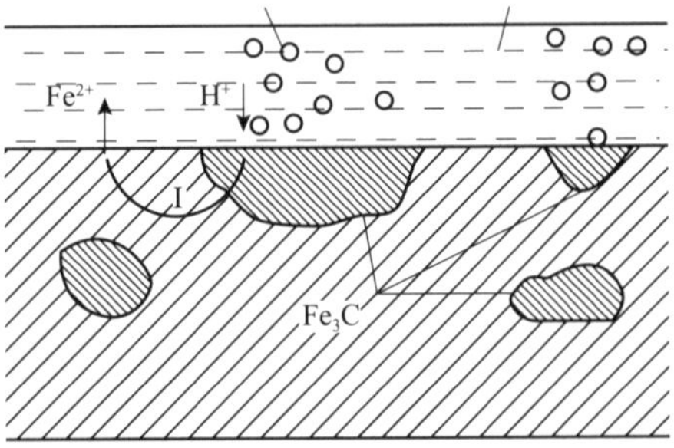

Figure 4-21 Conjugation reaction in Fe-Fe_3C microbattery

图 4-21 Fe-Fe_3C 微电池中的共轭反应

Conjugated system:

共轭体系：

$$Fe^{2+} + 2e^- \underset{\overleftarrow{i}_{1a}}{\overset{\overrightarrow{i}_{1k}}{\longleftrightarrow}} Fe2H^+ + 2e^- \underset{\overleftarrow{i}_{2a}}{\overset{\overrightarrow{i}_{2k}}{\longleftrightarrow}} H_2 \quad (4\text{–}19)$$

The dissolution rate of annealed carbon steel can be expressed as:

退火碳钢的溶解速度可以表示为：

$$i_a = \overleftarrow{i}_{1a} - \overrightarrow{i}_{1k} \quad (4\text{–}20)$$

The rate of hydrogen evolution is:

氢气的析出速度为：

$$i_k = \overrightarrow{i}_{2k} - \overleftarrow{i}_{2a} \quad (4\text{–}21)$$

The dissolution rate of metals in the system is equal to the evolution rate of hydrogen:

体系中金属的溶解速度与氢气的析出速度相等：

$$i_a = i_k = i_{corr} \quad (4\text{–}22)$$

The reaction coupling conditions of the two electrodes are:

两个电极反应耦合条件为：

$$V_{VH}-V_{eFe}>0 \quad (4\text{–}23)$$

$$V_{eFe}<V_{corr}<V_{eH} \quad (4\text{–}24)$$

The corrosion potential is the potential measured when the metal reaches a stable corrosion state without an applied current, which is the mixed potential of the anodic and cathodic reactions polarized by the self-corrosion current. The corrosion current corresponding to the corrosion potential not only represents the corrosion dissolution rate of the metal at the corrosion potential, but also represents the reduction reaction rate of the depolarizer in the solution at this time.

腐蚀电位是在没有外加电流时金属达到一个稳定腐蚀状态时测得的电位，它是被自腐蚀电流所极化的阳极反应和阴极反应的混合电位。腐蚀电位所对应的腐蚀电流既表示金属在腐蚀电位时的腐蚀溶解速度，又表示此时溶液中去极化剂还原反应的速度。

4.4 Passivation of metals

4.4.1 Passivation phenomenon

The concept of passivation originally came from Faraday's observations of changes in iron in nitric acid solutions. If a piece of industrial pure iron is put into nitric acid of different concentrations, it will be found that the dissolution rate of iron does not increase monotonically with the increase of nitric acid concentration, but when the nitric acid increases to a certain concentration, the dissolution rate of iron begins to drop sharply. If you continue to increase the concentration of nitric acid, the dissolution rate of iron will still be very low, as shown in Figure 4-22. Iron does not react in concentrated nitric acid at room temperature, but becomes stable. This anomaly is called passivation. That is, the metal loses its chemical activity under certain environmental conditions. And it was found that this stability can also be maintained for a certain period of time when the iron treated with concentrated nitric acid is put into dilute nitric acid (or water, steam and other media).

Other metals such as chromium, nickel, cobalt, molybdenum, zirconium, silver, titanium, etc. also have this passivation phenomenon. Other strong oxidants (such as potassium nitrate, etc.) can also cause passivation of some metals; while other non-oxidizing media can also cause some metals (such as magnesium in hydrofluoric acid, aluminum and niobium in hydrochloric acid) to passivate, Oxygen in solution or in the atmosphere is also a passivating agent. In addition, some metals with an

4.4 金属的钝化

4.4.1 钝化现象

钝化的概念最初来自法拉第对铁在硝酸溶液中变化的观察。若把一块工业纯铁放入不同浓度的硝酸中，就会发现铁的溶解速度并不随硝酸浓度的增长而单调增加，而是当硝酸增加到一定浓度后，铁的溶解速度开始急剧下降，当继续增加硝酸浓度，铁的溶解速度将仍然很低，如图4-22所示。铁在室温浓硝酸中不发生反应，而变得稳定，这一异常现象称为钝化。即在某些环境条件下金属失去了化学活性。而且发现，经过浓硝酸处理过的铁再放入稀硝酸（或水、水蒸气和其他介质）中，也能使这种稳定性保持一定时间。

其他金属如铬、镍、钴、钼、锆、银、钛等也同样具有这种钝化现象。其他强氧化剂（如硝酸钾等）也可引起一些金属产生钝化；而其他非氧化性介质也能使某些金属（如镁在氢氟酸中，铝和铌在盐酸中）产生钝化，溶液或大气中的氧也是一种钝化剂。除此以外，

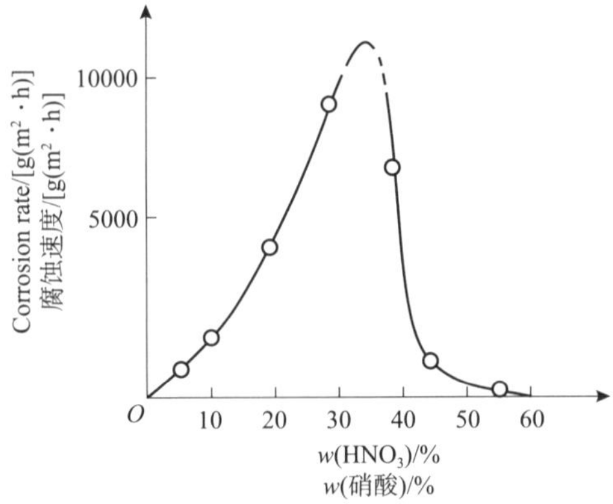

Figure 4-22 The relationship between the dissolution rate of iron and the mass fraction of HNO_3 (mass fraction is 25%)

图 4-22 铁的溶解速度与 HNO_3 质量分数的关系（质量分数为 25%）

activation-passivation transition can also be turned into a passivated state when an external cathodic polarization is applied.

Passivation has the following characteristics: the difficulty of metal passivation is related to the passivation agent, metal nature and temperature; after metal passivation, the potential rises sharply in the positive direction; the conversion between metal passivation and active state is often irreversible. Under certain conditions, the applied anodic current or local anodic current can also transform the metal from the active state to the passive state.

Before and after passivation, there are several significant changes in the metal or alloy:

(1) The corrosion rate of the metal is significantly reduced

When a metal is in a passivated state, its corrosion rate is very low. When changing from the active state to the passivation state, the corrosion rate will decrease by 10^4~10^8 orders of magnitude. This is mainly due to the formation of a very thin passivation film on the metal surface in the corrosion system. The thickness of the

一些具有活化-钝化转变的金属，当外加阴极极化时，也可使其转入钝化状态。

钝化具有以下特征：金属钝化的难易程度与钝化剂、金属本性和温度有关；金属钝化后电位往正方向急剧上升；金属钝态与活态之间的转换具有一定的不可逆性；一定条件下，利用外加阳极电流或局部阳极电流也可使金属从活态转变为钝态。

钝化前后，金属或合金有以下几个显著的变化：

（1）金属的腐蚀速度显著降低

金属处在钝化状态时，其腐蚀率非常低。当由活化态转入钝化态时，腐蚀速率将减小 10^4~10^8 数量级。这主要是由于腐蚀体系中的金属表面形成了一层极薄的钝化膜。钝化膜

passivation film is generally 3nm or thinner, and it is a layer of semiconductor oxide film with high resistance or an adsorption film of substances such as gases. The film contains a considerable amount of bound water, its nature is extremely fragile, and its passivation will change when removed from the metal surface or taken out from a corrosive environment.

的厚度一般为3nm或更薄，它是一层具有很高电阻的半导体氧化膜或者是气体等物质的吸附膜。膜内含有相当量的结合水，其性质极为脆弱，当从金属表面除去或从腐蚀环境中取出时，其钝化将发生变化。

（2) The electrode potential of the metal undergoes a sudden change, and the positive shift is obvious

（2）金属的电极电势发生突变，且明显正移

Passivation will move the potential of the metal in the positive direction by about 0.5~2V. For example, after iron passivation, the potential increases from −0.5~0.2V to 0.5~1.0V; after chromium passivation, the potential increases from −0.6~+0.4V to 0.8~1.0V. Passivation is therefore described as such: A reactive metal is passivated when its electrode potential becomes close to that of a less reactive noble metal.

钝化会使金属的电势朝正方向移动0.5~2V。例如，铁钝化后电势由−0.5~0.2V升高到0.5~1.0V；铬钝化后电势由−0.6~+0.4V升高到0.8~1.0V。因此有人这样描述钝化：当活泼金属的电极电势接近于较不活泼贵金属的电极电势时，活泼金属就钝化了。

4.4.2 Passivation characteristic curve of metal

4.4.2 金属的钝化特性曲线

Metal passivation caused by passivating agents, commonly referred to as “chemical passivation”. Anodic polarization can also cause passivation of the metal. Some metals are in a certain medium (usually does not contain Cl^- ions), when the applied anode current exceeds a certain value, the metal can be transformed from an active state to a passive state, which is called anode passivation or electrochemical passivation.

由钝化剂引起的金属钝化，通常称为“化学钝化”。阳极极化也可引起金属钝化。某些金属在一定的介质中（通常不含有Cl^-离子），当外加阳极电流超过某一定数值后，可使金属由活化状态转变为钝态，称为阳极钝化或电化学钝化。

The metal passivation process is relatively complex, and so far it cannot be expressed by strict mathematical analysis methods. At present, only graphical analysis methods are used to describe the passivation process.

金属钝化过程比较复杂，至今还不能用严格的数学解析方法表达，目前只是运用图形分析方法描述钝化过程。金属

Anode passivation of metals can be achieved by controlling the anode potential (potentiostatic method) or by controlling the anode current (galvanostatic method). The electrochemical behavior reflected by the metal anodic polarization curve measured by the potentiostatic method is relatively significant, and the passivation and activation laws of various metals can be easily analyzed, as shown in Figure 4-23(a). If the galvanostatic law is used, it is difficult to observe the change of the current in the passivation area with the potential, as shown in Figure 4-23(b). The *ABCD* curve can be obtained by normal measurement, and the *DFA* curve can be obtained by returning, but the curve shown in Figure 4-23(a) cannot be obtained.

的阳极钝化可以用控制阳极电势（恒电势法）或控制阳极电流（恒电流法）的方法来达到。用恒电势法测得的金属阳极极化曲线所反映的电化学行为比较显著，可以较方便地分析各种金属的钝化、活化规律，如图4-23（a）所示。若采用恒电流法则很难观察到钝化区内电流随电势的变化，如图4-23（b）所示。正常测定得到*ABCD*曲线，返程则得到*DFA*曲线，都无法得到如图4-23（a）所示的曲线。

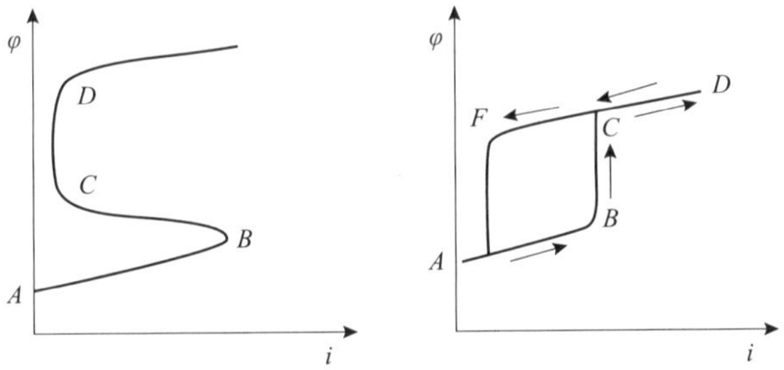

(a)Control the potential method
控制电势法

(b)Control current method
控制电流法

Figure 4-23 Anode passivation curves measured by different methods

图4-23 不同方法测得的阳极钝化曲线

The metal passivation curve with activation-passivation behavior measured by the controlled potential method, such as the schematic diagram of the anodic polarization curve of stainless steel. It reveals the characteristic points and characteristic areas of metal activation and passivation. It can be seen from the Figure 4-24 that from the open-circuit potential φ_0 of the metal, as the potential gradually increases, the current increases

用控制电势法测得的具有活化-钝化行为的金属钝化曲线，如不锈钢的阳极极化曲线示意。它揭示了金属活化、钝化的各特性点和特征区。由图4-24可知：从金属的开路电势φ_0起，随着电势逐渐升高，电流迅速增大，在*B*点达到最大

rapidly and reaches the maximum value at point *B*. The potential continues to rise, but the current begins to drop significantly. After reaching point *C*, the current maintains a small value, and the potential increases sharply in the *CD* potential range, and the current hardly changes with the potential. Beyond point *D*, the current increases again with the increase in potential. Therefore, this anodic polarization curve can be divided into several different segment.

值。电势继续升高，电流却开始大幅度下降，到达*C*点后，电流保持一个很小的数值，而且在*CD*电势范围内电势急剧增加，电流几乎不随电势而改变。超过*D*点后，电流又随电势升高而增大。因此，可将此阳极极化曲线划分成以下几个不同的区段。

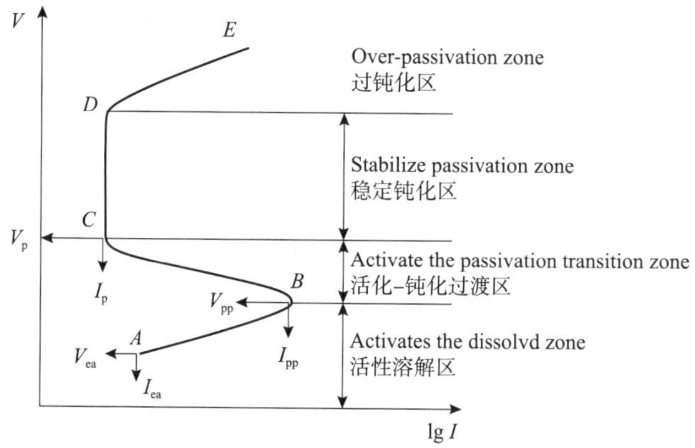

Figure 4-24 Schematic diagram of anodic polarization curve during metal passivation process

图 4-24 金属钝化过程阳极极化曲线示意

1. Active dissolution zone

The metal undergoes normal anodic dissolution, and the dissolution rate is controlled by activation polarization, V_{pp}—passivation potential (critical potential), and I_{pp}—called passivation current density.

2. Activation-passivation transition zone *BC* segment

The potential corresponding to point *B* is called the initial passivation potential V_{pp}, also called the passivation potential. The critical current density corresponding to point *B* is called passivation current density, which is expressed by I_{pp}. Because once the

1.活性溶解区AB段

金属进行正常的阳极溶解，溶解速度受活化极化控制，V_{pp}为钝化电位（临界电位），I_{pp}为致钝电流密度。

2.活化-钝化过渡区*BC*段

点*B*对应的电势称为初始钝化电势V_{pp}，也称致钝电势。点*B*对应的临界电流密度称为致钝电流密度，用I_{pp}表示。因为一旦电流密度超过I_{pp}，电势

current density exceeds I_{pp} and the potential is greater than V_{pp}, the metal begins to passivate, and the current density decreases sharply at this time. However, the *SC* segment is an activation-passivation transition region. In this potential range, the metal surface state changes sharply and is in an unstable state.

大于V_{pp}，金属就开始钝化，此时电流密度急剧降低。但*SC*段为活化–钝化过渡区，在此电势区间，金属表面状态发生急剧变化，并处于不稳定状态。

3. Stabilize passivation zone *CD*

When the potential reaches the point *C*, the metal turns into a completely passive state, and the potential at this point is usually referred to as the initial steady-state passivation potential V_p. The current density in the *CD* potential range is usually small, on the order of μA/cm^2, and hardly varies with potential. This tiny current density is called the passive current density I_p. The passivation current density is very small, reflecting that the dissolution rate of the metal in the passive state is very small.

3.稳定钝化区*CD*

当电势达到*C*点后，金属转入完全钝态，通常把该点的电势称为初始稳态钝化电势V_p。*CD*电势范围内的电流密度通常很小，在μA/cm^{2}数量级，而且几乎不随电势变化。这一微小的电流密度称为维钝电流密度I_p。维钝电流密度很小，反映了金属在钝态下的溶解速度很小。

4. *DE* section of overpassivation zone

After the potential exceeds point *D*, the current density starts to increase again. The potential at point *D* is called the overpassivation potential. The current density in this potential section increases again, usually due to the formation of soluble high-valent metal ions, such as stainless steel in this section due to the formation of high-valent chromium ions, causing the destruction of the passivation film and causing the metal to corrode again.

4.过钝化区*DE*段

电势超过*D*点后电流密度又开始增大。*D*点的电势称为过钝化电势。此电势区段电流密度又增大了，通常是由于形成了可溶性的高价金属离子，如不锈钢在此区段因有高价铬离子形成，引起钝化膜的破坏，使金属又发生了腐蚀。

4.4.3 Influence of cathodic polarization curve on metal passivation

4.4.3 阴极极化曲线对金属钝态的影响

The metal passivation due to the reduction of the oxidant (depolarizer) of the corrosive medium without any external polarization, i.e. the self-passivation of the metal. To achieve self-passivation of metals, the

没有任何外加极化的情况下，由于腐蚀介质的氧化剂（去极化剂）还原引起的金属钝化，即金属的自钝化。实现金

following two conditions must be met:

①The passivation potential V_{pp} of the metal must be higher than the oxidation-reduction equilibrium potential $V_{0,c}$ of the oxidant, that is, $V_{0,c}>V_{pp}$;

②Under the passivation potential V_{pp}, the current density of the cathodic reduction reaction of the oxidant must be greater than the passivation current density I_{pp} of the metal, that is, $I_c>I_{pp}$ under V_{pp}. Only in this way can the corrosion potential of the metal fall within the range of the anode passivation potential of the metal.

Because metal corrosion is the result of the conjugation reaction of cathodic and anodic in the corrosion system. For a metal corrosion system that may be passivated, such as a metal with activation-passivation behavior in a certain corrosive medium, whether the corrosion potential of the metal falls in the passivation zone depends not only on the range of the passivation zone on the anodic polarization curve The size also depends on the shape and location of the cathode curve. Graph showing the effect of cathodic polarization on passivation. If the cathodic polarization process in the figure is activation control, the polarization curve is a Tafel straight line, but there may be three different exchange currents, so the effect of cathodic polarization may appear in three different situations.

In the first case, there is only one intersection point *A* between the cathodic polarization curve *I* and the anodic polarization curve in Figure 4-25, which is in the activation zone. Point *A* corresponds to the corrosion potential and current of the corrosion system, such as the corrosion of titanium in dilute sulfuric acid or hydrochloric acid without air and the corrosion of iron

属的自钝化，必须满足下列两个条件：

①金属的致钝电势V_{pp}必须高于氧化剂的氧化-还原平衡电势$V_{0,c}$，即$V_{0,c}>V_{pp}$；

②在致钝电势V_{pp}下，氧化剂阴极还原反应的电流密度必须大于该金属的致钝电流密度I_{pp}，即在V_{pp}下$I_c>I_{pp}$。这样才能使金属的腐蚀电势落在该金属的阳极钝化电势范围内。

因为金属腐蚀是腐蚀体系中阴、阳极共轭反应的结果。对于一个可能钝化的金属腐蚀体系，如具有活化-钝化行为的金属在一定的腐蚀介质中，金属的腐蚀电势能否落在钝化区，不仅取决于阳极极化曲线上钝化区范围的大小，还取决于阴极曲线的形状和位置。图4-25表示阴极极化对钝化的影响。假若图中阴极极化过程为活化控制，极化曲线为Tafel直线，但可能有三种不同的交换电流，因此阴极极化的影响可能出现以下三种不同的情况。

第一种情况，图4-25中阴极极化曲线*I*与阳极极化曲线只有一个交点*A*，该点处于活化区。点*A*对应该腐蚀系统的腐蚀电位和电流，这种情况如钛在不含空气的稀硫酸或稀盐酸中的腐蚀及铁在稀硫酸中的

in dilute sulfuric acid. In the second case, the cathodic polarization curve 2 in Figure 4-25 and the anodic polarization curve have three intersection points, point B is in the activation zone, point *D* is in the passivation zone, and point *C* is in the transition zone, so the metal is in an unstable state. The metal may be in an activated state or in a passivation state, that is, the passivation is very unstable. In this case, if stainless steel is immersed in clesoxic acid, the passive film is damaged and cannot be repaired in time, causing metal corrosion. Points *B* and *D* are in the stable activation zone and passivation zone, respectively, corresponding to high corrosion rate and low corrosion rate. In the third case, in Figure 4-25, the cathodic polarization curve 3 and the anodic polarization curve intersect at point *E* in the passivation zone, and the metal or alloy of this type is in a stable passivation state. The metal will passivate spontaneously, so it is called self-passivation. Such is the case with iron in concentrated nitric acid such as stainless steel or titanium in an oxygenated acid solution.

腐蚀。第二种情况，图4-25中阴极极化曲线2与阳极极化曲线有三个交点，点*B*在活化区，点*D*在钝化区，点*C*处于过渡区，所以金属处于不稳定状态。金属可能处于活化态，也可能处于钝化态，即钝化很不稳定。这种情况如不锈钢浸入除去氧的酸中，钝化膜遭到破坏又得不到及时修补，使金属腐蚀。点*B*和*D*各处于稳定的活化区和钝化区，分别对应高的腐蚀速度和低的腐蚀速度。第三种情况，图4-25中阴极极化曲线3与阳极极化曲线交于钝化区的点*E*，这类体系金属或合金处于稳定钝态。金属会自发地钝化，所以称为自钝化。如不锈钢或钛处于含氧的酸溶液中，铁在浓硝酸中就属于此类情况。

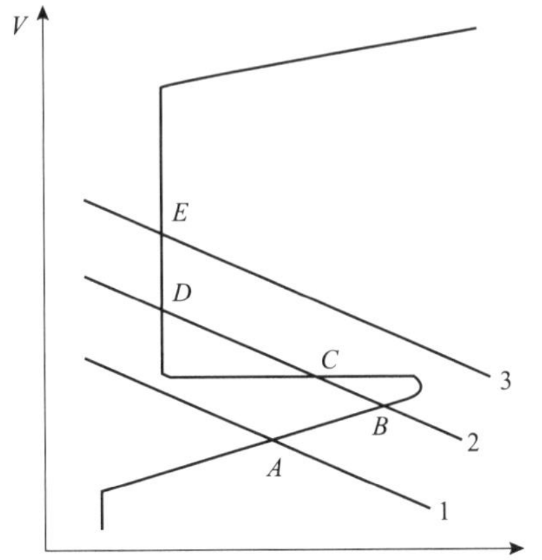

Figure 4-25 Schematic diagram of the intersection of different cathodic polarization curves and the anodic polarization curve of a passivating metal

图 4-25 不同的阴极极化曲线与钝化金属的阳极极化曲线的交点示意

4.4.4 Passivation theory

Metal passivation is an interfacial phenomenon, which does not change the properties of the metal body, but only changes the stability of the metal surface in the medium. It is a very complicated process for metals to change from an active state to a passive state, and there are still different views on its mechanism, and there is no complete theory that can explain all the passivation phenomena. At present, there are two theories that are considered to be able to explain most of the experimental facts satisfactorily, namely the phase-forming film theory and the adsorption theory.

1. Phase-forming film theory

Passivation is due to the formation of a dense solid product protective film with good coverage on the metal surface when the metal dissolves. This protective film exists as an independent phase, which either completely separates the metal from the electrolyte solution, or It strongly retards the progress of the anodic process, and as a result, the dissolution rate of the metal is greatly reduced, that is, the metal is transformed into a passive state.

2. Adsorption theory

The passivation of metals does not require the formation of solid product films, but it is sufficient to generate an adsorption layer of oxygen or oxygen-containing particles on the metal surface or part of the surface to passivate the metal. When these particles are adsorbed on the metal surface, the structure of the metal-solution interface is changed, and the activation energy of the anodic reaction is significantly increased, so the reactivity of the metal surface itself is reduced, that is, a

4.4.4 钝化理论

金属钝化是一种界面现象，它没有改变金属本体的性能，只是使金属表面在介质中的稳定性发生了变化。金属由活性状态变为钝态是一个很复杂的过程，目前对其机理还存在不同的看法，还没有一个完整的理论可以解释所有的钝化现象。目前认为能较满意地解释大部分实验事实的两种理论，即成像膜理论和吸附理论。

1.成像膜理论

钝化是由于金属溶解时，在金属表面形成了致密的、覆盖性能良好的固体产物保护膜，这层保护膜作为一个独立的相而存在，它或者使金属与电解质溶液完全隔开，或者强烈地阻滞了阳极过程的进行，结果使金属的溶解速度大大降低，亦即使金属转变为钝态。

2.吸附理论

金属钝化并不需要形成固态产物膜，而只需在金属表面或部分表面生成氧或含氧粒子的吸附层就足够使金属发生钝化。当这些粒子在金属表面吸附后，就改变了金属–溶液界面的结构，并使阳极反应的活化能显著升高，因而金属表面本身的反应能力降低了，亦即

passive state is presented.

3. Comparison of the two theories

These two passivation theories can better explain most of the experimental facts, and they are also supported by a large number of experimental facts, indicating that both theories partially reflect the nature of the passivation phenomenon. However, neither theory can A more comprehensive and complete explanation of various passivation mechanisms. The similarity between these two theories is that they both believe that the formation of a very thin passivation film on the metal surface hinders the dissolution of the metal, but the explanations for the surface film formation are different. The adsorption theory believes that as long as a two-dimensional film with a monomolecular layer is formed, the metal can be passivated; the phase-forming film theory believes that to protect the metal (insolubility), at least a three-dimensional film with a thickness of several molecular layers must be formed. The initially formed single-molecule adsorption film can only reduce the dissolution of the metal, and the thickened phase-forming film can achieve complete passivation. The difference between the two passivation theories involves the definition of passivation, adsorption film and phase-forming film, and there is not much essential difference. In fact, under different conditions in the passivation process of metals, the adsorption film and the phase-forming film can play the main roles, respectively.

Therefore, under different specific conditions, whether a phase-forming film or an adsorbed oxide film is formed on the surface, it may become the cause of passivation.

呈现出钝态。

3.两种理论的比较

这两种钝化理论都能较好地解释大部分实验事实，也都有大量的实验事实作为支持，说明这两种理论都部分反映了钝化现象的本质，然而无论哪一种理论都不能较全面、完整地解释各种钝化机理。这两种理论的相同之处是都认为由于在金属表面生成一层极薄的钝化膜而阻碍金属溶解，而对于表面成膜的解释，却各不相同。吸附理论认为，只要形成单分子层的二维膜就能导致金属产生钝化；而成像膜理论认为，要使金属得到保护（不溶解），至少要形成几个分子层厚的三维膜，而最初形成的单分子吸附膜只能减轻金属溶解，增厚的成像膜才能达到完全钝化。两种钝化理论之间的差别涉及对钝化、吸附膜和成像膜的定义问题，并无多大本质性区别。事实上，金属在钝化过程中的不同条件下，吸附膜和成像膜可分别起主要作用。

因此，在不同的具体条件下，表面无论形成成像膜还是吸附性氧化物膜，都有可能成为钝化的原因。

4.5 Hydrogen evolution corrosion and oxygen absorption corrosion

When the metal is corroded in acid, there is no other oxidant except H^+ in the acid solution, and the reduction reaction of hydrogen ions is the only cathodic reaction. Therefore, the study of hydrogen evolution reaction plays an important role in understanding the hydrogen evolution corrosion of metals.

In a neutral or alkaline medium, the hydrogen ion concentration is relatively low, so the hydrogen evolution equilibrium potential is relatively low. For some less active metals, the anodic dissolution equilibrium potential is relatively high, and the conjugation reaction of corrosion dissolution of these metals in neutral or alkaline media is often not the precipitation reaction of hydrogen, but the reduction reaction of dissolved oxygen, that is, the oxygen depolarization reaction promotes the continuous corrosion of the metal as the anode. This corrosion process is called oxygen-absorbing corrosion.

In the metal and electrolyte corrosion system, there is not only the dissolution of the anode, that is, the ionization process of the metal in the anode area, but also the process of reducing the corrosive agent in the cathode area. The types of cathodic processes are:

① Hydrogen precipitation reaction-H^+ ion discharge in acidic solution.

$$2H^+ + 2e^- \longrightarrow H_2 \tag{4-25}$$

② Reduction of oxygen-When there is dissolved oxygen in the solution, oxygen molecules are reduced in.

In neutral or alkaline solutions:

$$O_2 + 2H_2O + 4e^- \longrightarrow 4OH^- \tag{4-26}$$

4.5 析氢腐蚀与吸氧腐蚀

金属在酸中腐蚀时，设酸溶液中除H^+以外没有其他的氧化剂，氢离子的还原反应则是唯一的阴极反应。因此，析氢反应的研究对了解金属的析氢腐蚀起着重要作用。

在中性或碱性介质中，氢离子浓度较低，所以析氢平衡电势也较低。对某些不太活泼的金属，其阳极溶解平衡电势又较高，则这些金属在中性介质或碱性介质中的腐蚀溶解的共轭反应往往不是氢的析出反应，而是溶解氧的还原反应，即氧去极化反应促使作为阳极的金属不断被腐蚀。这种腐蚀过程称为吸氧腐蚀。

在金属与电解液腐蚀系统中，既要有阳极溶解，即阳极区金属的离子化过程，也要有在阴极区腐蚀剂被还原的过程。阴极过程的种类如下：

①氢的析出反应——在酸性溶液中H^+离子放电。

（4–25）

②氧的还原——溶液中有溶解氧时，氧分子被还原。

在中性或碱性溶液中：

（4–26）

In acidic solution:

在酸性溶液中：

$$O_2 + 2H^+ + 4e^- \longrightarrow H_2O \qquad (4\text{–}27)$$

③ Reduction of high valent metal ions, such as trivalent iron ions in Fe corrosion products:

③高价金属离子还原，如Fe锈蚀产物中的三价铁离子还原：

$$Fe(OH)_3 + e^- \longrightarrow Fe(OH)_2 + OH^- \qquad (4\text{–}28)$$

$$Fe_3O_4 + H_2O + 2e^- \longrightarrow 3FeO + 2OH^- \qquad (4\text{–}29)$$

④ Noble metal ion deposition, such as Cu^{2+} deposition on iron or aluminum:

④贵金属离子沉积，如Cu^{2+}在铁或铝上沉积：

$$Cu^{2+} + 2e^- \longrightarrow Cu \qquad (4\text{–}30)$$

⑤Reduction by oxidizing acid or some negative ions

⑤氧化性酸或某些负离子还原：

$$NO_3^- + 4H^+ + 3e^- \longrightarrow NO_2^- + 2H_2O \qquad (4\text{–}31)$$

⑥ Reduction of some organic compounds-more common in corrosion of chemical equipment, such as:

⑥某些有机化合物还原——多见于化工器械的腐蚀，如：

$$RO + 4H^+ + 4e^- \longrightarrow RH_2 + H_2O \qquad (4\text{–}32)$$

$$R + 2H^+ + 2e^- \longrightarrow RH_2 \qquad (4\text{–}33)$$

The electrons in the above cathode reaction process all come from the anode metal, which promotes the anode metal ions to continuously enter the solution. Among them, hydrogen precipitation reaction and oxygen reduction reaction are the most common cathode processes.

以上阴极反应过程的电子均来自阳极金属，从而促使阳极金属离子不断进入溶液，其中氢析出反应和氧还原反应是最常见的阴极过程。

4.5.1 Hydrogen evolution corrosion

4.5.1 析氢腐蚀

Hydrogen evolution corrosion is the metal corrosion caused by the hydrogen ion in the solution as a depolarizer, which discharges on the cathode and promotes the continuous dissolution of the metal anode.

析氢腐蚀是溶液中的氢离子作为去极化剂，在阴极上放电，促使金属阳极溶解过程持续进行而引起的金属腐蚀。

(1) Necessary conditions for hydrogen evolution corrosion

（1）析氢腐蚀的必要条件

The corrosion with hydrogen ion reduction reaction as the cathode process is called hydrogen evolution corrosion. The anode potential in the corrosion cell must be lower than the hydrogen evolution potential of the cathode.

以氢离子还原反应为阴极过程的腐蚀称为析氢腐蚀，腐蚀电池中的阳极电位必须低于阴极的析氢电位。

$$V_{anode} < V_H \qquad (4\text{–}34)$$

Hydrogen evolution potential refers to the difference between the equilibrium potential of hydrogen and the hydrogen overvoltage on the cathode under a certain cathode current density, i.e:

析氢电位是指在一定阴极电流密度下，氢的平衡电位和阴极上氢的超电压之差，即：

$$V_H = V_{eH} - \eta_H \quad (4\text{-}35)$$

Where: V_H ——hydrogen evolution potential under known cathode current density;

V_{eH} ——Equilibrium electrode potential of hydrogen;

η_H ——At this cathode current density, the hydrogen overvoltage.

式中 V_H ——已知阴极电流密度下的析氢电位；

V_{eH} ——氢的平衡电极电位；

η_H ——在该阴极电流密度下氢的超电压。

Electrode potential of hydrogen, V_{eH} can be measured experimentally or calculated according to nemst equation.

氢的电极电位，V_{eH} 可实验测定，也可根据能斯特方程式计算。

Whether a metal will undergo hydrogen evolution corrosion in a given corrosion medium can be judged by the above calculation. Generally speaking, hydrogen evolution corrosion occurs in low potentail metals, such as Fe, Zn, etc. In oxygen-free non-oxidizing acids and in very low potential metals, such as Mg, in neutral or alkaline solutions. However, for some highly passivative metals, such as Ti and Cr, the hydrogen evolution corrosion conditions can be met by thermodynamic calculation. However, since the passivation film is still stable in dilute acid, the actual potential is higher than the hydrogen evolution potential, so hydrogen evolution corrosion does not occur.

一种金属在给定的腐蚀介质中是否会发生析氢腐蚀，可通过上述计算来判断。一般来说，电势较低的金属，如Fe、Zn等在不含氧的非氧化性酸中，以及电势非常低的金属，如Mg，在中性溶液或碱性溶液中都发生析氢腐蚀，但是，对于一些强钝化性金属，如Ti、Cr，从热力学计算可满足析氢腐蚀条件，但由于钝化膜在稀酸中仍很稳定，实际电势高于析氢电势，因而不发生析氢腐蚀。

(2) Cathodic process of hydrogen evolution corrosion and hydrogen overvoltage

（2）析氢腐蚀的阴极过程和氢的超电压

The hydrogen ion cathodic depolarization reaction mainly consists of the following processes occuring in succession, as shown in Figure 4-26:

氢离子阴极去极化反应主要由下列几个过程连续发生（如图4-26所示）：

① Migration, convection and diffusion of hydrogen hydrate ions to the cathode surface:

①水化氢离子迁移、对流、扩散到阴极表面：

$$H_2O \cdot H^+ \rightarrow \text{阴极表面} \quad (4\text{-}36)$$

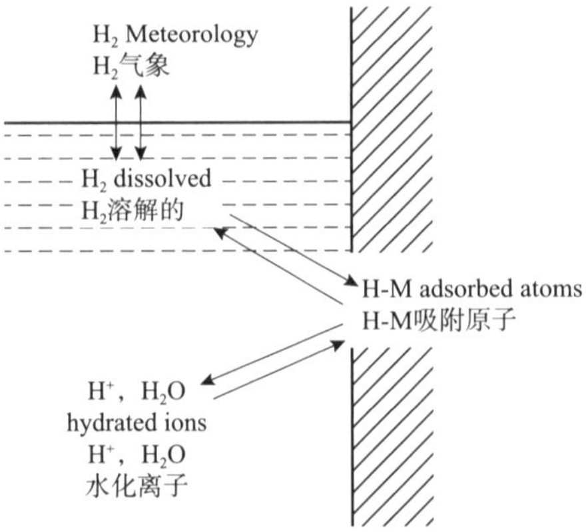

Figure 4-26 Schematic diagram of hydrogen electrode reaction
图 4-26 氢电极反应示意图

② The hydrated hydrogen ions are dehydrated, and the hydrogen ions are discharged into hydrogen atoms and adsorbed on the surface of the metal:

②水化氢离子脱水，氢离子放电成为氢原子并吸附在金属表面：

$$H_2O \cdot H^+ \longrightarrow H_2O + H^+ \tag{4-37}$$

$$H^+ + e^- \longrightarrow M - H_{吸附} \tag{4-38}$$

③ After complex desorption or electrochemical desorption, hydrogen atoms combine to form hydrogen molecules:

③复合脱附或电化学脱附后氢原子结合成氢分子：

$$(M - H_{吸附}) + (M - H_{吸附}) \xrightarrow{复合脱附} H_2 \tag{4-39}$$

$$(M - H_{吸附}) + (H_3O^+ + e^-) \xrightarrow{电化学脱附} H_2 \tag{4-40}$$

④ Hydrogen molecules on the electrode surface escape through diffusion and aggregation into hydrogen bubbles.

④电极表面的氢分子通过扩散、聚集成氢气泡逸出。

Among these steps, if a certain step is carried out slowly, the entire hydrogen depolarization reaction will be blocked, and the electrons transferred from the anode will accumulate at the cathode, causing the cathode potential to move in the negative direction, thus generating a constant overvoltage. The higher the overvoltage value of hydrogen, the smaller the depolarization effect of hydrogen, and the more difficult the hydrogen evolution corrosion process is.

这些步骤中，如果某一步骤进行得较缓慢，就会使整个氢去极化反应受到阻滞，由阳极迁移过来的电子就会在阴极积累，使阴极电势向负方向移动，从而产生一定的超电压。氢的超电压值越大，氢的去极化作用就越小，析氢腐蚀过程就越难发生。

A large number of experimental results show that hydrogen evolution is overvoltage,the following relationship exists between η_H and cathode current density i_c:

大量实验结果表明，析氢超电压 η_H 与阴极电流密度 i_c 之间存在下列关系：

$$\eta_H = a + b\lg i_c \quad (4\text{-}41)$$

Where: a——over potential under unit current density;

b——independent of the type of material, most metals are 0.116V.

As shown in Table 4-1, this rule is caused by the activation polarization of the hydrogen evolution reaction. Therefore, the control steps of the hydrogen depolarization reaction cannot be steps ① and ④. The slow discharge theory thinks that step ② is the control step of the whole hydrogen evolution process, and this step is the slowest, while the slow composite theory emphasizes that step ③ is the control step. The b value obtained according to the slow discharge theory is 116mV, which is almost the same as that measured on most metal electrodes. Therefore, the slow discharge theory is more universal. However, there are also a few metals (such as Pt, etc.), and it is more appropriate to explain the cause of hydrogen evolution overvoltage with the slow recombination theory.

式中　a——单位电流密度下的过电位；

b——与材料的种类无关，大多数金属为0.116V。

如表4-1所示，此规律由析氢反应的活化极化引起，因此，氢去极化反应的控制步骤不可能是步骤①和④。迟缓放电理论认为第②步是整个析氢过程的控制步骤，该步骤最慢，而迟缓复合理论则强调第③步为控制步骤。根据迟缓放电理论求得的b值为116mV，与大多数金属电极上实测的b值大致相同，因此迟缓放电理论更具有普遍意义。但也有少数金属（如Pt等），用迟缓复合理论解释其析氢超电压的成因更合适。

Table 4-1　Values of *a* and *b* for hydrogen evolution reaction of different metals(25℃)

表 4-1　不同金属析氢反应的 *a* 和 *b* 的值（25℃）

Metal 金属	Acidic solution 酸性溶液		Alkaline solution 碱性溶液	
	a/V	b/V	a/V	b/V
Pt	0.10	0.03	0.31	0.1
Pd	0.24	0.03	0.53	0.13
Au	0.40	0.12		
W	0.43	0.10		
Co	0.62	0.14	0.60	0.14
Ni	0.63	0.11	0.65	0.10
Mo	0.60	0.08	0.67	0.14
Fe	0.70	0.12	0.76	0.11
Mn	0.80	0.10	0.90	0.12
Nb	0.80	0.10		
Ti	0.82	0.14	0.83	0.14
Bi	0.84	0.12		
Cu	0.87	0.12	0.96	0.12
Ag	0.95	0.10	0.73	0.12

续表

Metal 金属	Acidic solution 酸性溶液		Alkaline solution 碱性溶液	
	a/V	*b*/V	*a*/V	*b*/V
Ge	0.97	0.12		
Al	1.00	0.10	0.64	0.14
Sb	1.00	0.11		
Be	1.08	0.12		
Sn	1.20	0.13	1.28	0.23
Zn	1.24	0.12	1.20	0.12
Cd	1.40	0.12	1.05	0.16
Hg	1.41	0.114	1.54	0.11
Ni	1.55	0.14		
1% C steel	1.56	0.11	1.36	0.25

(3) Factors affecting hydrogen evolution corrosion

In hydrogen evolution corrosion, when the metal or alloy is uniformly corroded in acid, if the impurity or alloy phase as the cathode has a low hydrogen evolution overvoltage, the corrosion rate is large; on the contrary, if the impurity or alloy phase has a high hydrogen evolution overvoltage, the corrosion rate is small.

Figure 4-27 shows hydrogen evolution overvoltage of different metals η_H with cathode current density. It can be seen from the figure that the hydrogen evolution overvoltage has a linear relationship with the logarithm of cathode current density, which is consistent with that described by Tafel equation. Generally, the constant *a* is related to the electrode material properties, surface conditions, solution composition and temperature, and its value is equal to the hydrogen overvoltage under the unit current density. The larger *a*, the greater the hydrogen overvoltage at *a* given current density. The constant *b* is independent of the electrode material, and the value of hydrogen evolution reaction on various metal cathodes is approximately the same, about 0.11~0.12V. It can be

（3）影响析氢腐蚀的因素

在析氢腐蚀中，金属或合金在酸中发生均匀腐蚀时，如果作为阴极的杂质或合金相具有较低的析氢超电压，则腐蚀速度较大；反之，若杂质或阴极相上的析氢超电压较大，则腐蚀速度较小。

图4-27显示了不同金属η_H随阴级电流密度的变化。由图可看出析氢超电压、与阴极电流密度的对数呈线性关系，与Tafel方程式所描述的相吻合。*a*越大，在给定电流密度下的析氢超电压越大。而常数*b*与电极材料无关，各种金属阴极上析氢反应的值大致相同，在0.11~0.12V。从图中可看出不同金属的析氢塔菲尔直线基本平行。

seen from the figure that Tafel straight lines of hydrogen evolution of different metals are basically parallel.

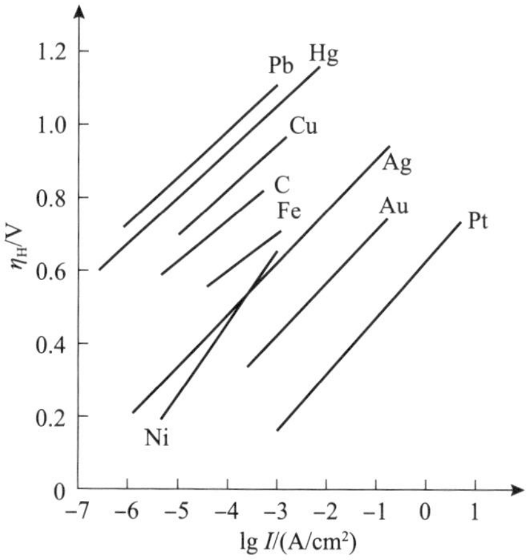

Figure 4-27 Relationship between hydrogen evolution overvoltage and current density on different metals

图 4-27 不同金属上的析氢超电压与电流密度的关系

According to the value of a, metals can be roughly divided into three categories, from which we can see the influence of metal materials on hydrogen evolution overvoltage:

根据a值的大小，可将金属大致分成三类，由此可看出金属材料对析氢超电压的影响：

High hydrogen overpotential metal: a value is between 1.0 and 1.5, such as Pb, Cd, Hg, Ti, Zn, Bi, Sn, etc.

高氢过电位金属：a值在1.0~1.5，如Pb、Cd、Hg、Ti、Zn、Bi、Sn等。

Medium hydrogen overpotential metal: a value is between 0.5 and 1.0, such as Fe, Co, Ni, Ag.

中氢过电位金属：a值在0.5~1.0，如Fe、Co、Ni、Ag等。

Low hydrogen overpotential metal: a value is between 0.1 and 0.5, such as Au, Pt, Pd, W.

低氢过电位金属：a值在0.1~0.5，如Au、Pt、Pd、W等。

Different metal materials have the same a, which is mainly caused by different exchange current density of hydrogen evolution reaction on different metals or different hydrogen evolution reaction mechanism. For example, metals with low hydrogen overvoltage, such as Pt and Pd, have great catalytic activity for hydrogen ion discharge, so that the exchange current density of

不同金属材料的a不同，这主要是因为不同金属上析氢反应的交换电流密度不同，或是析氢反应机理不同引起的。例如，低氢超电压金属（如Pt、Pd等）对氢离子放电有很大的催化活性，使析氢反应的交换

hydrogen evolution reaction is large, and the ability to adsorb hydrogen atoms is also strong. Thus, the slowest step in the reduction reaction of hydrogen on such metals is the composite desorption of adsorbed hydrogen atoms. The catalytic ability of high voltage metals for hydrogen ion discharge reaction is very weak, so the exchange current density is very small. Therefore, the slow discharge of hydrogen ions on these metals constitutes the control step of hydrogen depolarization process. For metals with moderate hydrogen overvoltage, such as Fe, Ni, Cu, etc., the slowest step in the process of hydrogen depolarization may be the electrochemical desorption reaction of adsorbed hydrogen.

电流密度很大，同时吸附氢原子的能力也很强，从而造成氢在这类金属还原反应过程中最慢的步骤为吸附氢原子的复合脱附。高氢超电压金属对氢离子放电反应的催化能力很弱，因而交换电流密度很小，因此，这类金属上氢离子的迟缓放电构成了氢去极化过程的控制步骤。对于中等氢超电压金属，如Fe、Ni、Cu等，氢去极化过程中最慢的步骤可能是吸附氢的电化学脱附反应。

$$(M-H_{吸附})+(H_3O^+ + e^-)\xrightarrow{电化学脱附}H_2 \tag{4-42}$$

The surface state of the electrode also affects the hydrogen evolution overvoltage. Therefore, for the same metal material, the hydrogen overvoltage on the rough surface is smaller than that on the smooth surface, because the true surface area on the rough surface is larger than that on the smooth surface.

电极表面状态对析氢超电压也有影响。因此相同的金属材料，粗糙表面的析氢超电压比光滑表面要小，这是因为粗糙表面的真实表面积比光滑表面要大。

If the solution contains platinum ions, they will precipitate on the corrosion metal Fe to form an additional cathode. Since the precipitation overvoltage of hydrogen on Pt is much smaller than that on Fe, the corrosion of Fe in acid is accelerated.

如果溶液中含有铂离子，它们将在腐蚀金属Fe上析出，形成附加阴极。由于氢在Pt上的析出超电压比在Fe上小得多，从而加速了Fe在酸中腐蚀。

On the contrary, if there is some surfactant in the solution, it will adsorb on the metal surface and hinder the precipitation of hydrogen, thus significantly improving the hydrogen evolution overvoltage. This surfactant can be used as a corrosion inhibitor to prevent metal corrosion in acid.

相反，如果溶液中含有某种表面活性剂，会在金属表面吸附并阻碍氢析出，从而显著提高析氢超电压。这种表面活性剂就可作为缓蚀剂，防止金属在酸中腐蚀。

The effect of pH value of solution on hydrogen evolution overvoltage is as follows: in acidic solution, hydrogen overvoltage increases with the increase of pH; In alkaline solution, the hydrogen overvoltage decreases with the increase of pH.

溶液pH值对析氢超电压的影响是：在酸性溶液中，析氢超电压随着pH增加而增大；在碱性溶液中，析氢超电压随着pH增加而减小。

The hydrogen overvoltage decreases with the increase of solution temperature. Generally, the hydrogen overvoltage decreases by about 2mV for every 1℃ increase in temperature.

溶液温度升高，析氢超电压降低。一般温度每升高1℃，析氢超电压约降低2mV。

The corrosion rate increases with the increase of cathode area.

阴极面积增加腐蚀速度加快。

Hydrogen evolution corrosion is mostly controlled by cathode or mixed cathode and anode, and the corrosion rate mainly depends on the hydrogen evolution overvoltage. Therefore, in order to reduce or prevent hydrogen evolution corrosion, we should try to reduce the cathode area and increase the hydrogen evolution overvoltage. For hydrogen evolution corrosion controlled by anode passivation, passivation shall be strengthened to prevent activation. The main ways to reduce and prevent hydrogen evolution corrosion are as follows:

析氢腐蚀多数为阴极控制或阴、阳极混合控制，腐蚀速度主要决定于析氢超电压的大小。因此，为减小或防止析氢腐蚀，应设法减小阴极面积，提高析氢超电压。对于阳极钝化控制的析氢腐蚀，则应加强钝化，防止活化。减小和防止析氢腐蚀的主要途径如下：

① Reduce or eliminate harmful impurities in metal, especially cathodic impurities with small hydrogen evolution overvoltage. The solution may precipitate noble metals on the metal, and the precipitation of noble metals provides an effective cathode. If the hydrogen evolution overvoltage of the precipitated noble metal is very small, it will accelerate the corrosion and should be removed.

①减少或消除金属中的有害杂质，特别是析氢超电压小的阴极性杂质。溶液可能在金属上析出贵金属，贵金属的析出提供了有效的阴极。如果此析出贵金属的析氢超电压很小，则会加速腐蚀，也应设法去除。

② Add components with high hydrogen overvoltage, such as Hg, Zn, Pb, etc.

②加入析氢超电压大的成分，如Hg、Zn、Pb等。

③ Add corrosion inhibitor to increase hydrogen evolution overvoltage. Such as acid washing inhibitor

③加入缓蚀剂，增大析氢超电压，如酸洗缓蚀剂若丁，

rodine, whose effective component is di-o-toluene thiourea.

其有效成分为二邻甲苯硫脲。

④ Reduce active anionic components, such as Cl^-, S^{2-}, etc.

④降低活性阴离子成分，如Cl^-、S^{2-}等。

4.5.2 Oxygen absorption corrosion

4.5.2 吸氧腐蚀

(1) Necessary conditions for oxygen absorption corrosion

In neutral or alkaline media, the concentration of hydrogen ions is often relatively low, so the equilibrium potential of hydrogen evolution is also relatively low. For some less active metals, the anodic dissolution equilibrium potential is relatively high. Therefore, the conjugate reaction of corrosion dissolution of these metals in neutral or alkaline media is usually not the precipitation reaction of hydrogen, but the reduction reaction of dissolved oxygen. The corrosion taking the reduction reaction of oxygen as the cathode process is called oxygen reduction corrosion or oxygen consuming corrosion. The necessary condition for oxygen absorption corrosion is that the potential of the metal is lower than that of the oxygen reduction reaction.

（1）吸氧腐蚀的必要条件

在中性介质或碱性介质中，氢离子浓度较低，所以析氢平衡电势也较低。对某些不太活泼的金属而言，其阳极溶解平衡电势又较高，因此这些金属在中性介质或碱性介质中腐蚀溶解的共轭反应通常不是氢析出反应，而是溶解氧还原反应。以氧还原反应为阴极过程的腐蚀，称为氧还原腐蚀或吸氧腐蚀。发生吸氧腐蚀的必要条件是金属电势比氧还原反应的电势低。

$$V_{\mathrm{M}}^{0} < V_{\mathrm{eO_2}}^{0} \tag{4-43}$$

Under the same conditions, $V_{\mathrm{eO_2}}^{0} \gg V_{\mathrm{eH}}^{0}$, the driving force of oxygen absorption corrosion is greater, so oxygen absorption corrosion is more likely to occur than hydrogen evolution corrosion.

相同条件下$V_{\mathrm{eO_2}}^{0} \gg V_{\mathrm{eH}}^{0}$，吸氧腐蚀的驱动力更大，所以吸氧腐蚀比析氢腐蚀更容易发生。

Oxygen consumption corrosion: metal corrosion caused by the continuous dissolution of anode metal due to the oxygen consumption reaction on the cathode.

耗氧腐蚀：阴极上耗氧反应的进行，促使阳极金属不断溶解而引起的金属腐蚀。

In neutral solution:

中性溶液中：

$$O_2 + 2H^+ + 4e^- \longrightarrow 2OH^- \quad (4-44)$$

In alkaline solution:

碱性溶液中：

$$O_2 + 2H_2O + 4e^- \longrightarrow 4OH^- \quad (4-45)$$

In acidic solution:

酸性溶液中：

$$O_2 + 4H^+ + 4e^- \longrightarrow 2H_2O \quad (4-46)$$

In nature, the solution is connected with the atmosphere, and oxygen is dissolved in the solution. The reduction potential of oxygen in neutral solution was 0.805V. It can be seen that as long as the potential of the metal in the solution is lower than this value, oxygen absorption corrosion may occur. Therefore, many metals can undergo oxygen absorption corrosion in neutral or alkaline solutions, in humid atmosphere, fresh water, seawater and humid soil, and even in acidic media. It is not difficult to see that compared with hydrogen evolution corrosion, oxygen absorption corrosion has more general and important significance.

自然界中，溶液与大气相通，氧气溶解在溶液中。在中性溶液中氧还原电势为0.805V。可见，只要金属在溶液中的电势低于这一数值，就可能发生吸氧腐蚀。所以，许多金属在中性溶液或碱性溶液中，在潮湿大气、淡水、海水、潮湿土壤中，都能发生吸氧腐蚀，甚至在酸性介质中也会有部分吸氧腐蚀。不难看出，与析氢腐蚀相比，吸氧腐蚀具有更普遍更重要的意义。

(2) Cathodic process of oxygen absorption corrosion and oxygen overvoltage

（2）吸氧腐蚀的阴极过程和氧的超电压

The cathodic depolarizer for oxygen absorption corrosion is dissolved oxygen in solution. With the progress of corrosion, oxygen consumed needs to be supplemented by oxygen in the air. Oxygen enters the solution from the air and migrates to the cathode surface for reduction reaction. The process includes:

吸氧腐蚀的阴极去极化剂是溶液中溶解的氧。随着腐蚀进行，消耗掉的氧，需空气中的氧来补充。如图4–28所示，氧从空气中进入溶液并迁移到阴极表面发生还原反应，此过程包括：

① Oxygen enters the solution through the air / solution interface;

①氧穿过空气/溶液界面进入溶液；

② Under the convection of the solution, oxygen migrates to the vicinity of the cathode surface;

②在溶液对流作用下，氧迁移到阴极表面附近；

③ Within the diffusion layer, oxygen diffuses to the cathode surface under the action of concentration

③在扩散层范围内，氧在浓度梯度作用下扩散到阴极

gradient;

④ The reduction reaction of oxygen molecules on the cathode surface is also called the ionization reaction of oxygen.

Among the four steps, steps ① and ② do not generally become the control step. In most cases, step ③ is the control step, and step ④ can become the control step in the corrosion medium with enhanced stirring or flow.

In acidic solution, the total reaction of oxygen reduction is:

表面；

④在阴极表面氧分子发生还原反应，也称氧离子化反应。

这四个步骤中，步骤①和②一般不成为控制步骤，多数情况下步骤③为控制步骤，在加强搅拌或流动的腐蚀介质中，步骤④可成为控制步骤。

在酸性溶液中，氧还原总反应为：

$$O_2 + 2H_2O + 4e^- \longrightarrow 4OH^- \quad (4\text{-}47)$$

(3) Oxygen ionization reaction process

（3）氧离子化反应过程

Formation of semivalent oxygen ions:

半价氧离子的形成：

$$O_2 + e^- \longrightarrow O_2^- \quad (4\text{-}48)$$

Formation of hydrogen dioxide:

二氧化一氢的形成：

$$O_2^- + H^+ \longrightarrow HO_2 \quad (4\text{-}49)$$

Formation of hydrogen dioxide ions:

二氧化一氢离子的形成：

$$HO_2 + e^- \longrightarrow HO_2^- \quad (4\text{-}50)$$

Hydrogen peroxide formation:

形成过氧化氢：

$$HO_2^- + H^+ \longrightarrow H_2O_2 \quad (4\text{-}51)$$

Forming hydroxide ions:

形成氢氧根离子：

$$H_2O_2 + 2e^- \longrightarrow 2OH^- \quad (4\text{-}52)$$

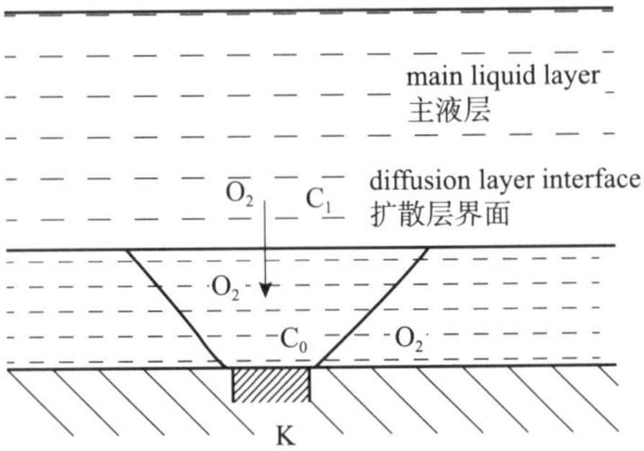

Figure 4-28 Oxygen transport to cathode

图 4-28 氧向阴极输送

ABC section: electrochemical polarization zone, $J_c < \frac{1}{2}J_d$ cathodic overpotential follows Tafel formula:

如图4–29所示，氧去极化过程的阴极极化曲线，*ABC*段：电化学极化区，$J_c < \frac{1}{2}J_d$ 阴极过电势服从Tafel公式：

$$\eta_{O_2} = a' + b'\lg J \tag{4-53}$$

a'——The electrochemical polarization overpotential of oxygen under potential current density depends on the electrode material;

b'——It is about 0.12V and has little relationship with the electrode material.

式中 *a'*——电位电流密度下氧的电化学极化过电位，取决于电极材料；

b'——约为0.12V，与电极材料关系不大。

BD section: mixed control area, $\frac{1}{2}J_d < J_c < J_d$ the expression of oxygen overpotential is:

*BD*段：混合控制区域，$\frac{1}{2}J_d < J_c < J_d$ 氧过电位的表达式为：

$$\eta_{O_2} = a' + b'\lg J_c + b'\lg\left(\frac{J_d}{J_d - J_c}\right) \tag{4-54}$$

EN segment: concentration polarization control area, the expression of oxygen overpotential is:

*EN*段：浓差极化控制区，氧过电位的表达式为：

$$\eta_{O_2} = \frac{RT}{nF}\ln\left(\frac{J_d}{J_d - J_c}\right) = b'\lg\left(\frac{J_d}{J_d - J_c}\right) \tag{4-55}$$

EFG section: oxygen and hydrogen combined depolarization zone $J_c \rightarrow J_d$.

*EFG*段：氧、氢联合去极化区$J_c \rightarrow J_d$。

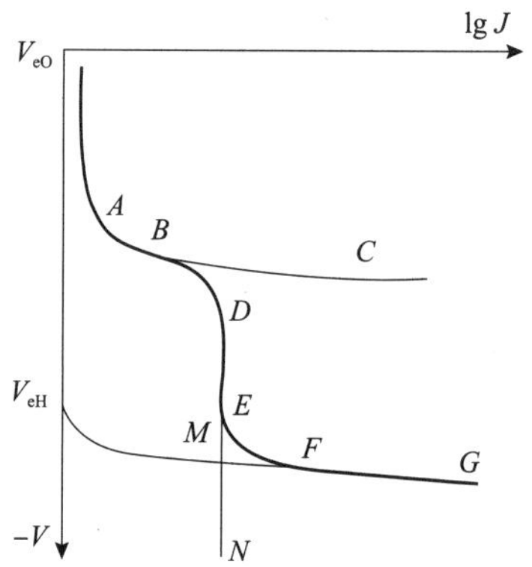

Figure 4–29 Tafel polarization curve of oxygen depolarization process
图4–29 氧去极化过程的阴极极化曲线

(4) Factors affecting oxygen absorption corrosion

If the oxygen absorption corrosion is controlled by the cathode and the oxygen supply speed is fast, the corrosion current is less than the limit diffusion current of oxygen. The corrosion rate of metal mainly depends on the activation overpotential of cathodic oxygen reduction reaction. Therefore, the cathode impurities, alloy composition and structure, micro cathode area, etc. in the metal will affect the oxygen absorption corrosion rate. However, in most cases, the oxygen supply rate is limited, and the oxygen absorption corrosion is controlled by the diffusion process of oxygen. In this case, the corrosion rate of metal is equal to the limiting diffusion current density of oxygen.

（4）影响吸氧腐蚀的因素

如果吸氧腐蚀受阴极控制，且供氧速度很快，腐蚀电流小于氧的极限扩散电流。金属的腐蚀速度主要取决于阴极氧还原反应的活化过电势。因而金属中的阴极杂质、合金成分和组织、微阴极面积等都将影响吸氧腐蚀速度。但大多数情况下，供氧速度有限，吸氧腐蚀受氧的扩散过程控制。这种情况下，金属的腐蚀速度就等于氧的极限扩散电流密度。

$$i_{corr} = i_{1,O_2} = nFD_{O_2}C_{O_2}^0 / \delta \qquad (4\text{–}56)$$

(5) Effect of dissolved oxygen concentration

With the increase of dissolved oxygen concentration, the limiting diffusion current density of oxygen increases, resulting in the increase of oxygen absorption corrosion rate, as shown in Figure 4-30. The initial potential of the cathode polarization curve 2 with high oxygen content is higher than that of the curve 1 with low oxygen content, i.e $\varphi_{O,2} > \varphi_{O,1}$, and the limiting diffusion current density is large, so the corrosion rate with high oxygen content is large, and the corrosion potential is also high. However, if the corrosion metal has passivation characteristics, when the oxygen concentration increases to a certain extent i_{1,O_2} reaches the passivation current density i_{pp} of the metal, the metal will change from active state to passive state instead, and the corrosion rate of oxygen depolarization will be significantly reduced. It can be seen that dissolved oxygen often has opposite double effects on metal corrosion,

（5）溶解氧浓度的影响

随着溶解氧浓度增大，氧的极限扩散电流密度增大，导致吸氧腐蚀速度增大，如图4–30所示。氧含量高的阴极极化曲线2比氧含量低的曲线1初始电势高，即 $\varphi_{O,2} > \varphi_{O,1}$，且极限扩散电流密度大，因而氧含量高的腐蚀速度大，腐蚀电势也较高，但如果腐蚀金属具有钝化特性，则当氧浓度增大到一定程度时，由于 i_{1,O_2} 达到了该金属的致钝电流密度 i_{pp}，则该金属反而由活化状态转为钝态，氧去极化的腐蚀速度将显著降低，可见，溶解氧对金属腐蚀有着相反的双重影响，这对研究具有钝化行

which is of great significance to study the corrosion of metal with passivation behavior in neutral solution.

为的金属在中性溶液中的腐蚀有重要意义。

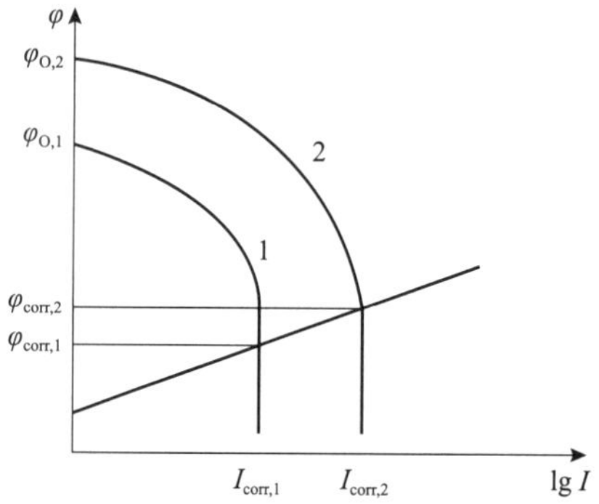

Figure 4-30 Effect of dissolved oxygen concentration on oxygen absorption corrosion

1—low oxygen concentration; 2—high oxygen concentration

图 4-30 溶解氧浓度对吸氧腐蚀的影响

1—氧浓度小；2—氧浓度大

(6) Effect of temperature

Generally speaking, the increase of the solution temperature is beneficial to the thermal movement of the solution and the improvement of the reaction speed of the interface electrode. As the temperature increases, the solution viscosity decreases, and the diffusion coefficient of dissolved oxygen increases. Therefore, the increase of temperature will accelerate the corrosion. But for the open system, when the temperature rises to a certain degree, especially near the boiling point, the oxygen solubility decreases sharply, and the corrosion rate decreases. Therefore, as shown in Figure 4-31, the corrosion rate of iron in water under the open system reaches the maximum near 80℃. For the closed system, the increase of temperature will increase the oxygen partial pressure in the gas phase, and there is a tendency to increase the oxygen solubility, which is roughly offset by the decrease of oxygen solubility in the solution due to the increase of temperature. Therefore, the corrosion rate will always increase with the increase of temperature.

（6）温度的影响

一般来说，溶液温度升高有利于溶液热运动及提高界面电极反应速度。温度升高，溶液黏度降低，从而溶解氧的扩散系数增大，故温度升高会加速腐蚀。但对于敞开体系，当温度升到一定程度，尤其接近沸点时，氧溶解度急剧降低，而使腐蚀速度减小。如图4-31所示，敞开体系下铁在水中的腐蚀速度在80℃附近达到最大值。对于封闭体系，温度升高使气相中氧分压增大，有增加氧溶解度的趋势，这与温度升高使氧在溶液中溶解度降低的作用大致抵消，因此腐蚀速度将一直随着温度升高而增大。

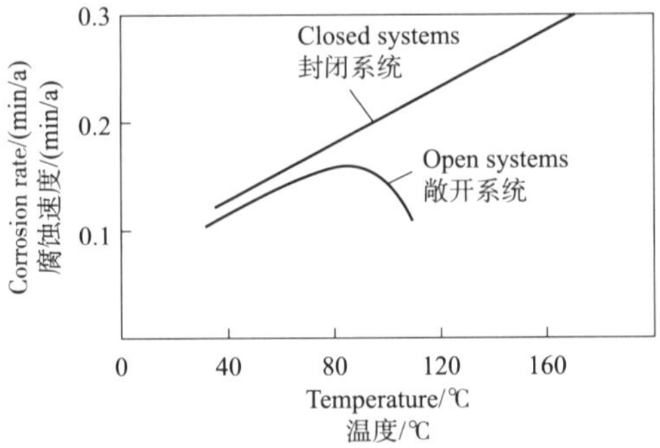

Figure 4-31 Effect of temperature on corrosion rate of iron in water

图 4-31 温度对铁在水中腐蚀速度的影响

(7) Effect of salt concentration

The amount of dissolved salt has a dual effect on the corrosion of metals. On the one hand, in neutral solution, when the salt concentration is low, the conductivity of the solution increases and the corrosion rate increases with the increase of the salt concentration. For example, when the mass fraction of NaCl increases to 3% (about equivalent to the NaCl content in seawater), the corrosion rate of iron reaches the maximum, as shown in Figure 4-32. On the other hand, with the further increase of NaCl concentration, the corrosion rate of iron decreases due to the significant decrease of oxygen solubility.

（7）盐浓度的影响

溶盐量对金属腐蚀有着双重作用。一方面，在中性溶液中，当盐浓度较低时，随着盐浓度增加，溶液电导率增大，腐蚀速度有所上升。例如，当NaCl质量分数增加到3%时（大约相当于海水中NaCl含量），铁的腐蚀速度达到最大值，如图4-32所示。另一方面，随着NaCl浓度进一步增加，由于氧溶解度显著降低，反而使铁的腐蚀速度下降。

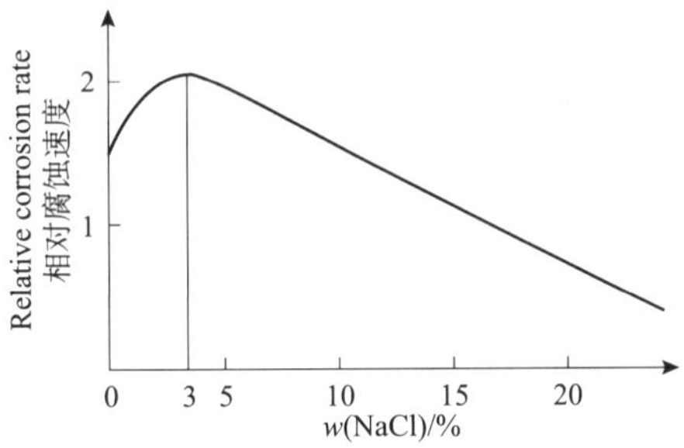

Figure 4-32 Effect of NaCl concentration on corrosion rate of iron in aerated solution

图 4-32 NaCl 浓度对铁在充气溶液中腐蚀速度的影响

(8) Effect of solution stirring and flow rate

Solution agitation or flow rate increase can reduce the thickness of diffusion layer, increase the limiting

（8）溶液搅拌和流速的影响

溶液搅拌或流速增加，可使扩散层厚度减小，氧的极限

diffusion current of oxygen, and thus increase the corrosion rate. As shown in Figure 4-33, in the laminar flow area, the corrosion rate increases slowly with the increase of the flow rate. When the flow rate increases to the speed at which turbulence begins to occur, that is, the critical velocity is V_{cri}, the turbulent liquid breaks through the almost static boundary layer close to the metal surface, and the protective film is damaged to a certain extent, so the corrosion rate increases sharply. In fact, the corrosion type has changed from uniform corrosion under laminar flow to wear corrosion under turbulent flow (turbulent corrosion). When the flow rate rises to a certain value, the anode polarization curve will not intersect with the concentration polarization part of the oxygen absorption reaction polarization curve, but will intersect with the activation polarization part. At this time, the corrosion rate is no longer controlled by the limit diffusion current of anode oxygen, and the corrosion type is no longer comprehensive corrosion, but turbulent corrosion. As the flow velocity further increases to the high flow velocity region, cavitation will be caused.

扩散电流增加，因而腐蚀速度增大。如图4–33所示，在层流区内，腐蚀速度随着流速增加而缓慢上升。当流速增加到开始出现湍流的速度，即临界速度为V_{cri}时，湍流液体击穿紧贴金属表面的、几乎静止的边界层，并使保护膜发生一定程度的破坏，因此腐蚀速度急剧增加。实际上，腐蚀类型已由层流下的均匀腐蚀，变为湍流下的磨损腐蚀（湍流腐蚀）。当流速上升到某一数值后，阳极极化曲线不会再与吸氧反应极化曲线的浓差极化部分相交，而与活化极化部分相交。这时，腐蚀速度不再受阳极氧极限扩散电流控制，腐蚀类型也不再是全面腐蚀，而变为湍流腐蚀。随着流速进一步提高到高流速区，将引起空泡。

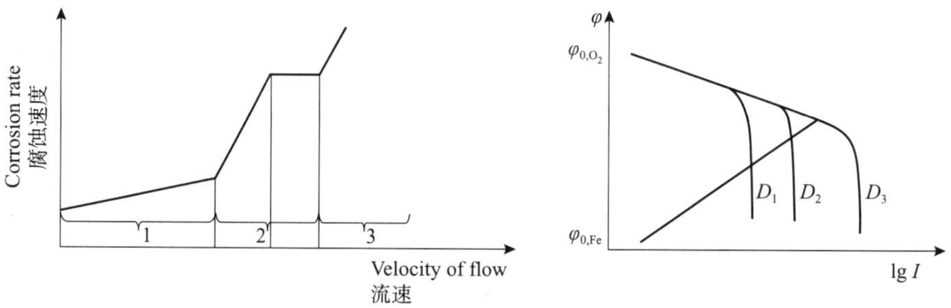

(a)Effect of flow rate on corrosion rate and type of corrosion (b)Oxygen absorption corrosion at different flow rates
(a)流速对腐蚀速率及腐蚀类型的影响 (b)不同流速下的吸氧腐蚀

Figure 4–33 Effect of flow rate on corrosion
图4–33 流速对腐蚀的影响

(9) Influence of the relative size of anode and cathode area on corrosion rate

The relative area ratio of anode and cathode has

（9）阴、阳极面积相对大小对腐蚀速度的影响

腐蚀电池中阴、阳极的相

a great influence on the corrosion rate of anode. If the anode area is small relative to the cathode, the anode will be rapidly corroded. This is because the corrosion current is concentrated on a small area.

对面积比对阳极的腐蚀速率有很大的影响。如果相对于阴极，阳极面积很小，则阳极将迅速被腐蚀。这是由于腐蚀电流集中于一个很小的面积上。

$$I_a=I_c=I_{corr} \quad (4\text{-}57)$$

Similarly, large cathodes may not be easily polarized and thus maintain a relatively high corrosion rate. The corrosion system of large cathode and small anode will accelerate the corrosion rate of metal anode. In the engineering practice, the structural system of large cathode and small anode shall be avoided as much as possible, and the structure of large anode and small cathode shall be designed as much as possible when there is no alternative.

同样，大阴极可能不易极化，因此保持较高的腐蚀速率。大阴极小阳极的腐蚀体系会加快金属阳极的腐蚀速度。工程实践中应尽量避免大阴极小阳极的结构体系，迫不得已的情况下尽量设计成大阳极小阴极的结构。

4.6 Influencing factors of electrochemical corrosion

4.6 电化学腐蚀影响因素

The influencing factors of electrochemical corrosion include metallurgical factors and environmental factors.

电化学腐蚀影响因素包括冶金因素对电化学腐蚀的影响，环境因素对电化学腐蚀的影响。

1. Oxidant and dissolved oxygen

In the non passivated metal corrosion system, the corrosion rate increases with the increase of oxidant concentration. For the metal corrosion system that is easy to passivate, the influence law is complex:

(1) If the presence of oxidant can not make the metal enter the passivation state, the corrosion rate of the activated metal increases with the increase of oxidant concentration.

(2) If the presence of oxidant can make the metal passivated stably, the corrosion rate of metal

1.氧化剂和溶解氧

非钝化金属腐蚀体系，腐蚀速度随着氧化剂浓度增加而增大。对易钝化金属腐蚀体系，影响规律较复杂：

（1）若氧化剂的存在不能使金属进入钝化状态，活化状态金属的腐蚀速度随着氧化剂浓度增加而增大。

（2）若氧化剂的存在能够使金属稳定钝化，金属腐蚀速

will decrease sharply, and will not change with the concentration of oxidant in a certain range.

度急剧减小，并且在一定的范围内不随氧化剂浓度而改变。

(3) If the oxidant concentration is too high or the activity is too large, the redox equilibrium potential of the oxidant is higher than the over passivation potential of the metal, so that the metal is in the over passivation state, and the corrosion rate of the metal increases with the increase of the oxidant concentration.

（3）若氧化剂浓度过高或活性太大，氧化剂氧化还原平衡电位高于金属过钝电位，使金属处于过钝化状态，金属腐蚀速度又随着氧化剂浓度增加而增大。

2. pH value of medium

2.介质的pH值

As shown in Figure 4-34, the metal corrosion velocity is related to the pH, in hydrogen evolution corrosion and oxygen absorption corrosion, with the decrease of pH value, the cathodic equilibrium potential V_{ec} moves forward and the corrosion rate increases.

如图4-34所示，金属腐蚀速度与pH值有关，析氢腐蚀和吸氧腐蚀中，随着pH值减小，阴极平衡电位V_{ec}正移，腐蚀速度增大。

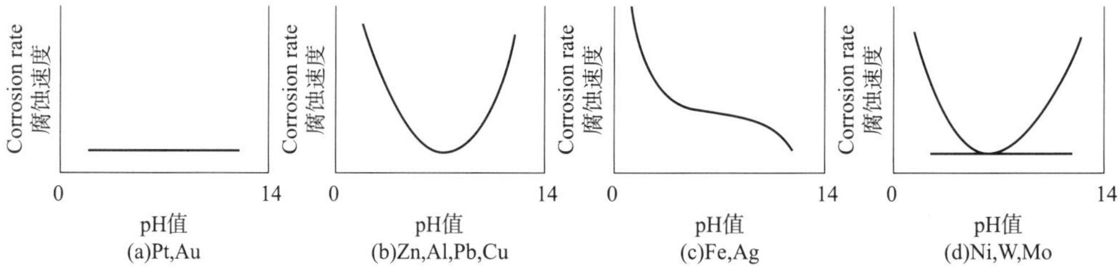

Figure 4-34 Elationship between corrosion rate of metal and pH value

图4-34 金属腐蚀速度与pH值的关系

For the metal that is easy to be passivated, the passivation tendency is generally greater with the increase of pH value. The influence of pH value on the corrosion rate of different metals is different, which mainly depends on the nature of metals.

对于易钝金属而言，随着pH值增加，钝化倾向越大。pH值对不同金属腐蚀速度的影响不尽相同，主要取决于金属的本性。

3. Salt and concentration in the medium

3.介质中的盐类及浓度

After hydrolysis of some salts, the pH value of the solution is changed, thus affecting the corrosion rate of the metal.

某些盐类水解后，改变溶液pH值，从而影响金属腐蚀速度。

The anions and cations ionized from the salts may act as depolarizers for the cathode or anode of the

盐类电离出的阴、阳离子可能作为腐蚀体系的阴极或

corrosion system, thus affecting the corrosion rate.

阳极去极化剂，从而影响腐蚀速度。

Some active ions, especially halogen anions, have a special destructive effect on the passivation film and promote local corrosion of metals.

某些活性离子，特别是卤素阴离子对钝化膜有特殊的破坏作用，促使金属发生局部腐蚀。

Some salts have no direct relationship with the corrosion reaction, but after dissolution, they increase the conductivity of the medium and accelerate the electrochemical corrosion process; At the same time, it reduces the solubility of oxygen and reduces the corrosion rate in some cases.

有的盐类本身与腐蚀反应没有直接关系，但溶解后增大了介质电导率，加速电化学腐蚀过程；同时，又减小了氧溶解度，在某些场合降低腐蚀速度。

Medium temperature: water temperature has a great influence on the corrosion process of steel caused by dissolved oxygen. In a closed system, the higher the water temperature, the faster the metal corrosion. This is because when the temperature rises, the diffusion speed of various substances in the aqueous solution increases and the resistance of the electrolyte aqueous solution decreases, which will accelerate the process of corroding the anode and cathode electrodes of the battery.

介质温度：水温对溶解氧引起的钢铁腐蚀过程有较大的影响。在封闭系统中，水的温度越高，金属腐蚀速度越快。这是因为温度升高时，各种物质在水溶液中的扩散速度加快和电解质水溶液的电阻降低，这些都会加速腐蚀电池阴阳两极的电极过程。

Medium flow: Generally speaking, the greater the flow rate of water, the faster the diffusion of various substances in water, thus accelerating the corrosion rate. In the open equipment where oxygen in the air enters the aqueous solution and causes corrosion, when the flow rate of water reaches a certain value, a large amount of oxygen will form a protective film on the metal surface, so the corrosion speed will slow down; However, when the flow rate of water is large, the protective film will be damaged and the corrosion rate will increase due to the mechanical scouring effect of water flow.

介质流动：一般来说，水流速越大，水中各种物质扩散速度也越快，从而使腐蚀速度加快。在空气中氧进入水溶液而引起腐蚀的敞口式设备中，当水流速达到一定数值时，多量的氧会使金属表面形成保护膜，所以腐蚀速度减慢；但当水流速很大时，由于水流的机械冲刷作用，保护膜遭到破坏，腐蚀速度又会增高。

4.7 References

[1] LI J J, ZHANG S, TANG Y P. Teaching design based on the cultivation of undergraduate research awareness: A case study of "The principle and application of galvanic batteries" [J]. Chemical Education, 2019, 40(24): 29-33.

[2] WANG B Q, ZHANG X L, YONG X Y. Numerical simulation study on galvanic corrosion of copper/steel pipes after coupling in seawater pipeline system of ships [J]. Journal of Chinese Society for Corrosion and Protection, 2022, 42(2): 200-210.

[3] GAO L J, LI H X. Study on corrosion behavior of A3 steel in soil with salt concentration difference battery [J]. Oil & Gas Storage and Transportation, 2000(2): 29–31+59–5.

4.7 参考文献

[1] 李佳佳，张拴，唐于平. 基于本科生科研意识培养的教学设计：以"原电池的原理及应用"为例[J]. 化学教育(中英文)，2019，40(24)：29–33.

[2] 王炳钦，张晓莲，雍兴跃. 舰船海水管系中紫铜/钢制管道耦接后接触腐蚀的数值模拟研究[J]. 中国腐蚀与防护学报，2022，42(2)：200–210.

[3] 高立君，李洪锡. A3 钢在土壤中盐浓差电池腐蚀行为的研究[J]. 油气储运，2000(2)：29–31+59–5.

Localized Corrosion 局部腐蚀 5

5.1 The corrosion morphology

5.1.1 General corrosion

If the corrosion is carried out on the entire surface of the metal, it is called total corrosion. Its characteratistic is corrosion is distributed throughout the metal surface, resulting in the reduction of the cross-section size of metal components until complete destruction.

The electrochemical process of total corrosion is characterized by the fact that the area size of the cathode and anode of the corrosion cell are very small and closely connected. Figure 5-1 illustrates the cross-section morphology of the oxide film of domestic E308L stainless steel electrode at 3000 h. The thickness of the oxide film of uniform corrosion products is less than 3μm.

5.1 腐蚀形态

5.1.1 全面腐蚀

如果腐蚀是在整个金属表面进行的，则称为全面腐蚀。其特征在于腐蚀分布在整个金属表面，结果使金属构件截面尺寸减小，直至完全破坏。

全面腐蚀的电化学过程特点：腐蚀电池的阴极和阳极面积尺寸非常微小且紧密相连。图 5－1 所示为3000h下国产E308L不锈钢焊条的截面氧化膜形貌，均匀腐蚀产物氧化膜的厚度在3μm以下。

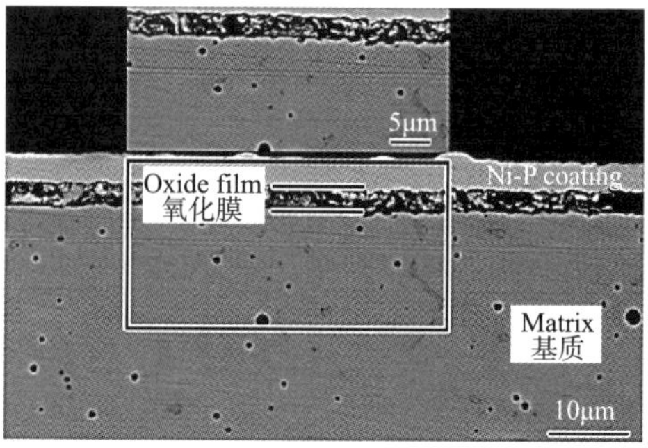

Figure 5-1 Morphology of cross section oxide film of domestic E308L stainless steel electrode at 3000 h

图 5-1 3000h 下国产 E308L 不锈钢焊条的截面氧化膜形貌

Overall corrosion often causes a large amount of loss of alloy materials, but will not constitute a sudden accident, the corrosion rate can be monitored/detected by linear polarization probe, resistance probe, etc. In addition, the engineering design and equipment manufacturing will allow sufficient corrosion margins to prevent the components from being damaged by comprehensive corrosion. For example, the corrosion margin C_2 should not less than 1mm in the national standards for pressure vessels. The method to control overall corrosion is shown in Figure 5-2: the reserved corrosion margin of the heat exchanger shell [1].

全面腐蚀往往造成合金材料的大量损失，但不会构成突发性事故，其腐蚀速率可通过线性极化探针、电阻探针等进行监/检测。另外，工程设计和装备制造均会预留足够的腐蚀裕量来防止部件因全面腐蚀而被破坏。例如，GB/T 150—2011《压力容器》中预留腐蚀裕量C_2不小于1mm。图5-2所示为控制全面腐蚀的方法：换热器壳体的预留腐蚀裕量[1]。

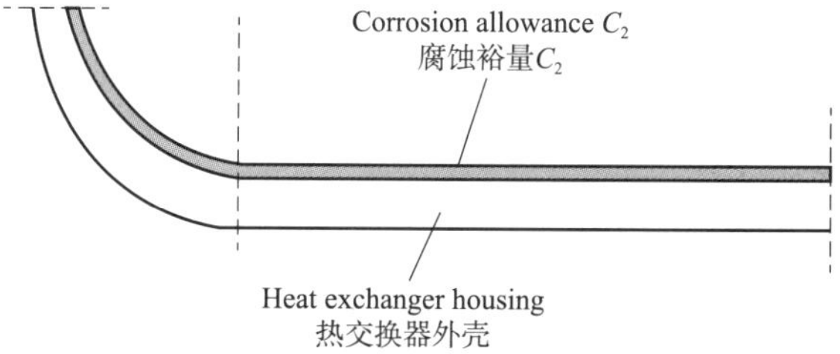

Figure 5-2 Corrosion allowance C_2 of heat exchanger shell in national standard

图 5-2 国家标准中换热器壳体的预留腐蚀裕量 C_2

5.1.2 Localized corrosion

5.1.2 局部腐蚀

If the corrosion is carried out on the surface of a part

如果腐蚀是在部分金属表

of the metal, it is called localized corrosion. Including galvanic corrosion, pitting corrosion, crevice corrosion, selective corrosion, intergranular corrosion, stress corrosion cracking, corrosion fatigue, wear corrosion and other corrosion forms.

面进行的，则称为局部腐蚀。包括电偶腐蚀、点蚀、缝隙腐蚀、选择性腐蚀、晶间腐蚀、应力腐蚀开裂、腐蚀疲劳、磨损腐蚀等腐蚀形态。

1. Characteristics of localized corrosion

1.局部腐蚀的特征

(1) Local corrosion usually exists the anode and cathode regions of macroscopic and microscopic corrosion cells that can be clearly identified and compared [2].

（1）局部腐蚀通常存在可以明确识别的和比较固定的宏观和微观腐蚀电池的阳极区和阴极区[2]。

(2) Usually, the anode area is relatively small, while the cathode area is relatively large in locally corroded corrosion cells.

（2）通常局部腐蚀的腐蚀电池中阳极面积相对较小，阴极面积相对很大。

(3) The electrochemical reaction process of local corrosion often has the characteristics of autocatalysis, which further creates the conditions for the occurrence of corrosion reaction, so that the development of local corrosion can be continuously accelerated.

（3）局部腐蚀的电化学反应过程具有自催化特征，进一步创造了腐蚀反应发生的条件，使得局部腐蚀得以持续地加速发展。

2. Hazards of localized corrosion

2.局部腐蚀的危害

Local corrosion accounts for a large proportion of the corrosion of chemical equipment, while overall corrosion only accounts for a small part. Figure 5-3 shows the statistics of the causes of chemical equipment damage in Mitsubishi Chemical Machinery Corporation in Japan in the past 10 years. Among them, stress corrosion and pitting corrosion are the main causes of chemical equipment damage, accounting for 45.6% and 21.6% [3].

局部腐蚀在化工设备腐蚀中占很大比例，全面腐蚀只占一小部分。图5-3所示日本三菱化工机械公司10年化工设备破坏原因统计。其中，应力腐蚀和孔蚀是化工设备破坏的主要原因，占比在45.6%和21.6%[3]。

5.1.3 References

5.1.3 参考文献

[1] BU H M, LI X W, QI J T, et al. Study on corrosion of alloys using microelectrode array technology [J]. Materials

[1] 卜红梅，李肖蔚，齐建涛，等. 利用微电极阵列技术研究合金

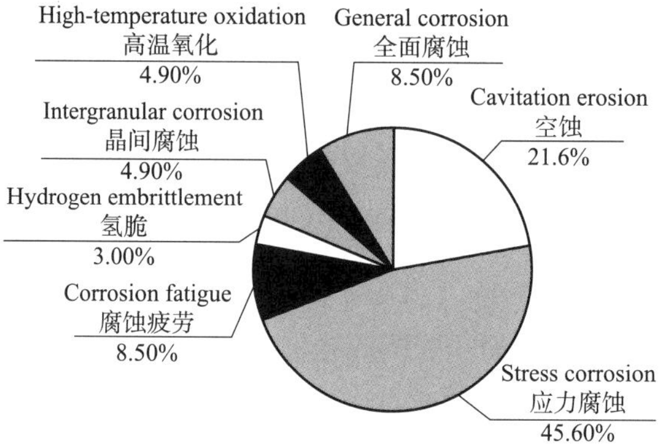

Figure 5-3 Cause statistics of chemical equipment damage of Mitsubishi Chemical Machinery Corporation in Japan in 10 years

图 5-3 日本三菱化工机械公司 10 年化工设备破坏原因统计

Reports, 2019, 33(23): 3963-3970.

[2] WANG S H. Research on application of complete sets of anti-corrosion technology of cold change equipment in refinery [D]. QingDao China University of Petroleum, 2009.

[3] Ghali E, Dietzel W, Kainer K U. General and localized corrosion of magnesium alloys : A critical review [J]. Journal of Materials Engineering & Performance, 2004, 13 (1): 7-23.

的腐蚀 [J]. 材料导报，2019，33（23）：3963-3970.

[2] 王世宏 . 炼厂冷换设备成套防腐技术应用研究 [D]. 青岛：中国石油大学（华东），2009.

[3]（文献同英文）

5.2 Galvanic corrosion

5.2 电偶腐蚀

5.2.1 Definition and mechanism of galvanic corrosion

5.2.1 定义及机理

Dissimilar metals in contact with each other or through other conductors, in the same medium, will cause local corrosion of the contact part. The dissolution rate of the metal with lower potential increases. Higher potential metal, the dissolution rate is reduced, this corrosion is called galvanic corrosion, or contact corrosion, bimetallic corrosion [1].

异种金属彼此接触或通过其他导体连通，处于同一介质中，会造成接触部分的局部腐蚀。其中电位较低的金属，溶解速度增大；电位较高的金属，溶解速度反而减小，这种腐蚀称为电偶腐蚀，或称为接触腐蚀、双金属腐蚀[1]。

In galvanic corrosion cells, the metal with lower

在电偶腐蚀电池中，腐蚀

corrosion potential is in contact with the metal with higher corrosion potential, resulting in anodic polarization, which increases the dissolution rate. However, the metal with more positive potential will undergo cathodic polarization due to contact with the metal with lower potential, resulting in a decrease in the dissolution rate, that is, it is protected by cathodic protection.

电位较低的金属由于和腐蚀电位较高的金属接触而产生了阳极极化，其结果使溶解速度增加；而电位较正的金属，由于和电位较低的金属接触而发生阴极极化，结果使溶解速度下降，即受到了阴极保护。

5.2.2 Influencing factors of galvanic corrosion and control measures

5.2.2 影响因素及控制措施

Galvanic sequence: in the galvanic sequence, the more distant the materials contact with each other, the more serious the galvanic corrosion; The closer the materials are to each other, the less the galvanic corrosion will occur[2].

电偶序：在电偶序中相距越远的材料互相接触，电偶腐蚀就越严重；相距越近的材料接触，发生电偶腐蚀的倾向就越小[2]。

Medium Conditions: Medium composition (the Cu-Fe pair will be reversed in NH_4^+); medium temperature (Cu-Zn pair reversed in hot water over 80℃); the pH value of the medium (Mg-Al pair from neutral or slightly acidic to alkaline change when the reversal); stirring or flowing medium; the resistivity of the medium (weak conductivity only at the junction; extensive severe distribution of conductive strong corrosion).

介质条件：介质成分（Cu-Fe偶对在NH_4^+中会逆转）；介质温度（Cu-Zn偶对在超过80℃热水中逆转）；介质的pH值（Mg-Al偶对由中性或微酸向碱性变化时的逆转）；介质的搅拌或流动；介质的电阻率（导电弱仅在结合部；导电强腐蚀严重分布面广）。

Area of effect: The area ratio of large cathode to small anode should be avoided. The following needs to be done: reasonable material selection (select the metal phase combination with close position in the electric couple sequence); reasonable structural design (including: try to avoid the structure of small anode and large cathode; different metals are insulated from each other; insert the third metal; design the anode parts to be easily replaceable or properly thickened to extend life, and so

面积效应：应避免大阴极和小阳极的面积比，需要做到以下几点：合理选材（选择在电偶序中位置靠近的金属相组合）；合理的结构设计（包括尽量避免小阳极、大阴极的结构；将不同的金属彼此绝缘；插入第三种金属；阴极保护将阳极部件设计成易于更换的，

on); application of coating protective layer; improve the corrosive environment; cathodic protection.

或适当增厚以延长寿命等）；应用涂料保护层；改善腐蚀环境；阴极保护。

5.2.3 Case analysis

Galvanic corrosion of zinc/steel couple in $MgCl_2$ aqueous solution: Tada et al. conducted galvanic corrosion tests on zinc/steel couple using 1M(1M=1mol/L) NaCl and 1M $MgCl_2$ solution to investigate the effect of magnesium ions on the galvanic corrosion behavior of zinc/steel[3]. Two types of zinc/steel couples were used to measure open-circuit potentials and currents. The open circuit potential (OCP) transient results of zinc/steel couple show that the corrosion potential of 1M $MgCl_2$ is more correct than that of 1M NaCl during Zn sacrificial dissolution. However, the galvanic open circuit potential increased earlier in 1M $MgCl_2$ solution, indicating that the sacrificial dissolution rate of Zn increased more in 1M $MgCl_2$ solution than in 1M NaCl solution, which is consistent with the current transient results. The reason for this result is that hydrogen evolution occurs in 1M $MgCl_2$, which enhances the cathodic reaction on the surface of the zinc/steel galvanic couple[4].

5.2.3 案例分析

锌/钢电偶在$MgCl_2$水溶液中的电偶腐蚀：Tada等采用1M（1M=1mol/L）NaCl和1M $MgCl_2$溶液对锌/钢电偶进行了电偶腐蚀试验，研究了镁离子对锌/钢电偶腐蚀行为的影响[3]。采用两种类型的锌/钢电偶测量开路电位和电流。锌/钢电偶的开路电位（OCP）瞬态结果表明：在Zn牺牲溶解过程中，1M $MgCl_2$的腐蚀电位相较于1M NaCl更正。但在1M $MgCl_2$溶液中电偶开路电位较早增加，表明1M $MgCl_2$溶液中Zn的牺牲溶解速率比1M NaCl溶液中增加更多，这与电流瞬态的结果一致。造成这种结果的原因是在1M $MgCl_2$中发生析氢反应，增强了锌/钢电偶表面的阴极反应[4]。

5.2.4 References

[1] MA L X, CHE R X. Improved galvanic sequence method for determining metal corrosion rate [J]. Journal of Dalian Railway Institute, 1995, 16(3): 97-99.

[2] JI Y, XU Z, ZHANG B, et al. Review of micro-scale and atomic-scale corrosion mechanisms of second phases in aluminum alloys [J]. Transactions of Nonferrous Metals Society of China, 2021, 31（11）: 3205-3227.

5.2.4 参考文献

[1] 马龙骧，车如心. 测定金属腐蚀速度的改进电偶序法[J]. 大连铁道学院学报，1995，16（3）：97–99.

[2]（文献同英文）

[3] HUANG H, BU F. Galvanic corrosion of electronic material copper coupled silver-coating in electronic systems [J]. Anti-Corrosion Methods and Materials, 2019, 66 (6): 730-45.

[4] Tada E, Katakami S, Nishikata A. Galvanic corrosion of Zn/steel couple in aqueous $MgCl_2$ [J]. Corrosion Science and Technology. 2017, 16 (4): 183-186.

5.3 Pitting

5.3.1 Definition and characteristics of pitting corrosion

Pitting corrosion, also known as corrosive pitting or pitted corrosion, is characterized by the localized formation of pits (small holes) on the metal surface that may extend into the metal interior. (Cl^-, Cu^{2+}, Fe^{3+}, Hg^{2+} aggravate pore-erosion; NO_3^-, OH^-, and SO_4^{2-} delay and inhibit pore erosion)

Morphological features: According to different geometric shapes, it can be divided into different types, as shown in Figure 5-4 [1].

[3]（文献同英文）

[4]（文献同英文）

5.3 点蚀

5.3.1 定义及其特征

孔蚀又称点蚀、坑蚀，是一种集中发生在某些点处并向金属内部发展的孔、坑状腐蚀。（Cl^-、Cu^{2+}、Fe^{3+}、Hg^{2+}加剧孔蚀；NO_3^-、OH^-、SO_4^{2-}延缓、抑制孔蚀）

形貌特征：根据不同的几何形状可进行不同的划分，如图 5-4 所示[1]。某炼化厂原油输送管线钢的点蚀形貌如图5-5所示。

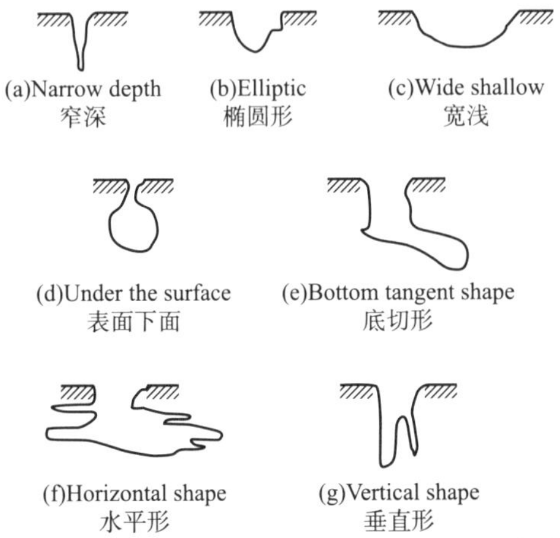

Figure 5-4 Morphology characteristics of pitting

图 5-4 点蚀的形貌特征

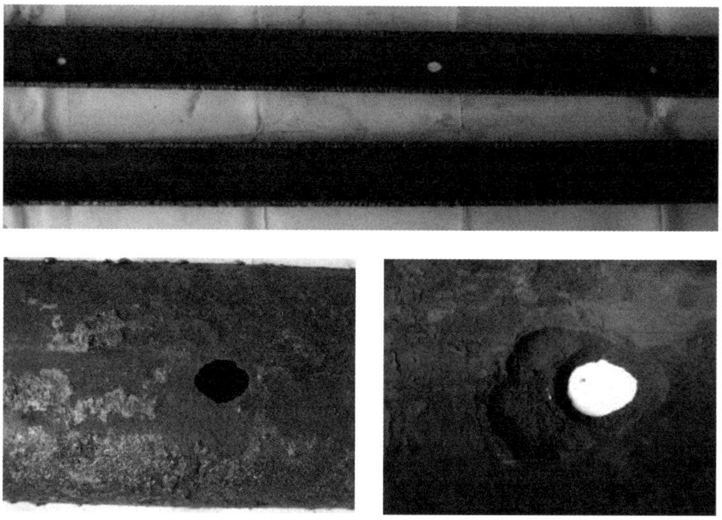

Figure 5-5 Pitting corrosion morphology of crude oil pipeline steel in a refining plant
图 5-5 某炼化厂原油输送管线钢的点蚀形貌

5.3.2 The mechanism of pitting

1. Nucleation of an etched pore

Pitting potential or rupture potential, breakdown potential (V_{br}).

Passivation film failure theory: The occurrence of holes is when corrosive anions (Cl^-) after passivation membrane adsorption, due to the small radius of Cl^- through the passivation membrane, after entering in the membrane polluted the passivation film, produced a strong induction ionic conduction, and the film on a fixed point become able to maintain high current density, and can make the cation mixed and disorderly movement and active, pitting occurs when the electric field at the membrane-solution interface reaches a critical value[2].

Adsorption theory: Pitting corrosion occurs due to the competitive adsorption of chloride ions and oxygen. Pitting occurs when the adsorption point of oxygen on

5.3.2 孔蚀的机理

1.蚀孔的形核

孔蚀电位或破裂电位、击穿电位（V_{br}）。

钝化膜破坏理论：小孔的发生是当腐蚀性阴离子（如Cl^-）在钝化膜上吸附以后，由于Cl^-半径小而穿过钝化膜，进入膜内后“污染了钝化膜”，产生了强烈的感应离子导电，于是此膜在一定点上变得能够维持高的电流密度，并能使阳离子杂乱移动而活跃起来，当膜–溶液界面的电场达到某一临界值时就发生点蚀[2]。

吸附理论：点蚀的发生是由于氯离子和氧的竞争吸附结果而造成的。当金属表面氧的

the metal surface is replaced by chloride ions.

吸附点被氯离子所替代时，就发生点蚀。

2. The development of pitting

2.孔蚀的发展

Block cell formation:

闭塞电池的形成如下。

Anodic reaction:

阳极反应：

$$Al \longrightarrow Al^{3+} + 3e^{-} \quad (5\text{-}1)$$

Cathodic reaction:

阴极反应：

$$O_2 + 2H_2O + 4e^{-} \longrightarrow 4OH^{-} \quad (5\text{-}2)$$

The overall reaction:

总的反应：

$$Al + 3O_2 + 6H_2O \longrightarrow 4Al(OH)_3 \longrightarrow 2Al_2O_3 + 6H_2O \quad (5\text{-}3)$$

Autocatalytic process of occluded cells[3]:

闭塞电池的自催化过程[3]如下：

① With the development of corrosion, Al^{3+} is concentrated in the corrosion pore. In order to maintain electrical neutrality, Cl^- is attracted by positive ions, and migrate and enrich into the pore.

①随着腐蚀的发展，蚀孔内Al^{3+}浓集，为保持电中性而受到正离子的吸引，Cl^-等向孔内迁移、富集。

② Al^{3+} hydrolyzes and acidifies the solution in the etching hole, which further accelerates the development of corrosion.

②Al^{3+}发生水解反应，并使蚀孔内溶液酸化，溶液酸化又进一步加速了腐蚀的发展。

Repassivation characteristics of pitting erosion[4]:

孔蚀的再钝化特点[4]如下：

With the development of pitting and the deepening of pitting, the pitting will slow down and stop development. Figure 5-6 shows the ring anode polarization curve.

随着孔蚀的发展，蚀孔的加深，也会发生孔蚀不断减速最终停止发展的现象。图5-6所示为环状阳极极化曲线。

① Above the pitting potential V_{br}, pitting will develop rapidly.

①在孔蚀电位V_{br}以上，孔蚀会迅速发生发展。

② When the potential is between V_{br} and V_p, the existing etch holes continue to develop, but no new etch holes will emerge.

②电位在V_{br}~V_p区间，已形成的蚀孔继续发展，但不会产生新的蚀孔。

③ When the potential is below V_p, the corrosion hole cannot occur, and the formed corrosion hole stops developing and enters the blunt state.

③电位在V_p以下，则蚀孔不能发生，已形成的蚀孔停止发展，进入钝态。

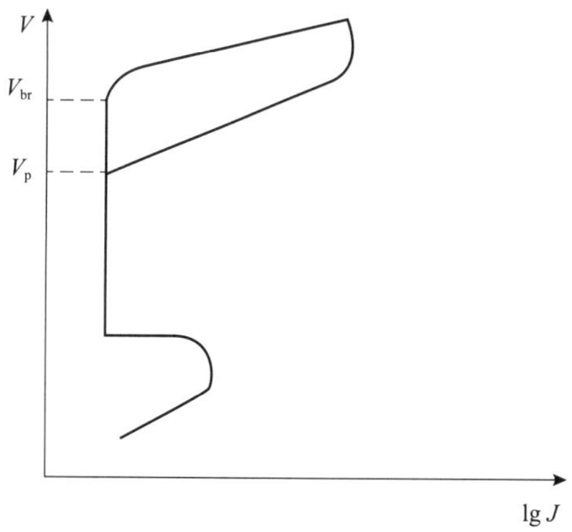

Figure 5-6 Polarization curve of ring anode

图 5-6 环状阳极极化曲线

The passivation characteristic curve of the metal is shown in Figure 5-7, which mainly includes four regions: active dissolution zone, activated passivation transition zone (unstable zone), passivation zone and over-passivation zone. Where, V_{pp} represents the passivation potential (critical potential), J_{pp} represents the obtuse current density, J_p represents the dimensional obtuse current density.

金属的钝化特性曲线如图5-7所示，主要包含四个区域，分别为活性溶解区、活化-钝化过渡区（不稳定区）、稳定钝化区及过钝化区。其中，V_{pp}代表钝化电位（临界电位）、J_{pp}代表致钝电流密度、J_p代表维钝电流密度。图5-8所示为铁形成钝化层的博克里斯模型。

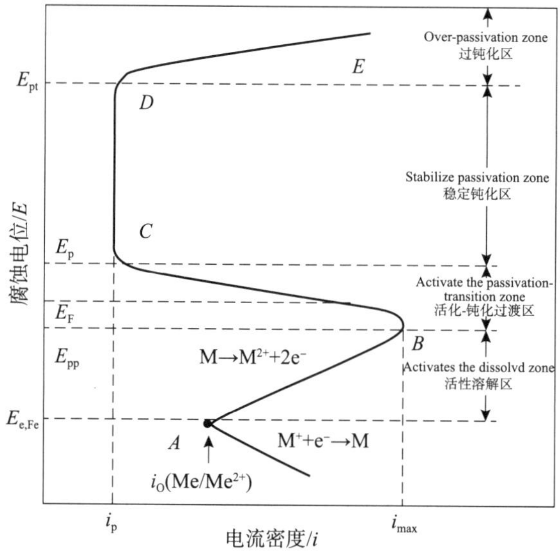

Figure 5-7 Passivation characteristic curve of metal

图 5-7 金属的钝化特性曲线

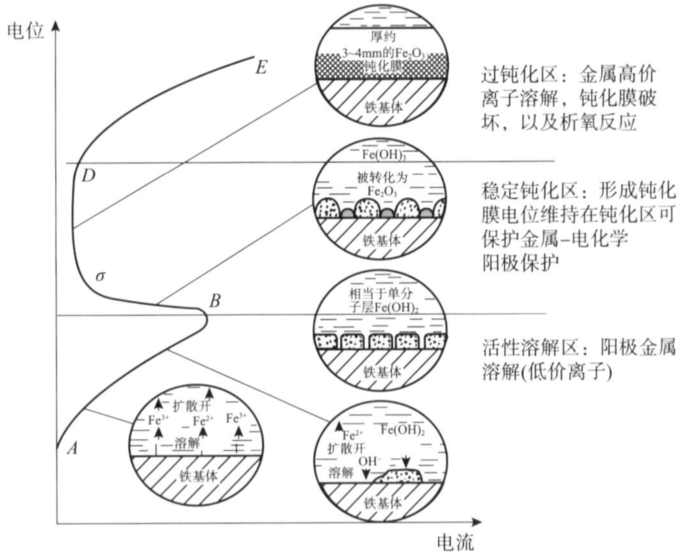

Figure 5-8 Bockless model of iron forming passivation layer

图 5-8 铁形成钝化层的博克里斯模型

5.3.3 Factors affecting pitting

Environmental factor: type of medium (stainless steel-Cl^-; $CuSO_4^{2-}$); medium concentration (occurs only when exceeding a certain concentration); the corrosion inhibition effect of other anions in the medium on stainless steel: $OH^- > NO_3^- > AC^- > SO_4^{2-} > ClO_4^-$; the influence of medium temperature; the effect of solution pH value: when pH >10, the pitting potential rises rapidly and the difficulty of pitting increases [5].

Metallurgical factors: stainless steel containing Mo, Ni and high Cr corrosion resistance; austenite is superior to ferrite and martensite; reduce C and S inclusions

5.3.3 影响孔蚀的因素

环境因素：介质的类型（不锈钢-Cl^-；Cu-SO_4^{2-}）；介质浓度（超过一定浓度才发生）；介质中其他阴离子的作用对不锈钢的缓蚀效果：$OH^- > NO_3^- > AC^- > SO_4^{2-} > ClO_4^-$；介质温度的影响；溶液pH值的影响：pH>10时孔蚀电位迅速上升，发生难度增加[5]。

冶金因素：不锈钢中含Mo、Ni及高Cr耐孔蚀；奥氏体优于铁素体、马氏体；减少C、S夹杂物。

5.3.4 Measures to prevent pitting

(1) Improve the medium conditions: reduce the Cl^- content in the solution, reduce the oxidant, reduce the temperature, improve the pH value and so on can reduce

5.3.4 防止孔蚀的措施

（1）改善介质条件：降低溶液中的Cl^-含量，减少氧化剂，降低温度，提高pH值等都

the occurrence of pitting corrosion.

(2) The selection of corrosion resistant alloy materials: choose austenitic stainless steel, duplex stainless steel and high purity ferritic stainless steel and other materials with good corrosion resistance.

(3) Electrochemical protection: using cathodic protection, cathode polarization makes the potential lower than V_{br}, so that the material is in a stable passivation zone.

(4) The surface of the alloy is passivated to improve the blunt stability of the material.

(5) Use of corrosion inhibitors: the use of corrosion inhibitors is most effective for closed systems. Common corrosion inhibitors used in stainless steel are nitrate, chromate, sulfate and alkali, the most effective being sodium nitrite. The amount of corrosion inhibitor should be suitable.

可以减少孔蚀的发生。

（2）选用耐孔蚀的合金材料：选用奥氏体不锈钢、双相不锈钢及高纯铁素体不锈钢等耐孔蚀性能好的材料。

（3）电化学保护：利用阴极保护，阴极极化使电位低于 V_{br}，使材料处于稳定钝化区。

（4）对合金表面进行钝化处理，提高材料钝态稳定性。

（5）使用缓蚀剂：对封闭系统使用缓蚀剂最有效。用于不锈钢的常用缓蚀剂有硝酸盐、铬酸盐、硫酸盐和碱，最有效的是亚硝酸钠。缓蚀剂的加入量要适宜。

5.3.5 References

[1] FAN X H, YU Y, ZHANG Z R, et al. Pitting corrosion and repassivation behavior of 316L austenitic stainless steel at different potentials [J]. Surface Technology, 2020, 49(7): 287-293, 318.

[2] RYAN M P, WILLIAMS D E, CHATER R J, et al. Why stainless steel corrodes [J]. Nature, 2002, 415(14): 770.

[3] SZKLARSKA-SMIALOWSKA Z. Mechanism of pit nucleation by electrical breakdown of the passive film [J]. Corrosion Science, 2002, 44(5): 1143.

[4] MARCUS P, MAURICE V, STREHBLOW H H. Localized corrosion (pitting): A model of passivity breakdown including the role of the oxide layer nanostructure [J]. Corrosion Science, 2008, 50(9): 2698.

[5] RODRIGUEZ A J, FRANCK C M. Radial Thermal Conductivity of all-Aluminum Alloy Conductors [J]. EEE Transactions on Power Delivery, 2015, 30(4): 1983-1990.

5.3.5 参考文献

[1] 樊学华，于勇，张子如，等. 316L 奥氏体不锈钢在不同电位下的点蚀和再钝化行为研究[J]. 表面技术，2020，49（7）：287–293，318.

[2]（文献同英文）

[3]（文献同英文）

[4]（文献同英文）

[5]（文献同英文）

5.4 Crevice corrosion

5.4 缝隙腐蚀

5.4.1 The concept of crevice corrosion

5.4.1 缝隙腐蚀的概念

Metal parts in the medium, due to the formation of a particularly small gap between metal and metal or metal and non-metal, so that the medium in the gap is in a stagnant state, causing accelerated corrosion of the metal in the seam, this local corrosion is called crevice corrosion. Figure 5-9 shows that AZ31 magnesium alloy is connected with 45 steels after chromate treatment. After a certain period of time, the microstructure of magnesium alloy occurs crevice corrosion [1].

金属部件在介质中，由于金属与金属或金属与非金属之间形成特别小的缝隙，使缝隙内的介质处于滞流状态，引起缝内金属的加速腐蚀，这种局部腐蚀称为缝隙腐蚀。图5-9所示为AZ31镁合金经铬酸盐处理后与45钢连接，经过一定时间后，镁合金组织发生缝隙腐蚀[1]。

Crevice corrosion conditions: main include connection of metal structures, metal and non-metal connection, deposits and attachments on the metal surface, and so on.

形成缝隙腐蚀条件：主要包括金属结构的连接、金属与非金属间的连接、金属表面的沉积物/附着物。

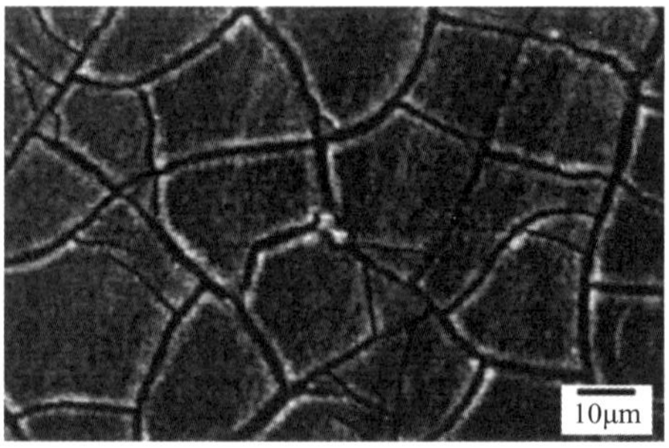

Figure 5-9 Crevice corrosion of AZ31 magnesium alloy connected to steel 45 after chromate treatment [2]

图 5-9 AZ31 镁合金经铬酸盐处理后与 45 钢连接发生的缝隙腐蚀 [2]

5.4.2 Characteristics of crevice corrosion

5.4.2 缝隙腐蚀的特征

Occurs on all metals and alloys, especially on corrosion resistant metals and alloys by passivation. The medium can be any corrosive solution, acidic or neutral, and the solution containing chloride ions is most

发生在所有金属与合金上，特别容易发生在靠钝化而耐蚀的金属与合金上。介质可以是任何侵蚀性溶液，酸性或

likely to cause crevice corrosion. Compared with pitting corrosion, crevice corrosion is more likely to occur for the same alloy. In the potential range between V_{br} and V_p, it can develop but not produce new corrosion pores. Crevice corrosion can both occur and develop in this potential region. The critical potential of the crevice corrosion is lower than the pitting potential.

中性，而含有氯离子的溶液最易引起缝隙腐蚀。与孔蚀相比，对同一种合金而言，缝隙腐蚀更易发生。在V_{br}~V_p区间的电位范围内，对孔蚀来说，可以发展但不产生新的蚀孔；而缝隙腐蚀在该电位区域内，既能发生又能发展。缝隙的临界电位要比孔蚀电位低。

5.4.3 Mechanism of crevice corrosion

5.4.3 缝隙腐蚀的机理

1. Initial stage

Oxygen depolarization corrosion is carried out uniformly in and out of the crevices. Due to stagnation, oxygen can only be diffused into the slit, making it difficult to replenish the oxygen in the slit after consumption, and the oxygen reduction reaction will soon terminate. And the oxygen outside the slit can be replaced at any time, so the REDOX reaction continues. The inside and outside of the slit constitute a macroscopic oxygen concentration cell.

1. 初期阶段

氧去极化腐蚀在缝隙内、外均匀进行。因滞流关系，氧只能以扩散方式向缝内传递，使缝内的氧消耗后难以得到补充，氧还原反应很快便终止。而缝外的氧随时可得到补充，所以氧化还原反应继续进行。缝内、外构成了宏观的氧浓差电池。

2. Latter stages

The formation of occluded cells marks the development of corrosion. At this time, it is difficult for metal cations in the slit to diffuse and migrate out. With the accumulation of Fe^{2+} and Fe^{3+}, excess positive charges are caused in the slit, which urges Cl^- outside the slit to migrate into the slit to maintain charge balance. The hydrolysis of metal chloride acidifies the medium in the seam and decreases the pH value, thus accelerating the dissolution of the anode.

2.后期阶段

闭塞电池的形成标志着腐蚀进入发展阶段。此时，缝内金属阳离子难以扩散、迁移出外，随着Fe^{2+}、Fe^{3+}的积累，缝内造成正电荷过剩，促使缝外的Cl^-迁移入内以保持电荷平衡。金属氯化物的水解使缝内介质酸化，pH值下降，因此加速了阳极溶解。阳极的加速溶解

The accelerated dissolution of the anode causes more Cl^- to move in, and the concentration of chloride increases. The hydrolysis of chloride further intensifies the acidification of the medium, thus forming an autocatalytic process, which accelerates the dissolution of the metal in the crack[3].

又引起更多的Cl^-迁入，氯化物浓度又增加，氯化物的水解又进一步加剧了介质的酸化，这样便形成一个自催化过程，使缝内金属的溶解加速进行下去[3]。

In conclusion, the formation of oxygen concentration cells promotes the onset of corrosion. However, the deepening and development of corrosion pits started from the formation of occluded cells, so acidification autocatalysis is the fundamental reason for accelerated corrosion.

总之，氧浓差电池的形成，对腐蚀的开始起促进作用。但蚀坑的深化和发展是从形成闭塞电池开始的，所以酸化自催化作用是造成腐蚀加速的根本原因。

5.4.4 Similarities and differences between pitting corrosion and crevice corrosion

5.4.4 孔蚀与缝蚀的异同点

1. Corrosion occurs conditions

Pitting corrosion originates from pitting core on metal surface and crevice corrosion originates from extra small crevices on metal surface. The former must occur in a medium containing an active anion, and the latter can occur even in a medium without an anion.

1. 腐蚀发生的条件

孔蚀起源于金属表面的孔蚀核，缝隙腐蚀起源于金属表面的特小缝隙。前者必须在含活性阴离子的介质中才会发生，后者即使在不含阴离子的介质中亦能发生。

2. Corrosion processes[4]

Pitting corrosion gradually forms occluded cells through corrosion, and then accelerates the corrosion, while crevice corrosion accelerates the corrosion by forming occluded cells quickly at the beginning of corrosion due to existing crevices.

2. 腐蚀过程[4]

孔蚀通过腐蚀逐渐形成闭塞电池，然后才加速腐蚀，而缝隙腐蚀由于事先已有缝隙，腐蚀刚开始便很快形成闭塞电池而加速腐蚀。

3. Complexity of reaction

Crevice corrosion is more likely to occur than pitting corrosion. In the range of $V_{br}\sim V_p$, for pitting

3. 发生难易程度

缝隙腐蚀比孔蚀更容易发生，在$V_{br}\sim V_p$区间。对孔蚀来

corrosion, the original pitting hole can develop, and no new pitting hole will occur. For crevice corrosion, new crevices may occur in addition to existing crevices.

说，原有的蚀孔可以发展，新的蚀孔不会产生。对缝隙腐蚀来说，除已形成的蚀坑可以扩展外，新的蚀坑仍会发生。

4. Corrosion morphology

4.腐蚀形态

The pitting hole of pitting corrosion is narrow and deep, while the pitting hole of crevice corrosion is relatively wide and shallow.

孔蚀的蚀孔窄而深，缝隙腐蚀的蚀坑相对广而浅。

5.4.5 Factors affecting crevice corrosion

5.4.5 影响缝隙腐蚀的因素

Geometrical shape: Gap width, gap depth and the area ratio between inside and outside the gap can affect crevice corrosion.

几何形状：缝隙宽度，缝隙深度及缝隙内、外面积比均会影响缝隙腐蚀。

Environmental factor: mainly includes: the concentration of oxygen in the solution (oxygen increases the corrosion speed); velocity of corrosive fluid (double influence); influence of temperature (double influence); pH value; Cl^- concentration in solution.

环境因素：主要包括溶液中氧浓度（氧增加腐蚀速度加快）；腐蚀液的流速（双重影响）；温度的影响（双重影响）；pH值；溶液中Cl^-浓度。

5.4.6 Measures to prevent crevice corrosion

5.4.6 防止缝隙腐蚀的措施

Measures to prevent crevice corrosion are as follow (Figure 5-10):

防止缝隙腐蚀的措施如下（图5-10）：

(1) Reasonable structural design.

（1）合理的结构设计。

(2) Electrochemical protection.

（2）电化学保护。

(3) Application of corrosion inhibitors: phosphate, chromate, nitrite on steel, brass, zinc.

（3）应用缓蚀剂：磷酸盐、铬酸盐、亚硝酸盐对钢、黄铜、锌。

(4) Clean up the non-metallic deposits on the metal surface in time.

（4）及时清理金属表面的非金属沉积物。

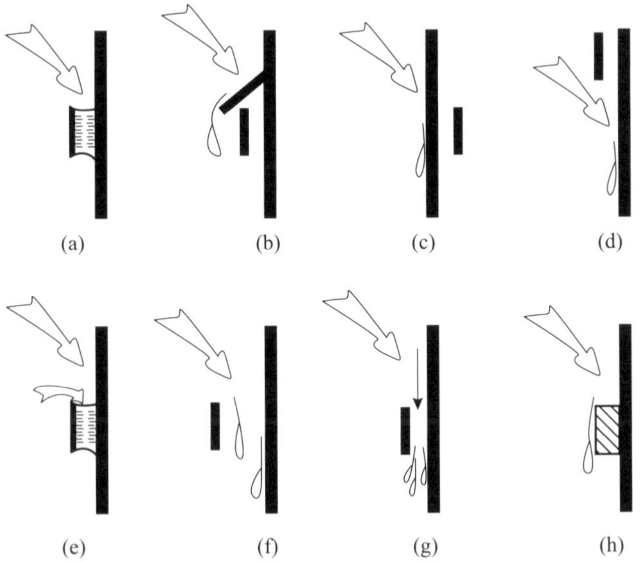

Figure 5-10 Measures to prevent crevice corrosion

图 5-10 防止缝隙腐蚀的措施

5.4.7 Case analysis

Critical crevice dissolution and IR Reduced corrosion model: Kennell[5] et al. proposed a mechanical model describing the dynamic concentration distribution and corrosion current in the corrosion gap. The model coupled the anode region in the crack with the cathode region on the bold surface, and combined the chemical erosion of the crack solution and the electric potential change on the crack and bold surface. The mathematical model is applied to predict the dynamic pH curves of fractures, and the results are in good agreement with the experimental data. Three crevice corrosion stages corresponding to the change of potential drop in crevices were determined(Figure 5-11).

5.4.7 案例分析

临界缝隙溶解和IR降低腐蚀模型：Kennell等[5]提出了一种描述腐蚀间隙内动态浓度分布和腐蚀电流的力学模型。该模型将裂缝内的阳极区与粗体表面的阴极区耦合在一起，同时结合了裂缝溶液的化学侵蚀以及裂缝和粗体表面电势变化的综合效应。将该数学模型应用于裂缝动态pH曲线的预测，结果与实验数据吻合较好。确定了与缝隙内电势降变化相对应的三个缝隙腐蚀阶段（图5-11）。

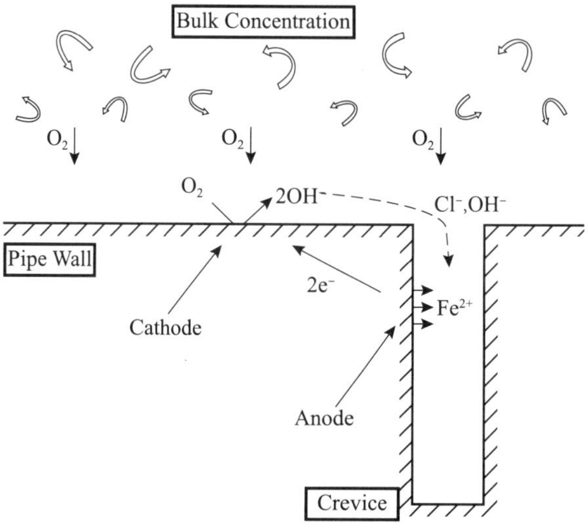

Figure 5-11 Idealized corrosion crack (not actual crack aspect ratio)
图 5-11 理想化腐蚀裂缝（非实际裂缝长宽比）

5.4.8 References

[1] HAN X K, QIN M, LI J R, et al. Research progress on corrosion behavior of stainless steel in seawater [J]. Materials Protection, 2017, 50(9): 75-81.

[2] YUAN B. Study on surface treatment of corrosion between AZ31 magnesium alloy and 45 steel [D]. Xi'an: Xi'an University of Technology, 2006.

[3] HU Q, ZhANG G, GUO X P. The crevice corrosion behaviour of stainless steel in sodium chloride solution [J]. Corrosion Science, 2011, 53: 4065-4072.

[4] LU B T, LUO J L, LU Y C. Effects of pH on lead-induced passivity degradation of nuclear steam generator tubing alloy in high temperature crevice chemistries [J]. Electrochim. Acta, 2013, 87: 824-838.

[5] KENNELL G F, EVITTS R W, HEPPNER K L. A critical crevice solution and IR drop crevice corrosion model [J]. Corrosion Science, 2008, 50(6): 1716-1725.

5.4.8 参考文献

[1] 韩小康，覃明，李佳润，等．不锈钢在海水中的腐蚀行为研究进展［J］．材料保护，2017，50（9）：75-81.

[2] 袁兵．表面处理 AZ31 镁合金与 45 钢连接腐蚀的研究［D］．西安：西安理工大学，2006.

[3]（文献同英文）

[4]（文献同英文）

[5]（文献同英文）

5.5 Selective corrosion

In the corrosion process of the alloy, the corrosion medium is not eroded according to the proportion of the alloy, but occurs selective dissolution of some of the components, so that the mechanical strength of the alloy decreases, this corrosion form is called composition selective corrosion, or called selective corrosion.

5.5.1 Dezincification of brass

1. Concept of dezincification corrosion of brass

Layered dezincification: Also known as uniform dezincification, the brass surface appears to be stripped away in layers.

Bolt dezincification: ALSO known as local dezincification (or zonal dezincification), corrosion holes ARE formed on the local surface of brass due to the dissolution of zinc and are sometimes covered with corrosion products[1].

2. Brass dezincification corrosion mechanism

Zinc-selective dissolution theory: This theory holds that only the zinc layer in the surface of the alloy acts as the anode for the corrosion battery to undergo selective dissolution. The zinc atoms in the slightly inside of the surface layer diffuse through the vacancies in the surface layer to the site of dissolution reaction and continue to be dissolved, leaving loose copper layer[2]. (Corrosion of copper in dilute hydrochloric acid)

Solution-deposition theory: This theory holds that copper and zinc enter the solution together as metal ions during corrosion. Since the precipitation potential of

5.5 选择性腐蚀

合金在腐蚀过程中，腐蚀介质不是按合金的比例侵蚀，而是发生了其中某种成分的选择性溶解，使合金机械强度下降，这种腐蚀形态称为成分选择腐蚀，或称为选择性腐蚀。

5.5.1 黄铜脱锌

1.黄铜脱锌腐蚀的概念

层状脱锌：又称均匀脱锌，黄铜表面像被一层层地剥走似的。

栓状脱锌：又称局部脱锌（或带状脱锌），在黄铜的局部表面，由于锌的溶解形成蚀孔，蚀孔有时被腐蚀产物覆盖[1]。

2.黄铜脱锌腐蚀的机理

锌选择溶解理论：这种理论认为仅是合金表层中的锌作为腐蚀电池的阳极发生选择性溶解，表面层稍里处的锌原子通过表面层上的空位经扩散抵达溶解反应的地点继续被溶解，留下疏松的铜层[2]。（铜在稀盐酸中的腐蚀）

溶解–沉积理论：这种理论认为腐蚀时铜和锌一起以金属离子的形式进入溶液，由于

copper ions is higher than the corrosion potential of the alloy, the copper ions will soon redeposit as pure copper on the surface of the brass near the dissolution site [3]. (Corrosion of copper in concentrated hydrochloric acid)

铜离子的析出电位比合金的腐蚀电位高，所以铜离子很快便在靠近溶解地点的黄铜表面上以纯铜的形式重新沉积[3]。（铜在浓盐酸中的腐蚀）

3. Influencing factors and control methods

3.影响因素与控制方法

(1) Dissolved oxygen in the medium has the effect of promoting dezinc, but even in the anoxic medium, the effect of dezinc can also be carried out. Because zinc is a very reactive metal, it can be reduced to hydrogen and hydroxide ions through the cathode of water in pure water.

（1）介质中的溶解氧有促进脱锌的作用，但即使在缺氧的介质中，脱锌的作用亦能够进行。因为锌是相当活泼的金属，在纯水中能够通过水的阴极还原生成氢气和氢氧根离子。

(2) The stagnant state of the solution, containing chloride ions, or loose scale layer or sediment on the surface of the brass can promote corrosion, otherwise reduce corrosion.

（2）溶液的滞流状态，含氯离子，或黄铜表面有疏松的垢层或沉积物都能促进腐蚀，反之则减轻腐蚀。

(3) The higher the zinc content of brass, the higher the dezincification tendency and corrosion rate. Under natural conditions, dezincification is mostly found on brass with zinc content higher than 15%.

（3）黄铜含锌量越高，其脱锌倾向和腐蚀速度越大。在自然条件下，多半是在含锌量高于15%的黄铜上发现脱锌。

(4) Adding a small amount of arsenic, antimony, tin, phosphorus in brass can reduce the sensitivity of dezincing. The main reason is that they play the role of “corrosion inhibitor”, forming a protective film on the brass surface and inhibiting the process of dezincification.

（4）黄铜中加入少量的砷、锑、锡、磷等可使脱锌敏感性下降。主要是它们起到“缓蚀剂”的作用，在黄铜表面形成保护性膜，抑制脱锌过程的进行。

5.5.2 Graphitization of gray cast iron

5.5.2 灰口铸铁石墨化

Graphite in gray cast iron is distributed in ferrite in lamellar form, and selective corrosion [4] of ferritic occurs in salt water, mineral water, soil and extremely

灰铸铁中的石墨以片层状分布于铁素体内，在盐水、矿水、土壤及极稀的酸性溶液中

dilute acid solution. As a result, graphite remains as cathode phase, which makes gray cast iron seem to turn into graphite as shown in Figure 5-12, so it is called graphitization corrosion of gray cast iron.

发生铁素体的选择性腐蚀[4]，作为阴极相的石墨残留下来，结果使灰铸铁变成石墨，如图5-12所示，故称为灰铸铁石墨化腐蚀。

The reason for the graphitization of gray cast iron IS as follows: in the corrosive medium, a high-efficiency corrosion cell with ferrite as anode and flake graphite as cathode is formed. After selective corrosion, it becomes a porous body composed of graphite, voids and rust, which makes cast iron lose its original metallic properties and properties.

灰铸铁石墨化产生的原因：在腐蚀介质中组成以铁素体为阳极、片状的石墨为阴极的高效腐蚀电池，经选择性腐蚀以后，成为石墨、空隙和铁锈构成的多孔体，使铸铁丧失原有的金属属性和性能。

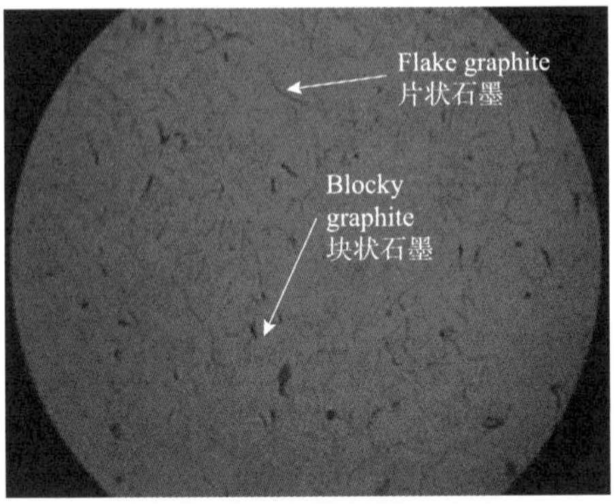

Figure 5-12 Graphitization of gray cast iron[5]

图5-12 灰铸铁石墨化腐蚀[5]

5.5.3 Case analysis

Interactions among elements during Cu-Zn alloying: Zhou[6] et al. studied the variation of element dissolution in Cu-Zn alloy with Zn content between 0 and 45 wt%. Atomic emission spectroelectrochemistry (AESEC) was used to directly monitor the release of Cu^{2+} and Zn^{2+} and the growth of oxides in potential kinetic experiments with time. The results show that the process of Cu

5.5.3 案例分析

Cu-Zn合金化过程中各元素间的相互作用：Zhou等[6]研究了Cu-Zn合金的元素溶解随着Zn含量质量分数在0~45wt%的变化规律。利用原子发射光谱电化学（AESEC）直接监测电位动力学实验中随着时间变化

dissolution occurs simultaneously with the formation of Cu_2O and the release of Cu^{2+}. The addition of Zn in Cu-Zn alloy does not significantly change the dissolution mechanism of Cu^{2+} and the dissolution rate of Cu^{2+} in solution is only dependent on potential. The formation of Cu(O) film hinders the dissolution of Zn, causesing the dissolution to shift towards the anode.

Cu^{2+}和Zn^{2+}的释放和氧化物生长。结果表明，Cu溶出过程是Cu_2O生成和Cu^{2+}释放同时发生的。Cu-Zn合金中Zn的加入未明显改变Cu^{2+}的溶解机制，并且溶液中Cu^{2+}的溶解速率仅依赖于电位。Cu（O）膜的形成阻碍了Zn溶解，使Zn溶解向阳极方向偏移。

5.5.4 References

［1］YANG S L, WU Y. Research progress of brass dezincification corrosion［J］. World Nonferrous Metals, 2011(4): 50-53.

［2］JIANG M J, GUO X C, YANG J. Characteristics and preventive measures of brass dezincification corrosion［J］. Journal of Logistics Engineering College, 2008, 24(4): 35-38, 61.

［3］ZHAO T X, CUI H J, HU X C, et al. Electron probe analysis of corrosion characteristics of brass dezincification［J］. Physical Testing, 2021, 39(3): 23-28.

［4］CHEN Z J. Application and analysis of alloying elements in gray cast iron cylinder block［C］//Innovative equipment technology to empower local economy—Proceedings of the 3rd National Local Mechanical Engineering Society Academic Annual Conference and Cross-Strait Machinery Science and Technology Forum, 2013: 212-222.

［5］CHAO S J, TSAI M H, YU R P, et al. Dezincification of brass water meters in a long-term study: effects of anions, alkalinity, and residual chlorine［J］. Environmental science-water research & technology, 2021, 7(9): 1666-1676.

［6］ZHOU P, ERNING J W, OGLE K. Interactions between elemental components during the dealloying of Cu-Zn alloys［J］. Electrochimica Acta, 2018, 293: 290-298.

5.5.4 参考文献

［1］杨胜利，吴语．黄铜脱锌腐蚀的研究进展［J］．世界有色金属，2011（4）：50-53.

［2］蒋明俊，郭小川，杨俊．黄铜脱锌腐蚀的特征及防止措施［J］．后勤工程学院学报，2008，24（4）：35-38，61.

［3］赵同新，崔会杰，胡晓春，等．黄铜脱锌腐蚀行为特征的电子探针分析［J］．物理测试，2021，39（3）：23-28.

［4］陈桢杰．灰铸铁缸体中合金元素的应用及分析［C］// 创新装备技术　给力地方经济——第三届全国地方机械工程学会学术年会暨海峡两岸机械科技论坛论文集，2013：212-222.

［5］（文献同英文）

［6］（文献同英文）

5.6 Intergranular corrosion

5.6 晶间腐蚀

5.6.1 Definition and conditions of intergranular corrosion

5.6.1 晶间腐蚀的定义及产生条件

Corrosion develops along the grain boundaries or other adjacent areas of metals or alloys, and the corrosion of the grains themselves is very mild. This type of corrosion is called intergranular.

Causes of intergranular corrosion:

The difference in chemical composition of grains and grain boundaries of metals or alloys leads to different electrochemical properties, thus having the tendency of intergranular corrosion. The corrosion medium can show the inhomogeneity of the electrochemical properties of grains and grain boundaries. In the corrosion system determined by the medium and metal, intergranular corrosion can occur when the dissolution current density of grain boundaries is much higher than that of grains themselves[1].

腐蚀沿着金属或合金的晶粒边界或其他的邻近区域发展，晶粒本身腐蚀很轻微，这种腐蚀称为晶间腐蚀，又称晶间腐蚀。

晶间腐蚀的原因：

金属或合金本身晶粒、晶界化学成分的差异，导致其电化学性质不同，从而具有晶间腐蚀的倾向。同时，腐蚀介质能够显示晶粒、晶界的电化学性质不均匀性，在由介质与金属所决定的腐蚀体系中，晶界的溶解电流密度远大于晶粒本身的溶解电流密度时，便可产生晶间腐蚀[1]，如图5-13所示。

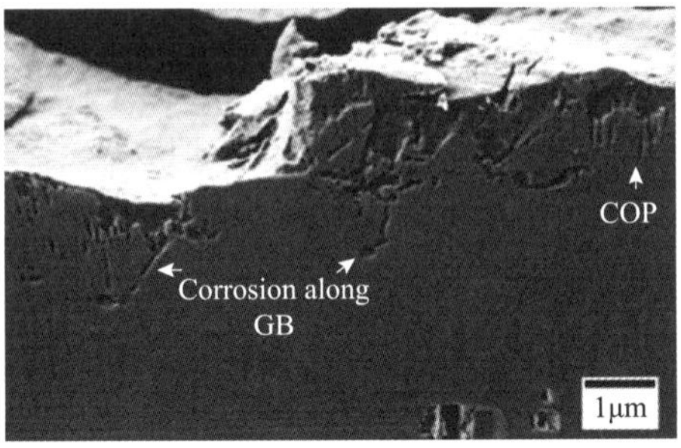

Figure 5-13 Morphology of intergranular corrosion

图 5-13 晶间腐蚀

5.6.2 Mechanism of intergranular corrosion

5.6.2 晶间腐蚀的机理

Dilution theory:

Due to the precipitation of new phase between

贫化理论：

由于晶间析出新相，造

grains, a component in the grain boundary zone is depleted, and a corrosion cell with the grain boundary zone as the anode and the grain body as the cathode is formed in the corrosion medium. The metal at the grain boundary is corroded and dissolved, leading to intergranular corrosion. As shown in Figure 5-14, the precipitation of Cr from stainless steel well proves the theory.Theory of selective dissolution of impurities or second phase in grain boundary zone [2].

成晶界区某一成分的贫化，在腐蚀介质中形成以晶界区为阳极、晶粒本体为阴极的腐蚀电池，晶界处的金属腐蚀溶解，引发晶间腐蚀。如图5－14所示，不锈钢Cr的析出很好地证明了该理论[2]。

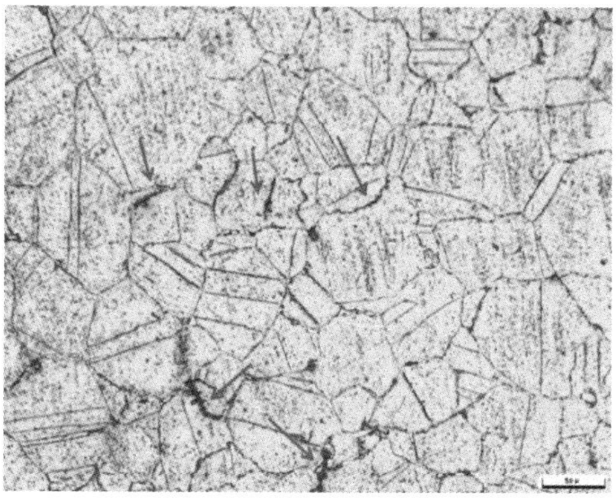

Figure 5-14 Intergranular corrosion of stainless steel
图5-14 不锈钢晶间腐蚀

Theory of selective dissolution of impurities or second phase in grain boundary zone:

The impurities at the grain boundary or the second phase distributed along the grain are preferentially corroded and dissolved, or the second phase itself constitutes the corrosion cell, in which a component is selectively dissolved, leading to the intergranular corrosion. This theory can well explain the mechanism of σ-phase intergranular precipitation corrosion and the intergranular corrosion of $MgZn_2$ and Mg_5Al_8 in ultra-low carbon stainless steel.

晶界区杂质或第二相选择性溶解理论：

晶界的杂质或沿晶分布的第二相优先发生腐蚀溶解，或者第二相本身就构成了腐蚀电池，其中某一组分发生了选择性溶解，导致晶间腐蚀。该理论较好地解释了超低碳不锈钢σ相沿晶析出腐蚀及$MgZn_2$、Mg_5Al_8等的晶间腐蚀机理。

5.6.3 Several special forms of intergranular corrosion

1. Peeling corrosion

Also known as denudation and lamination, it refers to the intergranular corrosion of materials with intergranular corrosion tendency after rolling or forging, under certain corrosion conditions, along the grain boundary direction parallel to the surface.

2. Welded corrosion

Some parts of the material with the tendency of intergranular corrosion in the heat affected zone of welding happen to be in the sensitive temperature range of precipitate phase. Under the action of corrosive medium, intergranular corrosion will occur in this part, which is also known as weld corrosion [3].

3. Knife-line attack

The knife line corrosion is the intergranular corrosion of the stabilized stainless steel after welding and reheating in a narrow range of several grain width on both sides of the adjacent weld. It is named because it looks like a knife edge.

5.6.4 Control measures for intergranular corrosion

It can be treated by resolution, stabilization, ultra-low carbon stainless steel, duplex steel (Ferrite and austenite), and so on [4].

5.6.5 Case analysis

Localized corrosion of AA2024-T351 aluminum alloy: Zhang et al [5]. investigated the relationship between intergranular corrosion and pitting corrosion of

5.6.3 几种特殊形式的晶间腐蚀

1.剥层腐蚀

剥层腐蚀又称剥蚀、层蚀，指具有晶间腐蚀倾向的材料经轧制或锻压加工后，在一定的腐蚀条件下，沿着与表面平行的晶界方向发生的晶间腐蚀。

2.焊缝腐蚀

具有晶间腐蚀倾向的材料，在焊接的热影响区内有的部位恰好处在析出沉淀相的敏感温度范围中，在腐蚀介质的作用下，这一部位就会发生晶间腐蚀，这种腐蚀又称焊缝腐蚀 [3]。

3.刀线腐蚀

刀线腐蚀是稳定化不锈钢经焊接并再次加热后，在紧邻焊缝两侧几个晶粒宽的狭窄范围内发生的晶间腐蚀，因形似刀刃而得名。

5.6.4 晶间腐蚀的控制措施

可通过重新固溶处理、稳定化处理、采用超低碳不锈钢、采用双相钢（铁素体和奥氏体）等方法 [4]。

5.6.5 案例分析

AA2024-T351铝合金的局部腐蚀：Zhang等 [5] 研究了AA2024-T351铝合金浸泡在质

AA2024-T351 aluminum alloy immersed in 3.5 wt.% NaCl solution. It was found that intergranular corrosion occurred first and pitting originated from the crack wall of intergranular corrosion. Propagating outward from grain boundaries, grain boundary pitting develops into the interior of selected grain boundaries. Because copper is enriched along pitting pit walls, crystalline pitting pits in AA2024 alloy have relatively irregular pit walls that are much finer in size than the well-defined hemicuboidal shape of crystal pits in pure aluminum(Figure 5-16).

量分数3.5wt% NaCl溶液中晶间腐蚀与点蚀的关系（图5-15）。发现晶间腐蚀首先发生并且点蚀起源于晶间腐蚀的裂隙壁。从晶界向外传播，晶界点蚀发展成选定的晶界内部。由于铜沿着点蚀坑壁富集，AA2024合金中的结晶点蚀坑具有相对不规则的坑壁，其尺寸比纯铝中清晰的半立方体形状的结晶坑要细得多（图5-16）。

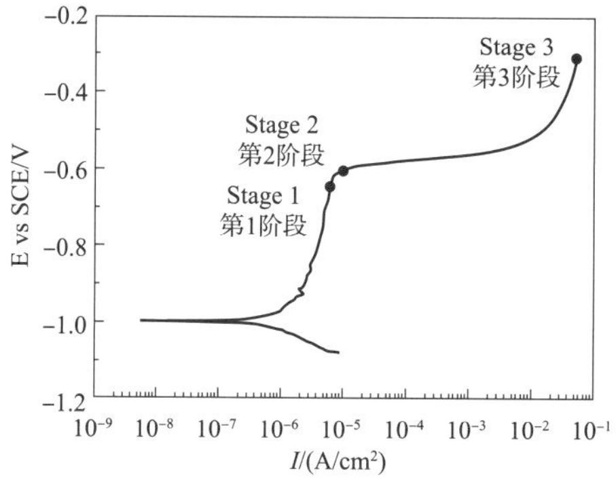

Figure 5-15　Potential-current density curves of AA2024-T351 aluminum alloy during dynamic polarization in 3.5% NaCl solution (deoxygenated) at room temperature

图5-15　室温下AA2024-T351铝合金在质量分数3.5% NaCl溶液（去氧）中动态极化过程的电位－电流密度曲线

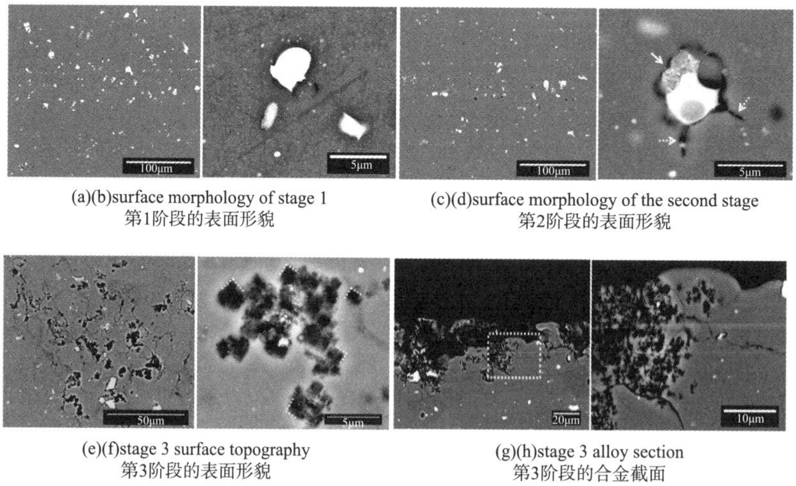

Figure 5-16　SEM micrographs of AA2024 alloy at each stage after anodic polarization

图5-16　AA2024合金阳极极化后各阶段的SEM图

5.6.6 References

[1] HU H B, TANG X Y. Research on mechanism and improvement application of intergranular corrosion of ferritic stainless steel [J]. Times Auto, 2021(7): 143-145.

[2] LI D P. Common corrosion analysis and preventive measures of chemical equipment [J]. Petrochemical Safety and Environmental Protection Technology, 2018, 34(4): 40-44, 63.

[3] PYO K H, Kim D. Intergranular corrosion of stainless steel [J]. Corrosion Science and Technology, 2018, 17(4): 183-192.

[4] BONFILS-LAHOVARY M-L De, JOSSE C, LAFFONT L, et al. Influence of hydrogen on the propagation of intergranular corrosion defects in 2024 aluminium alloy [J]. Corrosion science, 2019, 148: 198-205.

[5] ZHANG X, ZHOU X, HASHIMOTO T, et al. Localized corrosion in AA2024-T351 aluminium alloy: Transition from intergranular corrosion to crystallographic pitting [J]. Materials Characterization, 2017(130): 230-236.

5.6.6 参考文献

[1] 胡海波，汤旭炎．铁素体不锈钢晶间腐蚀的机理和改善应用研究[J]．时代汽车，2021（7）：143–145.

[2] 李东平．化工设备常见腐蚀分析及预防措施[J]．石油化工安全环保技术，2018，34（4）：40–44，63.

[3]（文献同英文）

[4]（文献同英文）

[5]（文献同英文）

5.7 Stress corrosion cracking

5.7 应力腐蚀开裂

5.7.1 Concept of stress corrosion cracking

Stress corrosion cracking, referred to as stress corrosion, is the phenomenon of metal material breaking under the joint action of tensile stress and specific corrosive medium [1].

5.7.1 应力腐蚀开裂的概念

应力腐蚀开裂简称应力腐蚀，它是在拉应力和特定的腐蚀介质共同作用下发生的金属材料的破断现象[1]。

5.7.2 Conditions under which stress corrosion occurs

Couditions under which stress corrosion occurs are as follows:

(1) Certain tensile stress.

(2) Sensitive materials.

(3) Specific environment.

Stainless steel, high strength steel, Cu, Al, Ti alloy, etc. The higher the strength level of the material, is the more prone it is to stress corrosion.

Table 5-1 shows the combinations of metal materials and environments prone to stress corrosion cracking.

5.7.2 应力腐蚀产生的条件

应力腐蚀产生的条件如下：

（1）一定的拉应力。

（2）敏感材料。

（3）特定环境。

不锈钢、高强钢、Cu、Al、Ti合金等；材料的强度水平越高，越易发生应力腐蚀。

表5-1所示为易产生应力腐蚀开裂的金属材料和环境组合。

Table 5-1 Combinations of metal materials and environments prone to stress corrosion cracking

表 5-1 易产生应力腐蚀开裂的金属材料和环境组合

Alloy 合金	Environment 环境	Alloy 合金	Environment 环境
Carbon steel and low alloy steel 碳钢及低合金钢	Caustic alkali solution 苛性碱溶液 Ammonia solution 氨溶液 Aqueous nitrate solution 硝酸盐水溶液 HCN 含 HCN 水溶液 Wet CO–CO_2–air 湿的 CO–CO_2– 空气 Carbonate and bicarbonate solutions 碳酸盐和重碳酸盐溶液 Aqueous solution containing H_2S 含 H_2S 水溶液 Seawater 海水 Marine atmosphere and industrial atmosphere 海洋大气和工业大气 CH_3COOH solution CH_3COOH 水溶液 CaC_2、$FeCl_3$ in water $CaCl_2$、$FeCl_3$ 水溶液 $(NH_4)_2CO_3$ $(NH_4)_2CO_3$ H_2SO_4–HNO_3 mixed acid solution H_2SO_4–NHO_3 混合酸水溶液	Austenitic stainless steel 奥氏体不锈钢	High temperature alkaline solution [$NaOH \cdot Ca(OH)_2 \cdot LiOH$] 高温碱液[$NaOH \cdot Ca(OH)_2LiOH$] Aqueous chloride solution 氯化物水溶液 Sea water, oceanic atmosphere 海水，海洋大气 Polysulfuric acid（$H_2S_nO_6$ n=2~5） 连多硫酸（$H_2S_nO_6$ n=2~5） High temperature and high pressure oxygenated high pure water 高温高压含氧高纯水 Concentrated boiler water 浓缩锅炉水 Water vapor 水蒸气（260℃） H_2SO_4（260℃） 260℃ H_2SO_4 Humid air（humidity 90%） 湿润空气（温度 90%） $NaCl+H_2O_2$ solution $NaCl+H_2O_2$ 水溶液 Hot NaCl 热 NaCl Wet magnesium chloride insulation 湿的氯化镁绝缘物 H_2S aqueous solution H_2S 水溶液

5.7.3 Characteristics of stress corrosion cracking

1. Several stages of stress corrosion cracking

(1) Crack initiation stage: crack or pitting stage caused by corrosion.

(2) Stable crack growth period: the time taken by the crack after nucleation until it reaches the critical size.

(3) The instability propagation period of the crack: after the crack reaches the critical size, the crack instability leads to fracture due to pure mechanical action.

2. Stress corrosion crack characteristics

There are three types of stress corrosion cracks: intergranular, transgranular and mixed. The main characteristics of stress corrosion cracks are as follows: the crack originates from the surface; The length and width of the crack are not proportional, and the difference is several orders of magnitude. The direction of crack propagation is generally perpendicular to the direction of main tensile stress. The cracks are generally dendritic.

3. Crack fracture

The stress corrosion fracture belongs to the brittle fracture on the macro scale and obvious traces of plastic flow can be observed on the micro scale of the fracture surface. Due to the action of corrosive medium, the color of fracture surface is dim, and the corrosion pits and secondary cracks are often seen on the microscopic fracture[2].

5.7.4 Stress corrosion cracking mechanism

1. Anodic dissolution theory

The stress concentration at the crack tip deforms the grain boundary, slip band, phase interface, etc., or pulls the crack apart to become a new active anode,

5.7.3 应力腐蚀开裂的特征

1.应力腐蚀开裂的几个阶段

（1）裂纹萌生期：由腐蚀引起裂纹或坑蚀阶段。

（2）裂纹稳定扩展期：裂纹成核后直至发展到临界尺寸所经历的时间。

（3）裂纹失稳扩展期：裂纹达到临界尺寸后，由纯力学作用裂纹失稳导致断裂。

2.应力腐蚀裂纹的特点

应力腐蚀裂纹有沿晶型、穿晶型和混合型三种类型。应力腐蚀裂纹的主要特点是：裂纹起源于表面；裂纹的长宽不成比例，相差几个数量级；裂纹扩展方向一般垂直于主拉应力方向；裂纹一般呈树枝状。

3.裂纹断口

应力腐蚀断裂从宏观上属于脆性断裂，在断裂面微观尺度上可观察到明显的塑性流变的痕迹。由于腐蚀介质的作用，断口表面颜色暗淡，显微断口上可见腐蚀坑及二次裂纹[2]。

5.7.4 应力腐蚀开裂机理

1.阳极溶解理论

裂纹尖端的应力集中使晶界、滑移带、相界面等处产生变形，或者将裂纹拉开，使之

forming an occluded corrosion cell with a large cathode at the crack tip and a small anode. Due to the autocatalysis of the occluded cell, the crack front has a very high dissolution rate[3].

成为新的活性阳极，形成裂纹尖端大阴极小阳极的闭塞腐蚀电池，由于闭塞电池的自催化作用使裂纹前沿具有非常大的溶解速度[3]。图5-17所示为304不锈钢的应力腐蚀裂纹。

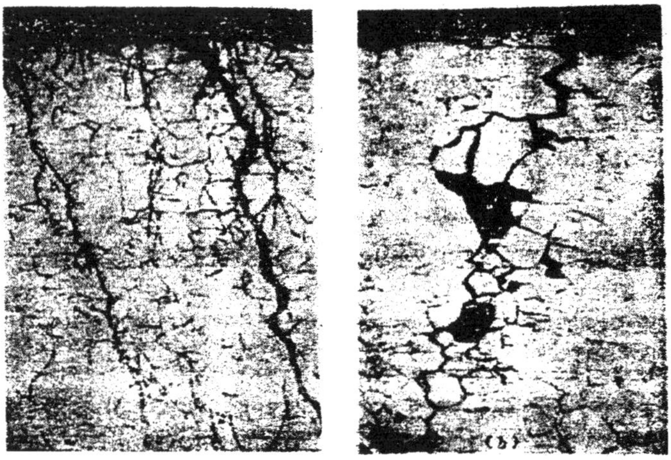

(a) transgranular crack in boiling 45% $MgCl_2$ solution
在沸腾45%$MgCl_2$溶液中的穿晶裂纹

(b) intercrystalline crack of softened stainless steel in polysulfuric acid solution at room temperature
软化不锈钢在室温连多硫酸溶液中的晶间裂纹

Figure 5-17 Stress corrosion crack of 304 stainless steel

图5-17 304不锈钢的应力腐蚀裂纹

2. Membrane rupture theory

The corrosion product film is damaged under the action of stress, exposing the active metal matrix, and forming a corrosion cell with a large cathode and a small anode with the unbroken film.

2.膜破裂理论

腐蚀产物膜在应力作用下引起破坏、裸露出活性金属基体，与膜未破裂处构成大阴极小阳极的腐蚀电池。

3. Hydrogen embrittlement theory

It is emphasized that hydrogen plays an important role in stress corrosion. The acidification of crack tip solution increases the concentration of hydrogen ion, and when hydrogen ion is reduced on the microcathode of crack tip, it becomes adsorbed hydrogen atom and diffuses into the metal, which causes the embrittlement of crack tip area, and the tensile stress causes the embrittlement of crack tip area and promotes crack expansion[4]. Figure 5-18 shows the stress corrosion cracking model.

3.氢脆理论

强调氢在应力腐蚀中起重要作用，裂纹尖端溶液酸化，氢离子浓度提高，在裂纹尖端微阴极上被还原以后，变成吸附氢原子向金属内部扩散，使裂尖区脆化，张应力使裂尖区脆断，促进裂纹扩张等[4]。图5-18所示为应力腐蚀开裂模型。

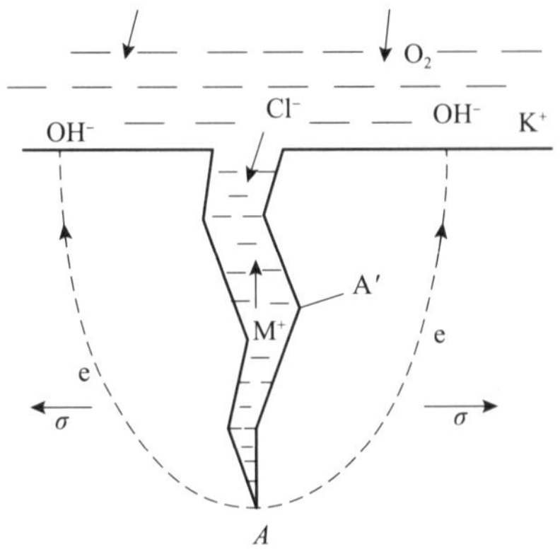

Figure 5-18 Stress corrosion cracking model

图 5-18 应力腐蚀开裂模型

5.7.5 Influencing factors and prevention measures

Reasonable material selection, control stress distribution, control of corrosive media and electrochemical protection are the promising methods to prevent stress corrosion cracking[5].

5.7.5 防止措施

可通过合理选材、控制应力分布、控制腐蚀介质、电化学保护来达到减缓应力腐蚀开裂的目的[5]。

5.7.6 References

[1] YANG Y, CHENG Y. Stress enhanced corrosion at the tip of near-neutral pH stress corrosion cracks on pipelines [J]. Corrosion, 2016, 72(8):1035-1043.

[2] MOSS T, KUANG W J, WAS, et al. Stress corrosion crack initiation in Alloy 690 in high temperature water [J]. Current Opinion in Solid State & Materials Science, 2018, 22(1): 16-25.

[3] FUJII T, TOHGO K, MORI Y, et al. Crystallographic and mechanical investigation of intergranular stress corrosion crack initiation in austenitic stainless steel [J]. Materials Science and Engineering A-Structural Materials Properties Microstructure and Processing, 2019, 751: 160-170.

[4] AALTONEN P, YAGODZINSKYY Y, SAUKKONEN T, et al. Role of excessive vacancies in transgranular stress corrosion cracking of pure copper [J]. Corrosion Reviews, 2015, 33(6): 487-500.

[5] MA H, ZHAO B, LIU Z, et al. Local chemistry-electrochemistry

5.7.6 参考文献

[1](文献同英文)

[2](文献同英文)

[3](文献同英文)

[4](文献同英文)

[5](文献同英文)

and stress corrosion susceptibility of X80 steel below disbonded coating in acidic soil environment under cathodic protection [J]. Construction and Building Materials, 2020, 243: 118203.

5.8 Corrosion fatigue

5.8.1 Concept of corrosion fatigue

Corrosion fatigue is caused by the failure of metal under the combined action of corrosive medium and alternating stress[1].

5.8.2 Characteristics of corrosion fatigue

Characteristics of corrosion fatigue are as follow:

(1) Corrosion fatigue has no definite combinatorial relationship between material and medium, and can occur in most water-soluble media.

(2) There is no lower limit critical stress value for fatigue corrosion.

(3) Most of the corrosion fatigue cracks are transgranular, and corrosion products are found in the fracture fatigue zone.

(4) Corrosion fatigue is not limited by the potential zone, and can occur in the activation zone or the passivation zone[2].

5.8.3 Corrosion fatigue mechanism

It is generally believed that under the action of alternating stress, plastic deformation, occurs in some parts of metal, resulting in uneven surface electrochemical properties, forming corrosion cells. The metal strain site is the anode, while the undeformed site is the cathode. The anode corrosion generates microcracks, which expand along the slip plane under the joint action of stress and corrosion medium, so

5.8 腐蚀疲劳

5.8.1 腐蚀疲劳的概念

金属在腐蚀介质和交变应力共同作用下引起的破坏称为腐蚀疲劳[1]。

5.8.2 腐蚀疲劳的特点

腐蚀疲劳的特点如下：

（1）腐蚀疲劳没有材料与介质之间明确限定的组合关系，可在大多数水溶性介质中发生。

（2）腐蚀疲劳无下限临界应力值。

（3）腐蚀疲劳裂纹多为穿晶型，断口疲劳区有腐蚀产物。

（4）腐蚀疲劳不受电位区段的限制，既可在活化区也可在钝化区发生[2]。

5.8.3 腐蚀疲劳机理

一般认为，金属在交变应力的作用下，某些部位先产生塑性变形，造成表面电化学性质的不均匀，形成腐蚀电池。金属应变部位为阳极而未变形处为阴极，阳极腐蚀生成微裂纹，裂纹在应力和腐蚀介质共同作用下沿滑移面扩展，故多

transgranular cracking occurs in most cases [3].

5.8.4 Factors affecting corrosion fatigue

1. Mechanical factor

(1) Stress cycle parameters:

f—— Stress cycle frequency;

R——Stress asymmetry coefficient.

(2) Fatigue loading mode:

Torsional fatigue > rotational bending fatigue > extrusion fatigue

(3) Stress concentration.

2. The influence of environmental factors

(1) Temperature: with the increase of temperature, the corrosion resistance fatigue performance of the material decreases, while the influence on the pure fatigue performance is small.

(2) The corrosion of the medium: the stronger the corrosion of the medium, the lower the corrosion fatigue strength [4].

产生穿晶开裂[3]。

5.8.4 影响腐蚀疲劳的因素

1. 力学因素

（1）应力循环参数：

f——应力循环频率；

R——应力不对称系数。

（2）疲劳加载方式：

扭转疲劳>旋转弯曲疲劳>挤压疲劳

（3）应力集中。

2. 环境因素的影响

（1）温度：随着温度升高，材料的耐腐蚀疲劳性能下降，而对纯疲劳性能的影响较小。

（2）介质的腐蚀性：介质的腐蚀性越强，腐蚀疲劳强度越低[4]。

图5-19所示为应力交变频率*f*对腐蚀疲劳的影响。

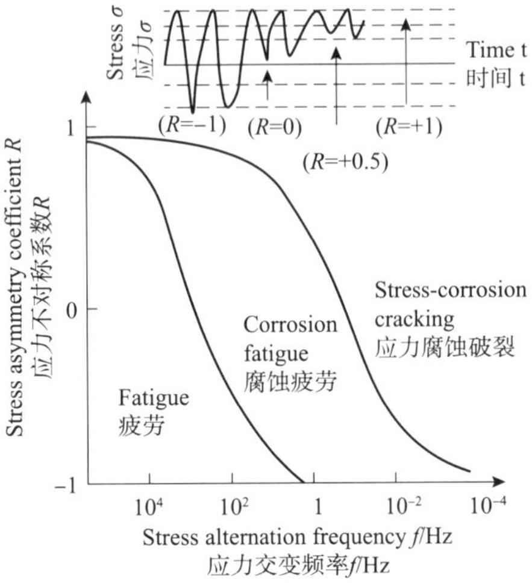

Figure 5-19 Influence of stress alternating frequency on corrosion fatigue

图5-19 应力交变频率 *f* 对腐蚀疲劳的影响

3. The influence of material factors

The corrosion resistance of materials is mainly related to the structure and surface state, which affects the corrosion resistance of materials.

3.材料因素的影响

材料的耐腐蚀疲劳性能主要与组织结构、表面状态等有关，进而影响材料的耐腐蚀性能。

5.8.5 Corrosion fatigue prevention measures

5.8.5 腐蚀疲劳防止措施

Corrosion fatigue prevention measures are as follow:

(1) Reasonable material selection.

(2) Stress reduction.

(3) Use corrosion inhibitor.

(4) Protective coating.

(5) Electrochemical protection.

腐蚀疲劳防止措施如下：

（1）合理选材。

（2）降低应力。

（3）采用缓蚀剂。

（4）保护涂层。

（5）电化学保护。

5.8.6 References

[1] LI B, DONG L H, WANG H D, et al. Research progress on corrosion fatigue of aerospace aluminum alloys [J]. Surface Technology, 2021, 50(7): 106-118.

[2] ZHANG Z Y, WU X Q, HAN E H, et al. Research status and progress of corrosion fatigue crack propagation behavior of nuclear power structural materials [J]. Journal of Chinese Society for Corrosion and Protection, 2022, 42(1): 9-15.

[3] YANG J, BAI R, CHEN B, et al. Hydrogel adhesion: a supramolecular synergy of chemistry, topology, and mechanics [J]. Advanced Functional Materials, 2019, 30 (2, SI).

[4] CHUKWUIKE V I, ECHEM O G, PRABHAKARAN S, et al. Laser shock peening (LSP): Electrochemical and hydrodynamic investigation of corrosion protection pre-treatment for a copper surface in 3.5 % NaCl medium [J]. Corrosion Science, 2020, 179.

5.8.6 参考文献

[1] 李斌，董丽虹，王海斗，等.航空航天铝合金腐蚀疲劳研究进展[J]. 表面技术，2021，50（7）：106–118.

[2] 张兹瑜，吴欣强，韩恩厚，等.核电结构材料腐蚀疲劳裂纹扩展行为研究现状与进展[J].中国腐蚀与防护学报，2022，42（1）：9–15.

[3]（文献同英文）

[4]（文献同英文）

5.9 Friction corrosion

Wear corrosion refers to the accelerated corrosion destruction of materials under the combined action of wear and corrosion. It can be divided into friction corrosion, turbulent corrosion and cavitation corrosion.

5.9.1 Friction corrosion

Friction corrosion, also known as fretting corrosion, friction oxidation, refers to the corrosion caused by vibration or sliding between surfaces in contact with loaded solids in a gas environment where oxygen exists. The presence of vibration or sliding, load and relative motion at the loaded interface can cause the interface to slip and deform, and the involvement of corrosive media can greatly increase the possibility of friction corrosion. Currently, wear-oxidation theory and oxidation-wear theory can explain the vibration corrosion[1].

Common preventive measures mainly include: increasing the load, so that there is no vibration or sliding between the contact surfaces of the assembly; Choose wear-resistant materials or plating suitable wear-resistant layers; Improve friction conditions, such as deoxygenation in the environment, add lubricants or reducing agents.

5.9.2 Turbulent corrosion

Turbulent corrosion also called impact corrosion (erosion), refers to the corrosive fluid and metal surface movement between the metal caused by accelerated

5.9 磨损腐蚀

磨损腐蚀是指在磨损和腐蚀的综合作用下材料发生的加速腐蚀破坏。主要包括摩振腐蚀、湍流腐蚀及空泡腐蚀。

5.9.1 摩振腐蚀

摩振腐蚀又称微动腐蚀、摩擦氧化，指在有氧气存在的气体环境中，加有载荷的固体相互接触的表面之间由于振动或滑动所产生的腐蚀。承受载荷的界面存在振动或滑动、载荷和相对运动使界面产生滑移和变形、有腐蚀介质参与作用时可大大增加摩振腐蚀的可能性。目前，磨损－氧化理论及氧化－磨损理论可对摩振腐蚀进行解释[1]。

常见的防止措施主要包括：增大载荷，使组合件的接触面之间不产生振动或滑动；选用耐磨材料或镀覆合适的耐磨层；改善摩擦条件，如环境中除氧，加入润滑剂或还原剂。

5.9.2 湍流腐蚀

湍流腐蚀又称冲击腐蚀（冲蚀），指腐蚀性流体与金属表面间相互运动而引起的金属

corrosion. It is influenced by the properties of the metal or alloy, the properties of the corrosive medium and the joint effect of other corrosion forms, which can be protected by reasonable material selection, correct design and optimize the medium.

加速腐蚀。受到金属或合金的性质、腐蚀介质的性质及其他腐蚀形式共同作用的影响。可通过合理选材、正确设计及优化介质进行防范。

5.9.3 Cavitation corrosion

Cavitation corrosion, also known as cavitation and cavitation, is a special form of impact corrosion. The reason is that the turbulence, bubbles and cavities generated when the fluid moves at high speed, the bubbles and cavities form and collapse at high speed, and the extreme pressure is generated at the moment of its collapse, which destroys the protective film on the metal surface and causes the metal to be strongly destroyed. Cavitation processes are mainly include ①the formation of bubbles on the metal surface film; ②the bubble burst to destroy the membrane; ③The exposed metal matrix corrodes and re-film; ④It is easy to form new bubbles there; ⑤bubble burst, the film is destroyed again; ⑥Bare area corrosion and re-film. This process is repeated and causes severe cavitation[2].

5.9.3 空泡腐蚀

空泡腐蚀又称空蚀、汽蚀，是冲击腐蚀的一种特殊形式。其产生的原因是流体高速运动时产生的湍流、气泡和空腔，气泡和空腔高速形成并破灭，在其破灭的瞬间产生极大压力，破坏了金属表面保护膜，使金属受到强烈的破坏。主要包括以下过程：①在金属表面膜上形成气泡；②气泡破灭使膜破坏；③暴露出的金属基体腐蚀并重新成膜；④在该处容易形成新的气泡；⑤气泡破灭，膜再次破坏；⑥裸区腐蚀并重新成膜。该过程反复进行，造成严重的汽蚀[2]。

5.9.4 References

[1] WU J P. Anti-corrosion mechanism and technical research of oil pipeline [J]. Science and Technology Innovation and Application, 2021, 11(29): 157-159.

[2] FEI J L, ZHANG T, YANG X L, et al. Analysis of erosion corrosion and leakage failure of steam system fittings [J]. HTM, 2019, 44(S1): 295-298.

5.9.4 参考文献

[1] 吴金萍 . 石油管道防腐机理与技术探究 [J]. 科技创新与应用，2021，11（29）：157–159.

[2] 费璟璐，张涛，杨晓蕾，等 . 蒸汽系统管配件冲刷腐蚀及泄漏失效分析 [J]. 金属热处理，2019，44（增刊 1）：295–298.

Natural Environment Corrosion
自然环境腐蚀
6

6.1 Atmospheric corrosion

Atmospheric corrosion is the most common type of metal corrosion. There are many factors that affect the speed of atmospheric corrosion, including the types and concentrations of impurities in the air, relative humidity, temperature and types of metals, etc.

The nature of atmospheric corrosion basically belongs to the scope of electrochemical corrosion. However, it is different from the corrosion in the electrolyte solution: when the thickness of the electrolyte solution film on the metal surface is very small, the electrolyte solution film is rich in oxygen, and the atmospheric corrosion is preferentially corroded by the oxygen depolarization process.

In terms of corrosion conditions, the main components of the atmosphere are water and oxygen, and the water vapor in the atmosphere is the main factor that determines the rate and course of atmospheric corrosion. Therefore, atmospheric corrosion can be divided into “dry”, “wet” and “wet” according to the degree of

6.1 大气腐蚀

大气腐蚀是最常见的金属腐蚀类型。影响大气腐蚀速度的因素很多，包括空气中杂质的种类和浓度、相对湿度、温度和金属的种类等。

大气腐蚀的性质基本上属于电化学腐蚀范围。然而，它与电解质溶液中的腐蚀不同：当金属表面电解质溶液膜的厚度非常小时，电解质溶液膜富含氧气，并且大气腐蚀优先被氧的去极化过程腐蚀。

从腐蚀条件看，大气的主要成分是水和氧，而大气中的水汽是决定大气腐蚀速度和历程的主要因素。因此，根据腐蚀金属表面的潮湿程度可把大气腐蚀分为“干的”“潮的”

wetness of the corroded metal surface. The degree of humidity is expressed by the thickness of the water film on the metal surface.

和“湿的”三种类型。潮湿程度通过金属表面水膜层的厚度来表示。

6.1.1 Types of atmospheric corrosion

6.1.1 大气腐蚀的类型

From the perspective of corrosion conditions, the main components of the atmosphere are water and oxygen, and the water vapor in the atmosphere is the main factor determining the speed and course of atmospheric corrosion. Therefore, atmospheric corrosion can be divided into three types: “dry”, “damp” and “wet” according to the moisture of the corroded metal surface. Moisture is measured by the thickness of the water film on the metal surface.

从腐蚀条件看，大气的主要成分是水和氧气，大气中的水蒸气是决定大气腐蚀速度和过程的主要因素。因此，大气腐蚀根据腐蚀金属表面的水分可分为“干”“潮”和“湿”三种。水分通过金属表面水膜的厚度来测量。

1. Dry atmospheric corrosion

1.干的大气腐蚀

Dry atmospheric corrosion is also called dry oxidation or low humidity corrosion, that is, there is no water film on the surface of the metal or the thickness of the water film is very small (the thickness of the water film is about 1~10nm) corrosion. This corrosion belongs to chemical corrosion.

干的大气腐蚀也称干的氧化或低湿度下的腐蚀，即金属表面没有水膜或者水膜的厚度极小（水膜厚度为1~10nm）情况下的腐蚀。这种腐蚀属于化学腐蚀。

2. Atmospheric corrosion of tides

2.潮的大气腐蚀

The atmospheric corrosion of the tides is the metal corrosion under the invisible water film (the thickness of the water film is about). This corrosion occurs when the air relative humidity is below 100%, as a result of capillary action, adsorption or chemical condensation to form a water film on the metal surface. The atmospheric corrosion of tides belongs to electrochemical corrosion.

潮的大气腐蚀是金属在肉眼不可见的水膜下（水膜厚度为10nm~1μm）进行的金属腐蚀。空气相对湿度在100%以下，由于毛细管作用、吸附作用或者化学凝聚作用在金属表面形成水膜，所以这种腐蚀是在超过临界相对湿度时发生的。潮的大气腐蚀属于电化学腐蚀。

3. Wet atmospheric corrosion

3.湿的大气腐蚀

Wet atmospheric corrosion is metal corrosion under

湿的大气腐蚀是金属在肉

the visible water film (ink thickness is about 1μm~1mm). The relative humidity of the air is close to 100%. Wet atmospheric corrosion is electrochemical corrosion.

眼可见的水膜下（水墨厚度为1μm~1mm）进行的金属腐蚀。空气相对湿度接近100%。湿的大气腐蚀属于电化学腐蚀。

The classification of atmospheric corrosion can be summarized as shown in Figure 6-1.

大气腐蚀的分类如图6-1所示。

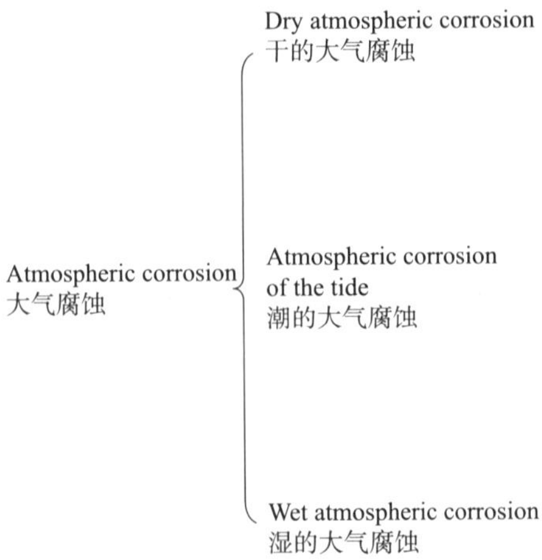

Figure 6-1 Classification of atmospheric corrosion

图6-1 大气腐蚀的分类

As shown in Figure 6-2, is the relationship between water film thickness and corrosion rate.

图6-2所示为水膜层厚度与腐蚀速度的关系。

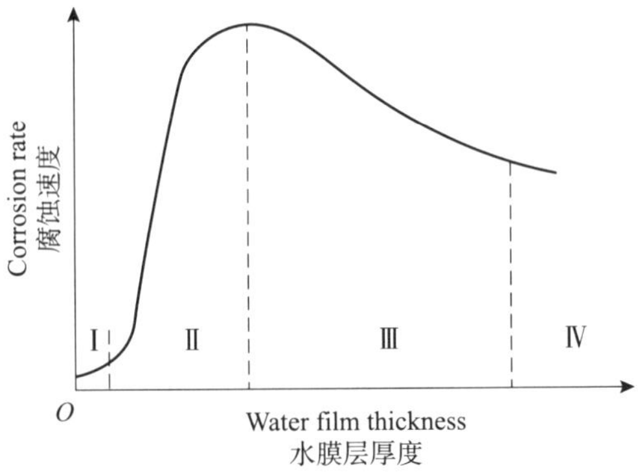

Figure 6-2 Relation between atmospheric corrosion velocity and the thickness of water film on metal surface

Ⅰ—Area of water film thickness δ=1~10nm; Ⅱ—δ=10nm~10μm;
Ⅲ—δ=1μm~1mm; Ⅳ—δ>1mm

图6-2 大气腐蚀速度与金属表面水膜层厚度之间的关系

Ⅰ—水膜厚度的区域 δ=1~10nm；Ⅱ—δ=10nm~10μm；Ⅲ—δ=1μm~1mm; Ⅳ—δ>1mm

In the figure 6-2, area Ⅰ is the adsorbed water film with several molecular layers thick on the metal surface, without forming continuous electrolyte, which is equivalent to dry atmospheric corrosion. Zone Ⅱ corresponds to the state of "tidal atmospheric corrosion". Due to the formation of the electrolyte film (tens or hundreds of water molecular layers thick), the electrochemical corrosion process starts, and the corrosion rate increases sharply. Zone Ⅲ is the visible liquid film layer (thickness of tens to hundreds of microns), which is equivalent to wet atmospheric corrosion. With the further increase of liquid film thickness, the diffusion of oxygen becomes difficult, so the corrosion rate decreases accordingly. In zone Ⅳ, the liquid film is thicker, which is similar to immersion in liquid.

图6-2中Ⅰ区为金属表面上有几个分子层厚的吸附水膜，未形成连续的电解液，相当于干的大气腐蚀。Ⅱ区对应于“潮的大气腐蚀”状态，由于电解液膜（几十个或几百个水分子层厚）的形成，开始电化学腐蚀过程，腐蚀速度急剧增加。Ⅲ区为可见的液膜层（厚度为几十至几百微米），相当于湿的大气腐蚀，随着液膜厚度进一步增加，氧扩散变得困难，因而腐蚀速度也相应降低。Ⅳ区为液膜更厚的情况，与浸泡在液体中类似。

According to the state of liquid film formation on metal surface, it can be divided into steady state and unsteady state liquid film. Steady state liquid film is continuous thin liquid film, while unsteady liquid film is usually dispersed droplets. The essence of thin liquid film atmospheric corrosion is that the water vapor in the air adsorbs on the metal matrix to form a thin electrolyte film, so as to promote the occurrence and development of electrochemical corrosion.

根据金属表面液膜形成的状态可分为稳态液膜和非稳态液膜，稳态液膜是连续的薄液膜，而非稳态液膜通常是分散的液滴。薄液膜大气腐蚀的本质就是空气中的水汽吸附在金属基体上形成薄电解质液膜，从而促进电化学腐蚀发生、发展。

At present, the electrochemical techniques suitable for atmospheric corrosion of thin liquid films mainly include scanning Kelvin technique (SKP), microdroplet electrode technique and filament beam electrode technique (WBE). SKP measures the work-function difference between the sample and the probe by vibrating capacitance probe, and then obtains the corrosion potential of the corroded metal surface. The Kelvin probe

目前，适用于薄液膜大气腐蚀的电化学技术主要包括扫描Kelvin技术（SKP）、微液滴电极技术及丝束电极技术（WBE）。SKP通过振动电容探针测量试样与探针间的功函数差，进而得到腐蚀金属表面的腐蚀电位。由于Kelvin探针可

is a powerful tool for studying atmospheric corrosion because it can measure the corrosion potential of metals under extremely thin liquid layers in the gas environment without contact. WBE is an integrated microelectrode array, which is composed of many microelectrodes. Since the area of each microelectrode is very small compared with the total area of the integrated electrode composed of all microelectrodes, it can be considered that the current and potential distribution on a single microelectrode is uniform. The data obtained by this method have good reproducibility and can be used to measure the corrosion potential, polarization resistance and corrosion current distribution of the metal under the liquid film. The microdroplet electrode is a three-electrode system composed of micro-sized auxiliary electrodes and reference electrodes to perform traditional polarization curve and AC impedance measurements, which is suitable for studying the local corrosion of metals under electrolyte droplets and can obtain relevant electrochemical information of micro-size.

不接触地测量气相环境中极薄液层下金属的腐蚀电位，是研究大气腐蚀的有力工具。WBE是一种集成微电极阵列，集成微电极由许多微电极组成，由于每一个微电极的面积相对于所有微电极组成的集成电极总面积非常小，可认为在单个微电极上的电流、电位分布是均匀的。由这种方法得到的数据的重现性较好，可用来测量液膜下金属的腐蚀电位、极化电阻及腐蚀电流的分布。微液滴电极是使用微观尺寸的辅助电极以及参比电极组成三电极体系，进行传统的极化曲线、交流阻抗等的测量，适用于研究金属在电解质液滴下的局部腐蚀，可得到微观尺寸的有关电化学信息。

6.1.2 Mechanism and process of atmospheric corrosion

6.1.2 大气腐蚀过程的机理和过程

The common atmospheric corrosion is mainly tidal atmospheric corrosion and wet atmospheric corrosion. A film of water will be adsorbed on the surface of the metal, thus forming the electrolyte film necessary for electrochemical corrosion. Therefore, the atmospheric corrosion of metals in wet and tidal atmospheres has an electrochemical nature.

一般常见的大气腐蚀都是以潮的大气腐蚀和湿的大气腐蚀为主。金属表面会吸附一层水膜，从而形成电化学腐蚀所必需的电解液膜。所以在湿的和潮的大气环境中，金属的大气腐蚀具有电化学本质。

As shown in Figure 6-3, the outer wall of the tank has been seriously corroded by the atmosphere.

如图6–3所示，储罐外壁已受到严重的大气腐蚀。

Figure 6-3　Atmospheric corrosion of the storage tank
图 6-3　储罐受到大气腐蚀

If only a layer of pure water is adsorbed on the surface of the metal, it will not cause serious corrosion, because the conductivity of pure water is poor. However, the liquid film formed on the metal surface will absorb impurities in the air, so that the liquid film contains water-soluble salts and corrosive gases, resulting in strong corrosion phenomenon.

如果金属表面仅仅吸附一层纯水时，不会造成严重的腐蚀，因为纯水的导电性较差。但是，金属表面形成的液膜会吸附空气中的杂质，使液膜中含有水溶性盐类及腐蚀性气体，从而造成强烈的腐蚀现象。

1. Formation of water film on metal surface

1.金属表面水膜的形成

On metal without oxide and on aluminum with oxide film, the thickness of the film is less than two water molecules at 90% relative humidity. At 60% relative humidity in the atmosphere, the thickness of the water vapor film is only one water molecule thick. However, when there is a hygroscopic attachment on the metal surface, even if very little, it will greatly increase the thickness of water molecules in the water vapor film, thus causing corrosion.

在不带氧化物的金属及在有氧化膜的铝上，90%相对湿度的大气下，水汽膜厚度不到两个水分子厚。60%相对湿度的大气下，水汽膜厚度只有一个水分子厚。但是，当金属表面存在吸湿性的附着物时，哪怕极少，也会极大地增加水汽膜的水分子厚度，从而引起腐蚀。

Water vapor film is an invisible liquid film with a thickness of 2 to 40 layers of water molecules. When the water vapor reaches saturation, condensation will occur

水汽膜是不可见液膜，其厚度为2~40个水分子层。当水汽达到饱和时，在金属表面会

on the metal surface, so that the metal surface to form a thicker layer of water, this layer is called wet film. A wet film is a visible liquid film with a thickness of 1~1000μm.

发生凝结现象，使金属表面形成一层更厚的水层，此层称为湿膜。湿膜是可见液膜，其厚度为1~1000μm。

(1) Formation of water vapor film

（1）水汽膜的形成

There are three main reasons for the formation of water vapor film: capillary condensation, adsorption condensation and chemical condensation.

水汽膜的形成主要有三种原因：毛细凝聚、吸附凝聚和化学凝聚。

① Capillary condensation: The smaller the curvature radius (r) of the liquid surface is, the smaller the saturation vapor pressure is, and the easier the water vapor condenses.

①毛细凝聚：液面的曲率半径（r）越小，饱和蒸汽压越小，水蒸气越容易凝聚。

Therefore, in Figure 6-4, the water vapor at the concave surface is the easiest to gather, followed by the flat surface, and the water vapor at the convex surface is not easy to gather.

如图6-4所示，凹形液面的水蒸气最容易聚集，平的其次，凸形液面的水蒸气不容易聚集。

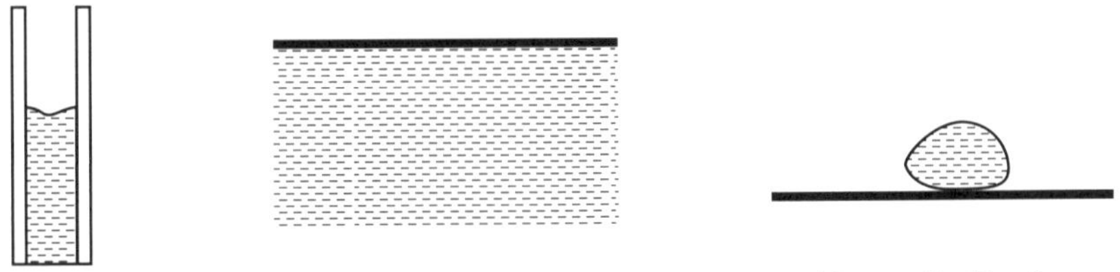

Figure 6-4 Influence of liquid level shape on saturated vapor pressure[1]

图6-4 液面形状对饱和蒸气压的影响[1]

The oxide film, the gap between the parts, corrosion products, the pores in the coating, the cracks of the material and the gaps under the ash and carbon particles falling on the surface of the plant, will form capillary condensation, promoting the atmospheric corrosion of these parts.

氧化膜、零件之间缝隙、腐蚀产物、镀层中的孔隙、材料的裂缝及落在金属表面的灰尘和碳粒下的缝隙等，会形成毛细凝聚，促进这些部位的大气腐蚀。

② Adsorption condensation: oxide film, cracks between parts, corrosion products, pores in the coating, cracks in the material and cracks falling on the surface of the plant dust and carbon particles, etc., will form

②吸附凝聚：氧化膜、零件之间缝隙、腐蚀产物、镀层中的孔隙、材料裂缝及落在金属表面的灰尘和碳粒下的缝隙

capillary condensation to promote the atmospheric corrosion of these parts.

等，会形成毛细凝聚，促进这些部位的大气腐蚀。

③ Chemical condensation: The material on the metal surface interacts chemically with the adsorbed water to form a salt solution. The vapor pressure on the salt solution is lower than that of pure water, so the condensation of water is easier. In this case, the moisture will condense at the humidity of 70%~80%, and the presence of electrolyte will greatly speed up the corrosion rate.

③化学凝聚：金属表面的物质与吸附的水分发生化学作用，形成盐溶液，盐溶液的水蒸气压力低于纯水的水蒸气压力，因此水的凝聚更加容易。在这种情况下，水分在湿度为70%~80%时就会凝聚，并且有电解质的同时存在，会极大地加快腐蚀速度。

(2) Formation of wet film

（2）湿膜的形成

When exposed to the outdoors, the surface of the metal tends to form a visible film of water about thick. If the temperature change is particularly dramatic in some areas, the relative humidity is less than 100%, it is easy to cause the condensation of water in the air. The greater the temperature difference, the lower the relative humidity of the condensation. The corrosion process in the case of wet film is similar to that in the case of immersion in water. The only difference is that in the case of wet film, because the thickness of water film is smaller, it is easier to obtain oxygen supply than that in the case of immersion in water, so the corrosion degree is more severe.

金属暴露在室外，其表面易形成1~10μm厚的可见水膜。若某些地区的气温变化特别剧烈，在相对湿度不足100%时，也容易造成空气中水分冷凝。温差越大引起凝露的相对湿度就越低。湿膜情况下发生的腐蚀与浸泡在水中的腐蚀历程类似，唯一不同的是，湿膜情况下，由于水膜厚度较小，比沉浸在水中的情况下更容易获得氧的补给，因而腐蚀程度更剧烈。

2. Electrochemical characteristics of atmospheric corrosion

2.大气腐蚀的电化学特征

The kinetics of corrosion process is closely related to electrode polarization, conduction and ion migration. The speed of the whole corrosion process depends on the process with the greatest resistance. During atmospheric corrosion, the cathodic process mainly depends on the depolarization of oxygen, and the reaction process is as follows:

腐蚀过程动力学问题与电极的极化、导电过程及离子迁离等密切相关。整个腐蚀过程的速度取决于所受阻力最大的过程。大气腐蚀时，阴极过程主要依靠氧的去极化作用，反应过程如下：

$$O_2 + 2H_2O + 4e^- \longrightarrow 4OH^-$$

The diffusion rate of oxygen determines the depolarization rate of oxygen on the cathode. Therefore, the rate of oxygen diffusion determines the speed of the corrosion process.

而氧的扩散速度决定阴极上氧的去极化作用的速度。因此，氧的扩散速度决定整个腐蚀过程的速度。

Figure 6-5 shows the cathodic polarization curve of copper in NaCl solution. As can be seen from the figure, the slope of the cathode polarization curve decreases with the thinning of the electrolyte film thickness. It indicates that the cathode current density increases gradually with the thinning of the film thickness at a certain potential.

图6-5所示为铜在0.1mol/L NaCl溶液下的阴极极化曲线。可见，随着电解液膜厚度减薄，阴极极化曲线的斜率也随之减小。说明随着膜厚减薄，在一定电位下阴极电流密度逐渐加大。

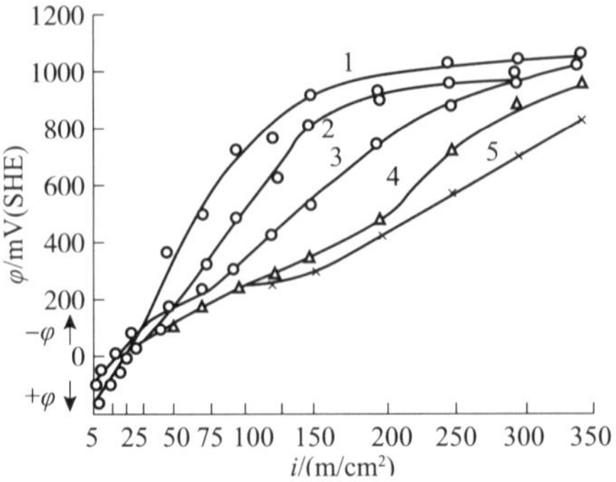

Figure 6-5 Influence of electrolyte film thickness on cathode polarization of copper in NaCl solution[1]

1—electrolyte; 2—liquid film thickness δ=300μm; 3—liquid film thickness δ=165μm; 4—liquid film thickness δ=100μm; 5—liquid film thickness δ=70μm

图6-5 电解液液膜厚度对铜在0.1mol/L NaCl溶液中阴极极化的影响[1]

1—电解液；2—液膜厚度 δ=300μm；3—液膜厚度 δ=165μm；4—液膜厚度 δ=100μm；5—液膜厚度 δ=70μm

In the atmosphere corrosion environment, oxygen in the air diffuses to the cathode surface through the liquid film. Therefore, the thinner the liquid film, the faster the transfer speed of oxygen and the faster the corrosion speed. However, when the liquid film is too thin, the water is not enough to support the oxygen reduction or hydrogen discharge of the cathode, and the polarization process of the cathode will be blocked.

在大气腐蚀环境下，空气中的氧通过液膜扩散到阴极表面，因此液膜越薄，氧的传递速度就越快，腐蚀速度就越快。但当液膜太薄时，水分不足以支持阴极的氧还原或氢放电，那么阴极的极化过程就会受到阻滞。

In short, the velocity of atmospheric corrosion

总之，大气腐蚀的速度

and the characteristics of the electrode polarization process vary with different atmospheric conditions: in the case of wet atmospheric corrosion (visible under water film or strongly wetting corrosion products), the corrosion velocity is mainly controlled by the cathode. However, the degree of cathodic control is weaker than that under full immersion, and the cathodic polarization process becomes easier and easier with the thinning of the electrolyte liquid film. In the atmospheric corrosion of tides, the corrosion rate is controlled by the anode, and with the thinning of the liquid film, the anodic polarization process becomes difficult, which will also lead to a significant increase in ohmic resistance. Some corrosion processes are controlled by the mixture of Yin and Yang, while for macroscopic battery contact corrosion, the corrosion is mostly controlled by ohmic resistance.

与电极极化过程的特征随着大气条件的不同而变化：在湿的（可见水膜下或水强烈湿润的腐蚀产物下）大气腐蚀时，腐蚀速度主要由阴极控制。但这种阴极控制程度已比全浸时减弱，并随着电解液液膜减薄，阴极极化过程变得越来越容易进行。在潮的大气腐蚀时，腐蚀速度则被阳极控制，并随着液膜减薄，阳极极化过程变得困难，这种情况也会使欧姆电阻显著增大。有的腐蚀过程同时受阴、阳两极混合控制，而对于宏观电池接触腐蚀来说，腐蚀却多半为欧姆电阻控制。

6.1.3 The main factors affecting atmospheric corrosion

The factors affecting atmospheric corrosion are more complex. With different climate and regions, the composition, humidity and temperature of the atmosphere are very different. In the main components of the atmosphere, the atmospheric corrosion has a greater impact on oxygen, water vapor and carbon dioxide: atmospheric corrosion has a strong role in promoting trace impurities, such as SO_2, H_2S, NH_3, NO_2, and a variety of suspended particles and dust.

1. The influence of atmospheric relative humidity

In an atmosphere with varying amounts of pollutants, metals have a critical relative humidity above which the rate of corrosion suddenly surges. Before

6.1.3 影响大气腐蚀的主要因素

影响大气腐蚀的因素比较复杂，因气候、地区不同，大气的成分、湿度、温度等有很大差别。在大气的主要成分中，对大气腐蚀有较大影响的是氧、水蒸气和二氧化碳：对大气腐蚀有强烈促进作用的微量杂质有SO_2、H_2S、NH_3和NO_2以及各种悬浮颗粒和灰尘。

1.大气相对湿度的影响

在含有不同数量污染物的大气中，金属都有一个临界相对湿度，超过这一临界值，腐

the critical value, the corrosion rate is very small or almost no corrosion. The occurrence of critical relative temperature indicates the formation of a lower layer of adsorbed electrolyte film on the metal surface. The existence of this layer of liquid film makes the metal change from chemical corrosion to electrochemical corrosion. Because of the sudden change in corrosion properties, the corrosion is greatly enhanced.

蚀速度就会突然猛增。在临界值之前，腐蚀速度很小或几乎不腐蚀。出现临界相对湿度，表明金属表面产生下一层吸附的电解液膜，这层液膜的存在使金属从化学腐蚀转为电化学腐蚀。由于腐蚀性质发生突变，因而腐蚀大大增强。

2. Influence of temperature and temperature difference

2.温度和温差的影响

In an atmospheric environment with high humidity, temperature plays a major role, and the corrosion rate generally accelerates with the increase of temperature. In an atmosphere where humidity is not very high, the temperature difference is more affected than temperature, because it affects not only the condensation of water vapor, but also the solubility of gases and salts in the condensed water film.

在湿度很高的大气环境中，温度起主要作用，随着温度升高，腐蚀速度加快。在湿度不是很高的大气中，温度差比温度的影响要大，它不但影响水汽凝聚，而且还影响凝聚水水膜中气体和盐类的溶解度。

3. Influence of acid, alkali and salt

3.酸碱盐的影响

The change of acid and alkalinity of the medium can significantly affect the content of depolarizer (such as H^+) and the stability of the metal surface film, thus affecting the size of the corrosion rate. Iron and magnesium, for example, corrode less rapidly in alkaline solutions than in neutral and acidic solutions because their oxides are insoluble in bases and form protective films on their surfaces. However, for some amphoteric metals (such as aluminum, zinc, and lead), they are unstable in both acid and base solutions, and their oxides are dissolved in both acids and bases.

介质的酸、碱性改变，能显著影响去极化剂（如H^+）含量及金属表面膜的稳定性，从而影响腐蚀速度的大小。例如铁和镁，由于它们的氧化物在碱中不溶解，使表面生成保护膜，所以它们在碱性溶液中的腐蚀速度比在中性和酸性溶液中要小。而对于一些两性金属（如铝、锌、铅）来说，在酸和碱溶液中都不稳定，它们的氧化物在酸、碱溶液中均溶解。

The effect of neutral salts on the corrosion rate of metals depends on many factors, including the solubility

中性盐类对金属腐蚀速度的影响取决于很多因素，其中

of corrosion products. If the corrosion products generated on the Yin and Yang poles of the metal surface are insoluble or even insoluble, the corrosion products will cover the metal surface to prevent the further occurrence of corrosion reaction, thus reducing the corrosion speed.

包括腐蚀产物的溶解度。如果在金属表面的阴、阳两极生成的腐蚀产物难溶甚至不溶，腐蚀产物就会覆盖在金属表面阻止腐蚀反应的进一步发生，从而降低腐蚀速度。

4. The influence of corrosive gases

4.腐蚀性气体的影响

The industrial atmosphere contains a large number of corrosive gases, such as SO_2, H_2S, NH_3, Cl_2, HCl. Among these pollution impurities, SO_2 is the most harmful to metal corrosion. As shown in Figure 6-6, copper is corroded in SO_2 and air containing water vapor. The curve shows that with the increase of the SO_2 concentration, the corrosion rate of copper increases firstly, then decreases and then increases. The *N* pitting corrosion rate reaches the slowest in the figure. The analysis shows that the composition of the pitting product is $CuSO_4$. At this point the corrosion product has an excess $Cu(OH)_2$ on the left and sulfuric acid on the right. It can be seen that the $Cu(OH)_2$ membrane at *N* is better protected.

工业大气中含有大量的腐蚀性气体，如SO_2、H_2S、NH_3、Cl_2、HCl等。在这些污染杂质中SO_2对金属腐蚀危害最大。图6-6所示为铜在含有SO_2和水蒸气的空气中腐蚀。曲线表明，随着SO_2浓度增加，铜的腐蚀速率呈先上升后下降再上升的趋势。可知*N*点腐蚀速率达到最慢。结果表明，该点腐蚀产物的成分为$CuSO_4$。在*N*点左侧腐蚀产物中含有过量的$Cu(OH)_2$，右侧却含有过量的H_2SO_4。可见在*N*点的膜$Cu(OH)_2$保护性较好。

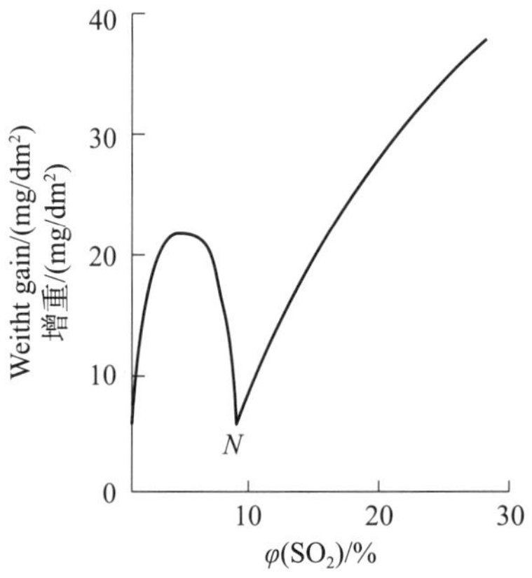

Figure 6-6 Copper corrosion in air containing SO_2 and 99% of different volume fractions [1]

图6-6 铜在含体积分数不同的SO_2 99% RH的空气中的腐蚀[1]

5. Solid particles, surface state and other factors

When all kinds of dust and sulfur dioxide in the air work with water, they speed up corrosion. Some solid particles, although not corrosive, will indirectly accelerate corrosion due to their ability to absorb corrosive gases. Although some solid particles are not corrosive or adsorptive, they can cause capillary condensation gap, which leads to the formation of electrolyte film on the metal surface and the formation of oxygen concentration battery, leading to crevice corrosion.

Compared with the smooth surface, the rough surface is easy to absorb dust, and the actual area of the dew and air is also larger, so the corrosion resistance is poor.

Due to serious problems such as air pollution, particulate matter in the air has greatly increased. Frequent occurrence of haze weather in winter and spring leads to greater challenges in air pollution prevention and control, and it is urgent to vigorously promote the precise control of air pollution. After dry settlement, haze particles will wear and dirty the surface of ground objects, increase the damage and cleaning cost, and improve the operation safety risk of devices and equipment. As a medium for adsorption and oxidation of gaseous pollutants, haze particles can also directly erode or corrode surface materials of surface objects after wet deposition. Haze pollution may cause acid rain in the south, while acid rain also has the potential to increase in the north after the reduction of mineral dust (dust storms). Some haze pollutants also become the nutrient

5. 固体颗粒、表面状态等因素的影响

当空气中各种灰尘和二氧化硫与水共同作用时，会加速腐蚀。一些虽然不具有腐蚀性的固体颗粒，由于具有吸附腐蚀性气体的作用，会间接地加速腐蚀。有些固体颗粒虽然不具有腐蚀性也不具有吸附性，但由于能造成毛细凝聚缝隙，促使金属表面形成电解液薄膜，形成氧浓差电池，导致缝隙腐蚀。

与光洁表面相比，加工粗糙的表面容易吸附灰尘，暴露于空气中的实际面积也更大，所以耐蚀性差。

由于空气污染等严重问题，造成空气中的颗粒物大大增加。冬春季雾霾天气频发大气氧化性增强，导致大气污染防治面临更大的挑战，急需大力推进对大气污染的精准治理。霾颗粒干沉降后会磨损、脏污地表物体表面，增加折损和清洁成本，提升装置设备的运行安全风险。霾颗粒作为吸附和氧化气态污染物的媒介，湿沉降后其酸性组分还可直接侵蚀或腐蚀地表物体的表面材料。霾污染在南方可能会造成酸雨，而北方在矿尘减少后（沙尘暴）酸雨也有增加的潜力。某些霾污

of microbial growth, such as mineral components. Therefore, in order to reduce the direct damage caused by haze pollution, it is required that the device and equipment increase airtightness and waterproofing, and clean the surface regularly[2].

染物还成为微生物滋生的营养物质，如矿物成分等。因此，为减少霾污染的直接损害，要求装置设备增加密闭性和防水性，并定期清洗表面[2]。

6.1.4 Measures to prevent atmospheric corrosion

6.1.4 防止大气腐蚀的措施

1. Alloying method

Through the alloying method in ordinary carbon steel to add some appropriate alloying elements to change the embroidery layer structure, the generation of protective embroidery layer, can significantly improve the atmospheric corrosion resistance of steel.

(1) Weathering steel

Corrosion resistance is better than the general structure of steel steel called weathering steel, generally contains phosphorus, copper, nickel, chromium, titanium and other metals, so that the metal surface to form a protective layer, to improve corrosion resistance. Its low temperature impact toughness is also better than the general structural steel. The standard is "Weathering steel for Welding structure" (GB/T 4171—2008).

(2) Hot-dip galvanizing

Hot dip zinc is to immerse the steel components after rust removal into the zinc liquid melted at a high temperature of about 600℃, so that the surface of the steel components is attached with a zinc layer. The thickness of the zinc layer is not less than 65μm for the 5mm thin plate, and not less than 86μm for the thick plate. Thus play the purpose of corrosion prevention. The advantages of this method are long durability, high degree of industrialization and stable quality. Therefore,

1.合金化法

通过合金化法在普通碳钢中加入某些适量合金元素改变锈层结构，生成具有保护作用的锈层，可显著改善钢的耐大气腐蚀性能。

（1）耐候钢

耐腐蚀性能优于一般结构用钢的钢材称为耐候钢，一般含有磷、铜、镍、铬、钛等金属，使金属表面形成保护层，以提高耐腐蚀性。其低温冲击韧性也比一般的结构用钢好。采用标准为GB/T 4171—2008《耐候结构钢》。

（2）热浸锌

热浸锌是将除锈后的钢构件浸入600℃左右高温熔化的锌液中，使钢构件表面附着锌层，锌层厚度对5mm以下薄板不小于65μm，对厚板不小于86μm，从而起到防腐蚀的目的。这种方法的优点是耐久年限长，生产工业化程度高，质量稳定。因而被大量用于受大

it is widely used in outdoor steel structures which are severely corroded by atmosphere and not easy to maintain. Such as a large number of transmission towers, communication towers and so on. In recent years, a large number of light steel structure system, such as profiled steel plate. Also more use of hot dip zinc corrosion. The first process of hot dipping zinc is pickling rust removal, followed by cleaning.

The incompleteness of these two processes will leave hidden dangers for corrosion prevention, so they must be treated thoroughly. For the steel structure designer, it is necessary to avoid designing the components with the mating surface, so as to avoid the phenomenon of yellow water flowing on the galvanized surface caused by incomplete acid cleaning or incomplete acid cleaning in the seams of the mating surface. Hot dip galvanizing is carried out at high temperature. For tubular members, both ends should be open. If both ends are closed, the air in the pipe will expand and the head plate will burst, resulting in safety accidents. If one end is closed, the zinc liquid can not flow smoothly and is easy to accumulate in the pipe[3].

2. The use of organic, inorganic coating and metal coating

Recently, the development of superhydrophobic surface has become one of the most effective means to solve the problem of atmospheric corrosion. From the perspective of the mechanism of atmospheric corrosion, the existence of liquid film/droplet is an important prerequisite for the occurrence of atmospheric corrosion, and superhydrophobic surface can effectively prevent the formation of liquid film/droplet on the surface. Droplet self-bouncing is a spontaneous behavior occurring

气腐蚀较严重且不易维修的室外钢结构中，如大量输电塔、通信塔等。近年来，大量出现的轻钢结构体系中的压型钢板等，也较多采用热浸锌防腐蚀。热浸锌的首道工序是酸洗除锈，然后是清洗。

这两道工序处理不彻底均会给防腐蚀留下隐患，所以必须处理彻底。对于钢结构设计者，应避免设计出具有相贴合面的构件，以免贴合面的缝隙中酸洗不彻底或酸液洗不净造成镀锌表面流黄水的现象。热浸锌是在高温下进行的。对于管形构件应让其两端开敞，若两端封闭会造成管内空气膨胀而使封头板爆裂，从而造成安全事故。若一端封闭则锌液流通不畅，易在管内积存[3]。

2.采用有机、无机涂层和金属镀层

近期超疏水表面的发展成为解决大气腐蚀问题极为有效的手段之一。从大气腐蚀机理看，液膜/滴的存在是发生大气腐蚀的重要前提，超疏水表面可有效阻止表面液膜/滴的形

超疏水表面在大气腐蚀中的应用

on a specific superhydrophobic surface. Droplet self-bouncing can effectively prevent the formation of liquid film/droplet on the surface, which may affect the protection performance of atmospheric corrosion on the surface. As shown in Figure 6-7, the nano-structure and micron-structure zinc oxide superhydrophobic surface prepared by Chen Xiaotong et al[2] on zinc sheets by the two-step method of hydrothermal synthesis and surface modification and by controlling the reaction concentration. This study revealed a new mechanism of atmospheric corrosion protection based on the self-bouncing effect of droplets on superhydrophobic surface, and confirmed the feasibility of applying the self-bouncing effect of droplets on superhydrophobic surface in atmospheric corrosion protection.

成。液滴自弹跳现象是发生在特定超疏水表面的一种自发行为，弹走的液滴能有效阻止表面液膜/滴的形成，进而可能对表面的大气腐蚀防护性能产生影响。如图 6-7 所示，陈晓彤等[2]以水热合成和表面修饰两步法，通过控制反应浓度，在锌片上制备的纳米结构和微米结构的氧化锌基超疏水表面。该研究揭示了基于超疏水表面液滴自弹跳效应的大气腐蚀防护新机制，证实了超疏水表面液滴自弹跳效应在大气腐蚀防护中应用的可行性。

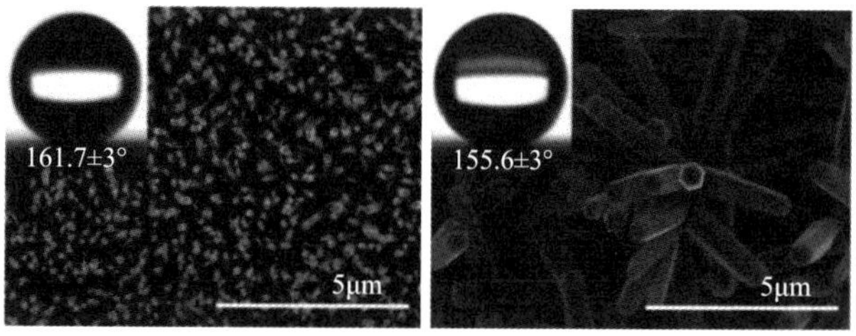

(a)~(b)Superhydrophobic surface structure and contact angle
超疏水表面结构和接触角

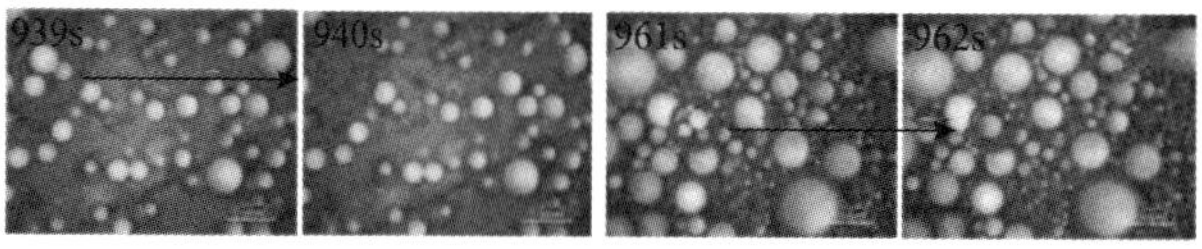

(c)~(d)Optical microscope of superhydrophobic surface
超疏水表面光学显微镜图

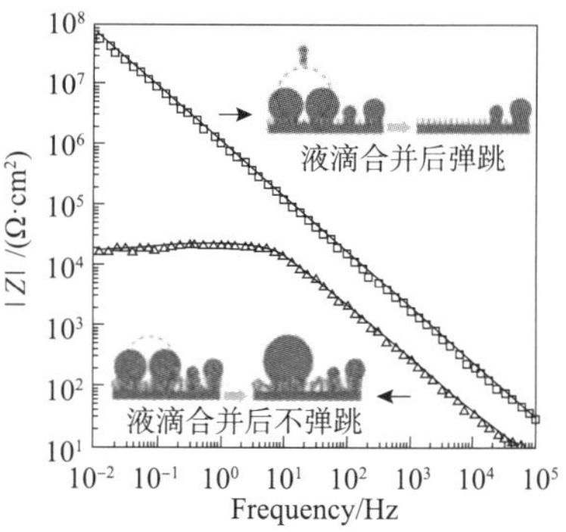

(e)Schematic diagram of atmospheric corrosion protection mechanism based on self-bouncing effect of droplets on superhydrophobic surface
基于超疏水表面液滴自弹跳效应的大气腐蚀防护机制示意

Figure 6-7 Hydrothermal synthesis and surface modification for two-step preparation of nanostructured and micron-structured ZnO-based superhydrophobic surfaces

图 6-7 水热合成和表面修饰两步法制备纳米结构和微米结构的氧化锌基超疏水表面

3. Corrosion inhibitor

A chemical or compound that prevents or slows the corrosion of a material when present in the environment (medium) in the appropriate concentration

3. 缓蚀剂

以适当的浓度和形式存在于环境（介质）中时，可防止或减缓材料腐蚀的化学物质或

and form. Therefore, corrosion inhibitors can also be called corrosion inhibitors. The amount used is small (0.1% to 1%), but the effect is significant. This method of protecting metals is called corrosion inhibitor protection. Corrosion inhibitor is used in neutral medium (boiler water, circulating cooling water), acid medium (hydrochloric acid in addition to pot scale, acid leaching solution for rust removal before plating) and gas medium (gas phase corrosion inhibitor).

复合物，因此缓蚀剂也称腐蚀抑制剂。其用量很小（0.1%~1%），但效果显著。这种保护金属的方法称为缓蚀剂保护。缓蚀剂用于中性介质（锅炉用水、循环冷却水）、酸性介质（除锅垢的盐酸、电镀前镀件除锈用的酸浸溶液）和气体介质（气相缓蚀剂）。

There are a variety of classification methods for corrosion inhibitors, which can be classified from different angles.

缓蚀剂有多种分类方法，可从不同角度对缓蚀剂进行分类。

(1) According to the chemical composition of the product, it can be divided into inorganic corrosion inhibitors, organic corrosion inhibitors, polymer corrosion inhibitors.

（1）根据产品化学成分，可分为无机缓蚀剂、有机缓蚀剂、聚合物类缓蚀剂。

① Inorganic corrosion inhibitors Inorganic corrosion inhibitors mainly include chromate, nitrite, silicate, molybdate, tungstate, polyphosphate, zinc salt and so on.

①无机缓蚀剂主要包括铬酸盐、亚硝酸盐、硅酸盐、钼酸盐、钨酸盐、聚磷酸盐、锌盐等。

② Organic corrosion inhibitors Organic corrosion inhibitors mainly include phosphonic acid (salt), phosphonocarboxylic acid, lucidyl benzothiazole, benzotriazole, sulfonated lignin and some other heterocyclic compounds containing nitrogen oxygen compounds.

②有机缓蚀剂主要包括膦酸（盐）、膦羧酸、巯基苯并噻唑、苯并三唑、磺化木质素等一些含氮氧化合物的杂环化合物。

③ Polymer corrosion inhibitors Polymer corrosion inhibitors as long as including polyethylene, POCA, poly aspartic acid and some other oligomers of polymer chemicals.

③聚合物类缓蚀剂主要包括聚乙烯类、POCA、聚天冬氨酸等一些低聚物的高分子化学物。

(2) According to the control position of corrosion inhibitor on electrochemical corrosion, it can be divided into anodic corrosion inhibitor, cathodic corrosion

（2）根据缓蚀剂对电化学腐蚀的控制部位分类，分为阳极型缓蚀剂、阴极型缓蚀剂和

inhibitor and mixed corrosion inhibitor.

① Anodic corrosion inhibitors are mostly inorganic strong oxidants, such as chromate, molybdate, tungstate, vanadate, nitrite, borate and so on. Their role is to interact with metal ions in the anodic area of the metal surface to generate oxide or hydroxide oxide film covering the anode to form a protective film. This prevents the metal from dissolving into water. The anodic reaction is controlled and the anode is passivated. Silicate can also fall into this category, it is also through the inhibition of the anodic process of corrosion reaction to achieve the purpose of corrosion inhibition.

Anode-type corrosion inhibitor requires a high concentration to make all the anodes passivated. Once the dose is insufficient, it will cause pitting corrosion on the parts that are not passivated.

② Cathodic corrosion inhibitor A chemical agent that inhibits the electrochemical cathodic reaction is called cathodic corrosion inhibitor.

Zinc carbonate, phosphate and hydroxide, calcium carbonate and phosphate are cathodic corrosion inhibitors. Cathodic corrosion inhibitors can react with water and the cathode area of the metal surface, and their reaction products are deposited on the cathode to form a film. With the thickening of the film, the reaction of the cathode to release electrons is blocked. In practical applications, because calcium ion, carbonate ion and hydroxide ion are naturally present in water, only soluble zinc salt or soluble phosphate need to be added to the water.

③ Mixed-type corrosion inhibitors, such as some organic corrosion inhibitors containing nitrogen, sulfur,

混合型缓蚀剂。

①阳极型缓蚀剂多为无机强氧化剂，如铬酸盐、钼酸盐、钨酸盐、钒酸盐、亚硝酸盐、硼酸盐等。它们的作用是在金属表面阳极区与金属离子作用，生成氧化物或氢氧化物氧化膜覆盖在阳极上形成保护膜。这样就抑制了金属向水中溶解。阳极反应被控制，阳极被钝化。硅酸盐也可归到此类，它也是通过抑制腐蚀反应的阳极过程来达到缓蚀的目的。

阳极型缓蚀剂要求有较高的浓度，以使全部阳极都被钝化，一旦剂量不足，将在未被钝化的部位造成点蚀。

②阴极型缓蚀剂抑制电化学阴极反应的化学药剂，称为阴极型缓蚀剂。

锌的碳酸盐、磷酸盐和氢氧化物，钙的碳酸盐和磷酸盐为阴极型缓蚀剂。阴极型缓蚀剂能与水中、金属表面的阴极区反应，其反应产物在阴极沉积成膜，随着膜增厚，阴极释放电子的反应被阻挡。在实际应用中，由于钙离子、碳酸根离子和氢氧根离子在水中是天然存在的，所以只需向水中加入可溶性锌盐或可溶性磷酸盐。

③混合型缓蚀剂，例如某些含氮、含硫或羟基的、具有

or hydroxyl groups and possessing surface activity, have two polar groups with opposite properties in their molecules. These inhibitors can adsorb on the clean metal surface to form a monolayer. They can form a film on both the anode and cathode. It prevents the diffusion of water and dissolved oxygen in water to the metal surface and plays a role in corrosion inhibition. Mercaptobenzothiazole, benzotriazole, hexadecylamine, etc. belong to this type of corrosion inhibitor.

表面活性的有机缓蚀剂，其分子中有两种性质相反的极性基团，能吸附在清洁的金属表面形成单分子膜，它们既能在阳极成膜，也能在阴极成膜。阻止水与水中溶解氧向金属表面扩散，起到缓蚀作用，巯基苯并噻唑、苯并三唑、十六烷胺等属于此类缓蚀剂。

(3) Classification according to the type of protective film

（3）根据生成保护膜的类型分类

In addition to neutralizing water treatment agents, the corrosion inhibition mechanism of most water treatment inhibitors is to form a metal protective film separating metal and water on the metal surface in contact with water to achieve the purpose of corrosion inhibition. Based on the type of protective film formed, corrosion inhibitors can be classified into oxide-film-type, deposition-film-type, and adsorption-film-type.

除中和性能的水处理剂外，大部分水处理用的缓蚀剂的缓蚀机理是在与水接触的金属表面形成一层将金属和水隔离的金属保护膜，以达到缓蚀目的。根据缓蚀剂形成的保护膜的类型，缓蚀剂可分为氧化膜型、沉积膜型和吸附膜型缓蚀剂。

① Oxide film type corrosion inhibitor chromate, nitrite, molybdate, tungstate, vanadate, orthophosphate, borate, etc. are all regarded as oxide film type corrosion inhibitor. Both chromate and nitrite are strong oxidants, which can react with metal without the help of dissolved oxygen in water, forming a dense oxide film on the anode area of metal surface. The others, either because of their weak oxidation ability or because they are not oxidants, need the help of oxygen to form an oxide film on the metal surface. Because these oxide film type corrosion inhibitors achieve corrosion inhibition by inhibiting the anodic process of the corrosion reaction, these anodic corrosion inhibitors can form oxides or chloride oxides

①氧化膜型缓蚀剂的铬酸盐、亚硝酸盐、钼酸盐、钨酸盐、钒酸盐、正磷酸盐、硼酸盐等均被看作氧化膜型缓蚀剂。铬酸盐和亚硝酸盐都是强氧化剂，无须水中溶解氧的帮助即能与金属反应，在金属表面阳极区形成一层致密的氧化膜。其余几种，或因本身氧化能力弱，或因本身并非氧化剂，都需要氧的帮助才能在金属表面形成氧化膜。这些氧化膜型缓蚀剂通过阻抑腐蚀反应

by the interaction between the anode and metal ions. The deposition covers the anode to form a protective film. Take chromate as an example, it forms $Cr(OH)_3$ and $Fe(OH)_3$ in the anodic reaction, and becomes a mixture of CrO_3 and Fe_2O_3 after dehydration (mainly γ-Fe_2O_3) forms a protective film on the anode. Therefore, it is sometimes referred to as anode corrosion inhibitor or dangerous corrosion inhibitor, because once the dosage is insufficient (when the dosage of 1L water is treated separately, the dosage is often as high as several hundred, even more than one thousand milligrams), it will cause pitting corrosion, which will make the corrosion problem that is not so serious, but become more serious. Chlorine ion, high temperature and high water flow rate will destroy the oxide film, so the concentration of corrosion inhibitor should be appropriately changed according to the process conditions during application. Silicate can also be roughly classified into this category, because it is mainly achieved by inhibiting the anodic process of corrosion reaction. However, it is not formed by the interaction with the metal iron itself, but possibly by the interaction of silicon dioxide and the corrosion products of iron, through the adsorption mechanism.

的阳极过程来达到缓蚀，它们能在阳极与金属离子作用形成氧化物或氯氧化物。沉积覆盖在阳极形成保护膜，以铬酸盐为例，它在阳极反应形成$Cr(OH)_3$和$Fe(OH)_3$，脱水后成为CrO_3和Fe_2O_3的混合物（主要是γ-Fe_2O_3）在阳极构成保护膜。因此又被称为阳极型缓蚀剂或危险型缓蚀剂，因为它们一旦剂量不足（单独缓蚀时，处理1L水，所需剂量高达几百、甚至过千毫克）就会造成点蚀，使本来不太严重的腐蚀问题，反而变得更加严重。氯离子、高温及高的水流速都会破坏氧化膜，故在应用时，要根据工艺条件，适当改变缓蚀剂浓度。硅酸盐也可粗略地归到这类缓蚀剂中，因为它也是通过阻抑腐蚀反应的阳极过程来达到缓蚀的。但是，它不是通过与金属铁本身，而可能是由二氧化硅与铁的腐蚀产物相互作用，以吸附机制来成膜的。

②The carbonate, phosphate and hydroxide of zinc and calcium carbonate and phosphate are the most common precipitation-film corrosion inhibitors. They are also called cathodic corrosion inhibitors because they are formed by the reaction of zinc and calcium cations with carbonate, phosphate and hydroxide anions in water and in the cathodic area of the metal surface. The cathodic corrosion inhibitor can react with the relevant ions in

②沉淀膜型缓蚀剂锌的碳酸盐、磷酸盐和氢氧化物，钙的碳酸盐和磷酸盐是最常见的沉淀膜型缓蚀剂。由于它们是由锌、钙阳离子与碳酸根、磷酸根和氢氧根阴离子在水中、于金属表面阴极区反应而沉积成膜，所以又被称为阴极型缓

the water, and the reaction products are deposited on the cathode to form a film; Take zinc salt as an example, it produces $Zn(OH)_2$ precipitation at the cathode, which acts as a protective film. The combination of zinc salt and other corrosion inhibitors can play a synergistic role. In the presence of orthophosphate, $Zn_3(PO_4)_2$ or $(Zn, Fe)_3(PO_4)_2$ precipitates out and adheres tightly to the metal surface, and the corrosion inhibition effect is better. In practical applications, because calcium ions, carbonate and hydroxide are naturally present in water, it is generally only necessary to add soluble zinc salts (such as zinc nitrate, zinc sulfate or zinc chloride to provide zinc ions) or soluble phosphates (such as sodium orthophosphate or sodium polyphosphate that can be hydrolyzed into sodium orthophosphate to provide phosphate radicals) to the water. These soluble zinc salts and soluble phosphates are usually called deposition-film corrosion inhibitors or cathodic corrosion inhibitors. In this way, soluble phosphates (including polyphosphates) are both oxide film type corrosion inhibitors and deposition film type corrosion inhibitors. In addition, some organic compounds containing phosphorus, such as organophosphate (salt), organophosphate ester and organophosphate carboxylic acid, can also be included in this type of corrosion inhibitor, which is related to their final hydrolysis to orthophosphate. Because the precipitation-type corrosion inhibition film is not directly combined with the metal surface, and is porous, it often appears that the adhesion on the metal surface is not good, and the corrosion inhibition effect is not as good as the oxidation type film.

蚀剂。阴极型缓蚀剂能与水中有关离子反应，反应产物在阴极沉积成膜；以锌盐为例，它在阴极部位产生$Zn(OH)_2$沉淀，起保护膜的作用。锌盐与其他缓蚀剂复合使用可起增效作用，在有正磷酸盐存在时，则有$Zn_3(PO_4)_2$或$(Zn, Fe)_3(PO_4)_2$沉淀出来并紧紧黏附于金属表面，缓蚀效果更好。在实际应用中，由于钙离子、碳酸根和氢氧根在水中是天然存在的，一般只需向水中加入可溶性锌盐（如硝酸锌、硫酸锌或氯化锌，提供锌离子）或可溶性磷酸盐（如正磷酸钠或可水解为正磷酸钠的聚合磷酸钠，提供磷酸根），因此，通常就把这些可溶性锌盐和可溶性磷酸盐称为沉积膜型缓蚀剂或阴极型缓蚀剂。这样，可溶性磷酸盐（包括聚合磷酸盐）既是氧化膜型缓蚀剂，又是沉积膜型缓蚀剂。另外，一些含磷的有机化合物，如有机磷酸（盐）、有机磷酸酯和有机磷羧酸，也可归到这类缓蚀剂中，可能与其最终能水解为正磷酸盐不无关系。由于沉淀型缓蚀膜未和金属表面直接结合，而且是多孔的，往往出现在金属表面附着不好的现象，缓蚀效果不如氧化膜型。

③ The adsorbed film type corrosion inhibitors are mostly organic corrosion inhibitors. They have polarity genes and can be adsorbed by the surface charge of metal to form a monolayer in the entire anode and cathode area, thus preventing or slowing down the corresponding electrochemical reaction. For example, some organic compounds containing nitrogen, sulfur or hydroxyl, which have surface activity, have two groups with opposite properties in their molecules; Hydrophilic and lipophilic groups. The molecules of these compounds are adsorbed on the metal surface with hydrophilic groups (for example, amino groups) to form a dense hydrophobic film to protect the metal surface from water corrosion. Tallow amine, hexadecane amine and octadecane amine, which are called "membrane amines", are common adsorbed membrane corrosion inhibitors in water treatment. Mercaptobenzothiazole, benzotriazole and methylbenzotriazole are ideal corrosion inhibitors for non-ferrous metals (especially copper). Although they react with copper metal itself to form a film, they are different from the typical oxide film type corrosion inhibitors mentioned above. They are not formed by oxidation, but by forming a complex with copper ions on the metal surface to form a film by chemical adsorption. When the metal surface is clean or active, such corrosion inhibitor can form an adsorption film with satisfactory corrosion inhibition effect. However, if there are corrosion products or scale deposits on the metal surface, it is difficult to form a corrosion inhibitor film with good effect. At this time, a small amount of surfactant can be properly added to help such corrosion inhibitor film.

③吸附膜型缓蚀剂多为有机缓蚀剂，它们具有极性基因，可被金属表面的电荷吸附，在整个阳极和阴极区域形成一层单分子膜，从而阻止或减缓相应电化学的反应。如某些含氮、含硫或含羟基的、具有表面活性的有机化合物，其分子中有两种性质相反的基团，亲水基和亲油基。这些化合物的分子以亲水基（如氨基）吸附于金属表面，形成一层致密的憎水膜，保护金属表面不受水腐蚀。牛脂胺、十六烷胺和十八烷胺等被称为“膜胺”的胺类，就是水处理中常见的吸附膜型缓蚀剂。巯基苯并噻唑、苯并三唑和甲基苯并三唑等，是有色金属（尤其是铜）的理想缓蚀剂。它们虽然与铜金属本身作用成膜，但与上述典型的氧化膜型缓蚀剂不同，不是通过氧化，而是通过与金属表面的铜离子形成络合物，以化学吸附成膜的。当金属表面为清洁或活性状态时，此类缓蚀剂能形成缓蚀效果令人满意的吸附膜。但如果金属表面有腐蚀产物或有垢沉积的情况下，就很难形成效果良好的缓蚀膜，此时可适当加入少量表面活性剂，以帮助此类缓蚀剂成膜。

Because the corrosion inhibition mechanism of corrosion inhibitor is film formation, the key to successful corrosion inhibition is to quickly form a dense and solid film on the metal surface. In order to be quick, the concentration of corrosion inhibitor in water should be high enough, and after the formation of the film, it should be reduced to the concentration that can only repair the damage of the film; In order to compact, the metal surface should be very clean. For this reason, chemical cleaning, oil removal, dirt removal and scale removal on the metal surface before film formation is an essential step.

由于缓蚀剂的缓蚀机理在于成膜，故迅速在金属表面形成一层密而实的膜，是获得缓蚀成功的关键。为了迅速成膜，水中缓蚀剂浓度应足够高，等膜形成后，再降至只对膜的破损起修补作用的浓度；为了成膜密实，金属表面应十分清洁，为此，成膜前对金属表面进行化学清洗除油、除污和除垢是必不可少的步骤。

4. Control the environment

It is mainly to control the relative humidity of sealed metal containers or non-metal containers and filled with inert nitrogen or to remove the air, so as to make the parts isolated from the peripheral medium, so as to avoid corrosion, and make the non-residual parts mildew and aging. The methods include ammonia storage method, oxygen absorption agent method and dry air storage method.

4.控制环境

控制环境主要是控制密封金属容器或非金属容器内的相对湿度和充以惰性氮气或抽去空气，以使制件与外围介质隔离，从而避免锈蚀，并使非金属件防霉、防老化。其方法有充氨封存法、吸氧剂法和干燥空气封存法等。

6.1.5 References

[1] GAO W, LIU X Y, CHENG T T. Influence of smog pollution on near-ground atmospheric environment [C] //Proceedings of the Annual Conference of Science and Technology of Chinese Society of Environmental Sciences (Volume I), 2019:917-922.

[2] CHEN X T,WANG P, ZHANG D. Application of superhydrophobic surface in atmospheric corrosion protection [C] //Abstract of the 10th National Corrosion Conference, 2019:254.

6.1.5 参考文献

[1] 高伟，刘小英，成天涛．雾霾污染对近地面大气环境的影响［C］// 中国环境科学学会科学技术年会论文集（第一卷），2019：917–922.

[2] 陈晓彤，王鹏，张盾．超疏水表面在大气腐蚀防护中的应用［C］// 第十届全国腐蚀大会摘要集，2019：254.

6.2 Seawater corrosion

The oceans account for about 70% of the earth's area and are closely related to human life. Ships are sailing in the sea. A variety of marine development projects are being built in large quantities. Submarine cables and oil pipelines are increasing on the seabed. The production and domestic water of factories in coastal areas can not be separated from seawater. With the development of coastal transportation, industrial production and national defense construction in China, metal structures are increasingly threatened by seawater and marine atmospheric corrosion. Therefore, it is of great significance to study the corrosion and protection of marine environment. The marine corrosion area is divided into five areas: atmospheric area, splash area, tidal range area, full immersion area and sea mud area. Among them, the marine splash zone is the most severe zone in the marine corrosion area due to the high alternating frequency of dry and wet, sufficient oxygen supply, high salt particle content, ultraviolet rays from the sun and other factors [1,2]. Figure 6-8 shows the marine biological corrosion.

6.2 海水腐蚀

海洋约占地球面积的70%，与人类生活有着密切的联系。舰船在海洋中航行，各种各样的海洋开发工程正在大量兴建，海底电缆、输油管道在海底日益增多，沿海地区工厂的生产和生活用水更离不开海水。随着我国沿海交通运输、工业生产和国防建设的发展，金属结构物受到海水和海洋大气腐蚀的威胁也越来越突出。因此，研究海洋环境的腐蚀及其防护有重要意义。海洋腐蚀区域划分为大气区、浪溅区、潮差区、全浸区和海泥区五个区域。其中，海洋浪溅区由于干湿交替频率高、供氧充分、含盐粒子量高、来自太阳的紫外线等因素，是海洋腐蚀区域中最苛刻的区带[1,2]。海洋生物腐蚀如图6-8所示。

Figure 6-8 Marine biological corrosion

图6-8 海洋生物腐蚀

6.2.1 Factors of seawater corrosion

Seawater is rich in natural electrolytes. Seawater contains almost all the chemical elements on the earth, and the composition is very complex. In addition to containing a large number of salts, seawater also contains dissolved oxygen, marine organisms and putrid organic matter. The temperature, flow rate and pH value of seawater have great influence on seawater corrosion[3,4].

1. Salts and concentration

Seawater contains a lot of salts and is a strong electrolyte solution. The presence of chloride ions in the humid and hot marine atmosphere is the most corrosive to stainless steel[5]. The conductivity of seawater is about 2.3~3.0S/m. There is little difference in the salt content of the sea areas in the world, ranging from 3.3% to 3.8%. However, there is a large deviation in the salt content of inland sea water (such as Qinghai's salt lake and Caspian Sea). It can be seen from Table 6-1 that the content of chloride ion in seawater is very high, accounting for 55.04% of the total salt content, making seawater highly corrosive[6]. The commonly used structural metals and alloys are corroded by seawater. The concentration of NaCl in seawater is approximately equivalent to 0.5 mol/L of NaCl solution.

In addition to these main components, there are a small amount of ozone, free iodine and bromine and a small amount of other elements in the seawater. The diluted seawater at the sea outfall is highly corrosive, although the electrolyte itself is not corrosive. Ordinary seawater is usually saturated with carbonate, but the

6.2.1 影响海水腐蚀性的因素

海水是丰富的天然电解质，海水中几乎含有地球上所有化学元素的化合物，成分很复杂。除含有大量盐类外，海水中还含有溶解氧、海洋生物和腐败的有机物。海水的温度、流速与pH值等都对海水腐蚀有很大的影响[3, 4]。

1.盐类及浓度

海水含有大量盐类，是较强的电解质溶液，除湿热海洋大气环境中氯离子的存在外，对不锈钢腐蚀性最强[5]。海水电导率在2.3~3.0S/m。世界上各海区的含盐量差别不大，均在3.3%~3.8%。但内海海水的含盐量有较大偏差（如青海的盐湖和里海）。由表6-1可以看出，海水中氯离子含量很高，占总盐量的55.04%，使海水具有较大的腐蚀性[6]。常用的结构金属和合金均受海水的侵蚀。海水中NaCl浓度近似地相当于0.5 mol/L NaCl溶液。

除这些主要的成分外，海水中还有少量的臭氧、游离的碘和溴及少量的其他元素。出海口处的稀释海水，尽管电解质本身的侵蚀性不大，却有较大的腐蚀性。普通的海水通

carbonate cannot reach saturation in diluted seawater, so it is not easy to generate protective carbonate scale. Moreover, in diluted seawater, the activity of marine organisms decreases or disappears, so it is not easy to generate a protective layer of contaminating organisms. The salt content in water directly affects the conductivity and oxygen content of water, so it must have an impact on corrosion. With the increase of salt content in water, the conductivity of water increases while the oxygen content decreases, so there will be a maximum corrosion rate at a certain salt content. The salt content of seawater is just close to the salt content corresponding to the maximum corrosion rate.

常被碳酸盐饱和，但在稀释海水中碳酸盐达不到饱和，不易生成保护性碳酸盐型水垢。而且，在稀释的海水中，海生物的活性降低或消失，因而不易生成沾污生物的保护层。水中含盐量直接影响水的电导率和含氧量，因此必然对腐蚀产生影响。随着水中含盐量增加，水的电导率增加而含氧量降低，所以在某一含盐量时将存在一个腐蚀速度的最大值。海水的含盐量刚好接近腐蚀速度最大时所对应的含盐量。

Table 6-1 Main salt content in seawater

表 6-1 海水中主要的盐类含量

Component 成分	100g seawater salt grams 100g 海水含盐克数	% of total salt 占总盐的量 /%	Component 成分	100g seawater salt grams 100g 海水含盐克数	% of total salt 占总盐的量 /%
NaCl	2.213	77.8	KSO_4	0.0863	2.5
$MgCl_2$	0.3807	10.9	$CaCO_3$	0.0123	0.3
$MgSO_4$	0.1658	4.7	$MgBr_2$	0.0076	0.2
$CaSO_4$	0.1260	3.6	合计	3.5	100

2. Dissolved oxygen

Dissolved oxygen in seawater is an important factor in seawater corrosion, because the corrosion of most metals in seawater is controlled by oxygen depolarization. The sea water surface is always in contact with the atmosphere, and the contact surface area is very large. The sea water is also constantly stirred by waves and has severe natural convection. Therefore, the oxygen content in seawater is usually high. It can be considered that the surface of seawater is oxygen saturated. When the mass fraction of seawater is 3%, the oxygen content

2.溶解氧

海水中含有的溶解氧是海水腐蚀的重要因素，因为绝大多数金属在海水中的腐蚀受氧的去极化作用控制。海水表面始终与大气接触，而且接触表面积非常大，海水还不断地受到波浪的搅拌作用并有剧烈的自然对流。所以，通常海水中含氧量较高。可以认为，海水表层已为氧饱和。当海水含质

of seawater at 20℃ is 8mg/L or 5.6cm^3/L. With the increase of salt concentration and temperature in seawater, the dissolved oxygen in seawater will decrease. The table lists the relationship between dissolved oxygen, salt concentration in seawater and temperature. It can be seen from Table 6-2 that the higher the salt concentration and temperature, the smaller the oxygen solubility.

量分数量为3%，在20℃时海水氧含量为8mg/L或5.6cm^3/L。随着海水中盐浓度增大和温度升高，海水中溶解氧量将下降。表6-2列举出溶解氧量、海水中盐的浓度与温度之间的关系。可见，盐的浓度和温度越高，氧的溶解度越小。

Table 6-2　Solubility of oxygen in seawater(cm^3/L)

表6-2　氧在海水中的溶解度（cm^3/L）

Temperature/℃ 温度/℃	W（SALT）/% W（盐）/%					
	0.0	1.0	2.0	3.0	3.5	4.0
0	10.30	9.65	9.00	8.36	8.04	7.72
10	8.02	7.56	7.09	6.63	6.41	6.18
20	6.57	6.22	5.88	5.52	5.35	5.17
30	5.57	5.27	4.95	4.65	4.50	4.34

From sea level to 80 meters deep, the content gradually decreases and reaches the lowest value. This is because marine animals need to consume oxygen, and the decomposition of animal carcasses falling from the upper layer of the sea also needs to consume oxygen. However, the oxygen replenished by convection is not enough to offset the consumed oxygen, so an anoxic layer appears. From 80 m to 100 m deep, the dissolved oxygen content starts to rise again and is close to the oxygen concentration in the sea surface. This is due to the low temperature and high pressure of deep sea water. The oxygen concentration of surface seawater is usually different from the oxygen content of seawater in the sea area. When marine animals are attached to the surface of ships or offshore structures, the surface seawater above them will be anoxic, and the amount of CO is very

自海平面至80m深，含量逐渐减少并达到最低值。这是因为海洋动物要消耗氧气，从海水上层下降的动物尸体发生分解时也要消耗氧气。然而，通过对流形式补充的氧不足以抵消消耗了的氧，所以出现了缺氧层。从80m再降至100m深，溶解氧量又开始上升，并接近海水表层的氧浓度。这是由于深海海水温度较低、压力较高的缘故。表面海水的氧浓度通常与所在海域海水中的氧含量不同。当船舶或海上构筑物表面附着了海洋动物，其上方的表面海水将缺氧，而CO量

high. If they are covered with marine plants, the oxygen concentration can be higher than the oxygen content in the sea.

很高，如果它们表面长满了海洋植物时，则氧浓度可高出海中的氧含量。

3. Temperature

3.温度

Sea water temperature varies with latitude, season and depth. The closer to the equator (the smaller the latitude), the higher the temperature of seawater, and the greater the corrosion rate of metals. The deeper the sea water and the lower the temperature, the smaller the corrosion rate. When the seawater temperature increases by 10 ℃, the chemical reaction rate increases by about 14%, and the metal corrosion rate in seawater will double. However, the solubility of oxygen in seawater decreases by about 20% after the temperature rises, which causes the reduction of metal corrosion rate. In addition, the temperature change also affects the biological activity of seawater and the calcareous scale deposit. Due to the seasonal change of temperature, the corrosion rate of iron, copper and their alloys is relatively high in hot seasons.

海水温度随着纬度、季节和深度的不同而发生变化。越靠近赤道（纬度越小）海水的温度越高，金属腐蚀速度也大。而海水越深、温度越低，则腐蚀速度越小。海水温度每升高10℃,化学反应速度提高约14%，海水中的金属腐蚀速度将增大1倍。但是，温度升高后氧在海水中的溶解度约降低20%，引起金属腐蚀速度的减小。此外，温度变化还给海水的生物活性和石灰质水垢沉积层带来影响。由于温度的季节性变化，铁、铜和它们的多种合金在炎热的季节里腐蚀速度较大。

4. pH value

4. pH值

The pH value of seawater is between 7.2 and 8.6, close to neutral. With the increase of seawater depth, the pH value gradually decreases. The pH of seawater may vary slightly due to photosynthesis. In the daytime, plants consume CO_2, which affects the pH value. The CO_2 in seawater at sea level exchanges with CO_2 in the atmosphere, thus changing the CO_2 content. The pH value of seawater is far less influential than the oxygen content on the corrosion rate. The pH value in seawater mainly affects the deposition of calcareous scale, thus affecting the corrosivity of seawater. Although the pH value of surface seawater is higher than that of

海水pH值在7.2~8.6，接近中性。海水深度增加，pH值逐渐降低。海水pH值可因光合作用而稍有变化。白天，植物消耗CO_2，影响pH值。海面处海水中的CO_2与大气中的CO_2相交换，从而改变CO_2含量。海水pH值远没有含氧量对腐蚀速度的影响大。海水pH值主要影响钙质水垢沉积，从而影响海水的腐蚀性。尽管表层海水

deep seawater, because the oxygen content of surface seawater is higher than that of deep seawater, the corrosion of surface seawater to steel is greater than that of deep seawater. With the increase of seawater pressure in the deep sea area, the equilibrium of reaction $CaCO_3 \rightleftharpoons CaO + CO_2$ will generate CO_2. As a result, the pH value decreases and it is not easy to generate protective carbonate scale. The corrosion products under acidic conditions are looser and easier to fall off than those under neutral conditions, which weakens the protection of the matrix and increases the corrosion rate[7].

pH值比深处海水高，但由于表层海水含氧量比深处海水高，所以表层海水对钢的腐蚀性比深处海水大。深海区海水压力增加，反应$CaCO_3 \rightleftharpoons CaO + CO_2$的平衡向生成$CO_2$的方向进行，因而pH值减小，不易生成保护性碳酸盐水垢，酸性条件腐蚀产物比中性条件下的腐蚀产物疏松、易脱落，对基体的保护作用减弱，使腐蚀速度增大[7]。

5. Flow rate

The corrosion of many metals is greatly related to the seawater flow rate. In particular, there is a critical flow rate for common metals such as iron and copper. When the flow rate is exceeded, metal corrosion is significantly accelerated, but passive metals are more resistant to corrosion in high-speed seawater[8]. The movement of sea water is easy to dissolve into the air and promote the diffusion of dissolved oxygen to the metal surface, so the depolarization of oxygen will be strengthened after the increase of flow rate, which will accelerate the metal corrosion. The relationship between seawater flow rate and corrosion rate of carbon steel is shown in the table 6-3. The corrosion rate of steel piles immersed in seawater is different in different parts. Near the waterline, especially at 0.3~1.0 m above the water surface, due to the impact of sea waves, the oxygen supply is particularly sufficient and the corrosion products are constantly taken away, so the corrosion rate at this place is 3~4 times higher than that at the fully immersed part.

5.流速

许多金属发生腐蚀时与海水流速有较大关系，尤其是铁、铜等常用金属存在一个临界流速，超过此流速时金属腐蚀明显加快，但钝态金属在高速海水中更能抗腐蚀[8]。海水运动易溶入空气，并且促使溶解氧扩散到金属表面，所以流速增大后氧去极化作用加强，使金属腐蚀速度加快。海水流速与碳钢腐蚀速度之间的关系见表6-3。浸泡在海水中的钢桩，其各部位的腐蚀速度不同。水线附近，特别是在水面以上0.3~1.0 m的位置由于受到海浪冲击，供氧特别充分且腐蚀产物不断被带走，因此该处的腐蚀速度要比全浸部位大3~4倍。

Table 6-3 Relationship between corrosion rate of carbon steel and seawater flow rate

表 6-3 碳钢腐蚀速度与海水流速的关系

Seawater Velocity/ (m/s) 海水流速 / (m/s)	0	1.0	3.0	4.0	6.0	7.5
Corrosion Rate/ [(mg/cm^2) · d] 腐蚀速度 / [(mg/cm^2) · d]	0.3	1.1	1.6	1.8	1.9	1.95

6. Marine life

When the fresh metal surface is immersed in seawater for several hours, a layer of biological slime is attached to the surface, which is convenient for marine organisms to parasitize. Immobilized organisms have the greatest impact on metal corrosion. They cover the surface with clay and firmly attach to the surface of metal components. Then they grow up quickly and are fixed. Marine organisms produce crevices on the surface attachment of ships or offshore structures, which are easy to induce crevice corrosion. In addition, the physiological action of microorganisms will produce corrosive substances such as ammonia, carbon dioxide and hydrogen sulfide, and the action of sulfate reducing bacteria will produce oxygen, which can accelerate the corrosion of metals. The mixture of anaerobic bacteria and aerobic bacteria will produce synergistic effect to make metal corrosion more serious. Aerobic bacteria can cause steel corrosion under aerobic conditions and produce local corrosion products, which forms an anoxic environment, thus promoting the rapid growth of anaerobic bacteria. The interaction of the two bacteria can accelerate metal corrosion.

6.海洋生物

金属新鲜表面浸入海水中数小时后表面即附着一层生物黏泥，便于海洋生物寄生。对金属腐蚀影响最大的是固着生物，它们以黏泥覆盖表面，并牢牢地附着在金属构件表面，随后很快长大，并固定不动。海洋生物在船舶或海上构筑物表面附着处产生缝隙，容易诱发缝隙腐蚀。另外，微生物的生理作用会产生氨、二氧化碳和硫化氢等腐蚀性物质，硫酸盐还原菌的作用则产生氧，这些都能加快金属腐蚀。厌氧菌和好氧菌混合会产生协同效应，使金属腐蚀更加严重。好氧菌在有氧状态下造成钢铁腐蚀产生局部腐蚀产物，形成缺氧环境，从而促进了厌氧菌迅速生长。两种菌的相互作用会加速金属腐蚀。

Marine biological contamination is also an important factor affecting the performance of marine facilities. Additional contamination loads can overload buildings in the sea, such as making buoys lose buoyancy. For ships

海洋生物沾污也是影响海洋设施性能的重要因素，额外的沾污负荷能使海水中建筑物过载，如使浮标失去浮力。对于在

sailing at sea, the serious contamination of the hull will affect the motion performance of the ship, resulting in excessive fuel consumption.

海洋航行中的船只，船身的严重沾污会影响船体航行的运动性能，造成燃料的过量消耗。

6.2.2 Characteristics of seawater corrosion

6.2.2 海水腐蚀特点

Seawater is a typical electrolyte solution, and the basic rules of electrochemical corrosion are applicable to the corrosion of metals in seawater. The electrochemical process of seawater corrosion has its own characteristics, which can be summarized as follows:

① The pH value of seawater is 7.2~8.6, which is close to neutral and contains a large amount of dissolved oxygen. Therefore, except for particularly active metals, such as Mg and its alloys, the corrosion process of most metals and alloys in seawater is a depolarization process of oxygen, and the corrosion rate is controlled by cathodic polarization.

② The concentration of Cl^- in seawater is high. For steel, iron, zinc, cadmium and other metals, when they are subjected to electrochemical corrosion in seawater, the blocking effect of the anodic process is very small. Increasing the resistance of the anodic process has no significant effect on reducing seawater corrosion. It is difficult to significantly slow down the corrosion rate of seawater if ordinary carbon steel structures are replaced by stainless steel. The reason is that stainless steel is easily damaged by pitting in seawater. Only by improving the stability of the passive film on the alloy surface, such as adding alloy element aluminum, can the damage of Cl^- to the passive film be reduced and the corrosion resistance of the material in seawater be improved. In addition, alloys based on titanium, germanium, tantalum, niobium,

海水是典型的电解质溶液，有关电化学腐蚀的基本规律对于海水中金属腐蚀都是适用的。海水腐蚀时电化学过程的特征，可归纳为以下几方面：

①海水pH值在7.2~8.6，接近中性，并含有大量溶解氧，因此除特别活泼的金属，如Mg及其合金外，大多数金属和合金在海水中的腐蚀过程都是氧的去极化过程，腐蚀速度由阴极极化控制。

②海水中Cl^-浓度高，对于钢、铁、锌、镉等金属来说，它们在海水中发生电化学腐蚀时，阳极过程的阻滞作用很小，增加阳极过程阻力对减轻海水腐蚀的效果并不显著。如将一般碳钢制造结构件改用不锈钢，很难达到显著减缓海水腐蚀速度的目的。其原因是不锈钢在海水中易由点蚀而遭到破坏。只有通过提高合金表面钝化膜的稳定性，如添加合金元素铝等，才能减轻Cl^-对钝化膜的破坏作用，改进材料在海水中的耐蚀性。另外，以金属钛、锗、钽、铌等为

etc. can also maintain stable passivity in seawater.

③ Water is a good conductive medium with low resistance, so it can not only corrode the battery in the micro way, but also corrode the battery in the macro way. The galvanic corrosion caused by the contact of dissimilar metals in seawater has an important destructive effect on metals. The electrode potential of most metals or alloys in seawater is not a constant value, but changes with the dissolved oxygen content in water, seawater flow rate, temperature, metal structure and surface state and other factors. The table shows the potential of some common metals in aerated flowing seawater (for saturated calomel electrode). When different metals contact in seawater, the corrosion of metals with low potential will be accelerated, while the corrosion of metals with high potential will be slowed down. The flowing speed of seawater, the type of metal and the area of anode and cathode are all factors that affect galvanic corrosion. For example, in static or low flow seawater, due to galvanic corrosion, the corrosion rate of carbon steel increases only in proportion to the area of the cathode electrode, and has little relationship with the nature of the cathode metal it contacts. The corrosion rate of carbon steel is controlled by oxygen depolarization. When the seawater flow rate is high and oxygen depolarization has not become the main control factor of corrosion, the cathodic metal polarization property in contact with carbon steel will have a significant impact on the corrosion rate. When carbon steel and copper form a couple, the corrosion rate increases much more than when carbon steel and titanium contact, because cathode titanium is more easily polarized than copper.

基础的合金也能在海水中保持稳定的钝态。

③水是良好的导电介质，电阻较小，因此在海水中不仅有微观腐蚀电池的作用，还有宏观腐蚀电池的作用。在海水中由于异种金属接触引起的电偶腐蚀对金属有重要的破坏作用，大多数金属或合金在海水中的电极电势不是一个恒定的数值，而是随着水中溶解氧含量、海水流速、温度及金属结构与表面状态等多种因素的变化而变化。表6–4为一些常用金属在充气流动海水中的电势（对饱和甘汞电极）。海水中不同金属间相接触时，将导致电势较低的金属腐蚀加速，而电势较高的金属腐蚀减缓。海水的流动速度、金属种类及阴、阳极电极面积的大小都是影响接触腐蚀的因素。例如，在静止或流速不大的海水中，碳钢由于电偶腐蚀，其腐蚀速度增加程度仅与阴极电极面积大小成比例，而与所接触的阴极金属本身几乎没有关系。碳钢的腐蚀速度由氧去极化控制。而当海水流速很大，氧去极化已不成为腐蚀的主要控制因素时，与碳钢接触的阴极金属极化性能将对腐蚀速度带来明显的影响。碳钢与铜组成电偶时引起腐蚀速度增大的程度要比

碳钢与钛相接触时大得多，原因是阴极钛比铜更容易极化。

④ Metals in seawater are prone to local corrosion damage. In addition to the above mentioned galvanic corrosion, the common damage forms include pitting corrosion, crevice corrosion, turbulence corrosion and cavitation corrosion.

④海水中金属易发生局部腐蚀破坏，除上面提到的接触腐蚀外，常见的破坏形式还有点蚀、缝隙腐蚀、湍流腐蚀和空泡腐蚀等。

⑤ The composition of seawater and the concentration of salt in different regions are not different, so geographical factors are not important in seawater corrosion.

⑤不同地区的海水组成及盐的浓度差别不大，因此地理因素在海水腐蚀中并不重要。

Table 6-4 Potential of some metals in aerated flowing seawater

表 6-4 一些金属在充气流动海水中的电势

Metal 金属	Potential /V 电势 /V	Metal 金属	Potential /V 电势 /V
Magnesium 镁	−1.5	Aluminium Brass 铝黄铜	−0.27
Zinc 锌	−1.03	Copper Nickel Alloy (90/10) 铜镍合金（90/10）	−0.26
Aluminium 铝	−0.79	Copper Nickel Alloy (80/20) 铜镍合金（80/20）	−0.25
Cadmium 镉	−0.70	Copper Nickel Alloy (70/30) 铜镍合金（70/30）	−0.25
Steel 钢	−0.61	Nickel 镍	−0.14
Lead 铅	−0.50	Silver 银	−0.13
Tin 锡	−0.42	Titanium 钛	−0.10
Brass 黄铜	−0.30	18–8 Stainless Steel (Passive) 18–8 不锈钢（钝态）	−0.08
Copper 铜	−0.28	18–8 Stainless Steel 18–8 不锈钢（活化态）	−0.53

6.2.3 Corrosion protection methods

6.2.3 防止海水腐蚀的措施

The air-sea exchange interface area in the middle of the marine environment is often in a humid and oxygen-rich state, and there is strong and frequent wave impact.

在海洋环境中间的海 – 气交换界面区，常处于潮湿多氧状态，还有较强而频繁的海浪

The steel structure in the spray splash area is in a dry and wet alternate state, combined with high salt content, seawater scouring, sufficient oxygen supply, sufficient sunshine, temperature rise and other factors, which comprehensively lead to the most serious corrosion of steel structures in the environment. This section introduces the anti-corrosion measures of materials in the marine environment[9].

冲击作用，在浪花飞溅区钢结构处于干湿交替状态，加上高含盐量、海水冲刷、供氧充分、日照充足、温度上升等因素，综合导致了钢结构在该环境中腐蚀最为严重，本节介绍了材料在海水环境中的防腐措施[9]。

1. Reasonable material selection

1.合理选材

Table 6-5 shows the corrosion resistance of some metals in seawater. Titanium and nickel chromium aluminum alloys have the best corrosion resistance, while cast iron and carbon steel are poor. Copper based alloys such as aluminum bronze and copper nickel alloys also have better corrosion resistance. Although stainless steel is resistant to uniform corrosion, it is prone to pitting.

表6-5所示为部分金属在海水中的耐蚀性。钛及镍铬铝合金的耐蚀性最好，铸铁和碳钢较差，铜基合金如铝青铜、铜镍合金也较耐蚀。不锈钢虽耐均匀腐蚀，但易产生点蚀。

2. Electrochemical protection

2.电化学保护

Cathodic protection[10] is one of the common methods to prevent seawater corrosion, but it is only effective in the full immersion area. Sacrificial anodes can be installed on metal structures at the bottom of the ship or in sea water, or impressed current cathodic protection can be used.

阴极保护[10]是防止海水腐蚀常用的方法之一，但只在全浸区才有效。可在船底或海水中金属结构上安装牺牲阳极，也可采用外加电流的阴极保护法。

3. Coating protection

3.涂层保护

The most common method to prevent seawater corrosion is to use oil knee layer or antifouling coating to prevent biological contamination. The antifouling coating is a kind of coating containing toxic substances such as CuO, HgO, organic tin and organic lead. After being coated on the metal surface, it can be dissolved and diffused in the sea water to distribute toxicity to resist and kill marine organisms that stay on the metal surface, which can reduce or prevent crevice corrosion caused by marine organisms.

防止海水腐蚀最普通的方法是采用油漆层，或采用防止生物沾污的防污涂层。这种防污涂层是一种含有CuO、HgO、有机锡及有机铅等毒性物质的涂料。涂在金属表面后，在海水中能溶解扩散，以散发毒性来抵抗并杀死停留在金属表面的海洋生物，这样可减少或防止因海洋生物造成的缝隙腐蚀。

Table 6-5　Seawater corrosion resistance of metal materials

表 6-5　金属材料耐海水腐蚀性能

Alloy 合金	Corrosion rate of full immersion area/(mm /a) 全浸区腐蚀率 /（mm /a）		Corrosion rate in tidal area/(mm /a) 潮汐区腐蚀率 /（mm /a）		Impact corrosion property 冲击腐蚀性能
	Avg 平均	Max 最大	Avg 平均	Max 最大	
Mild steel (no scale) 低碳钢（无氧化皮）	0.12	0.40	0.3	0.5	Inferior 劣
Mild steel (with oxide scale) 低碳钢（有氧化皮）	0.09	0.90	0.2	1.0	Inferior 劣
Common cast iron 普通铸铁	0.15		0.4		Inferior 劣
Copper (cold rolled) 铜（冷轧）	0.04	0.08	0.02	0.18	Not good 不好
Brass (containing 10% Zn by mass) 黄铜（含质量分数 10%Zn）	0.04	0.05	0.03		Not good 不好
Brass (70Zn–30Zn) 黄铜（70Zn– 30Zn）	0.05				Satisfied 满意
Brass (22Zn–2Al–0.02As) 黄铜（22Zn–2Al–0.02As）	0.02	0.18			Good 良好
Brass (20Zn–1Sn–0.02As) 黄铜（20Zn–1Sn–0.02As）	0.04				Satisfied 满意
Brass (60Cu–40Zn) 黄铜（60Cu– 40Zn）	0.06	脱 Zn	0.02	脱 Zn	Good 良好
Bronze (Sn mass fraction 5%–0.1P) 青铜（Sn 质量分数 5%–0.1P）	0.03	0.1			Good 良好
Aluminum bronze (5% by mass of Al and 2% by mass of Si) 铝青铜（Al质量分数5%和Si质量分数2%）	0.03	0.08	0.01	0.05	Good 良好
Copper nickel alloy (70Cu–30Ni) 铜镍合金（70Cu–30Ni）	0.003	0.03	0.05	0.3	质量分数 0.15%Fe，良好 质量分数 0.45%Fe，优秀
Nickel 镍	0.02	0.1	0.4		Good 良好
Monel［66Ni–31Cu–4 (Fe+Mn)］ 镍合金［66Ni–31Cu–4（Fe+Mn）］	0.03	0.2	0.5	0.25	Good 良好
Inconel (80Ni–13Cr) 因科镍合金（80Ni–13Cr）	0.05	0.1			Good 良好
Hastelloy (5Ni–19Mo–17Cr) 哈氏合金（5Ni–19Mo–17Cr）	0.001	0.001			Excellent 优秀
Cr_{13} Cr_{13}		0.28			Satisfied 满意
Cr_{17} Cr_{17}		0.20			Satisfied 满意
$Cr_{18}Ni_9$ $Cr_{18}Ni_9$		0.18			Good 良好
Cl_{28}–Ni_{20} Cl_{28}–Ni_{20}		0.02			Good 良好
Zn（mass fraction 99.5%Ti） Zn（质量分数 99.5%Ti）	0.028	0.03			Good 良好
Ti Ti	0.00	0.00	0.00	0.00	Excellent 优秀

可怕的海水腐蚀

6.2.4 Reference

[1] LI Z, QI Q J, MA Z H, et al. Influencing factors of iron and steel corrosion in humid and hot marine atmospheric environment [J]. Baosteel Technology, 2015(1): 77-80.

[2] XU Y Z. Research on key technologies for monitoring local corrosion of inner wall of marine pipeline [D]. Dalian University of Technology, 2019.

[3] FAN J, LI T, SUN M S. Effect of pH on corrosion behavior of automotive steel in harsh marine atmospheric environment [J]. Journal of Inner Mongolia University of Science and Technology, 2019, 38(1): 50-53.

[4] ZHOU Q P. Study on Erosion-Corrosion coupling damage mechanism of carbon steel used in offshore engineering [D]. Dalian University of Technology, 2021.

[5] LI T, ZHANG Y W, LV Y. Research progress on corrosion and protective measures of hull materials [J]. Guangzhou Chemical Industry, 2015, 43(6):33-35, 71.

[6] HOU S Z. Research and application of cathodic protection technology [J]. Total Corrosion Control, 2018, 32(10):39-44, 65.

[7] MENG H, HU X, NEVILLE A. A systematic erosion-corrosion study of two stainless steels in marine conditions via experimental design [J]. Wear, 2007, 263(1–6): 355-362.

[8] HOSSAIN K, EASA S M, LACHEMI M. Evaluation of the effect of marine salts on urban built infrastructure [J]. Building & Environment, 2009, 44(4): 713-722.

[9] CASTAÑO J G, BOTERO C A, RESTREPO A H, et al. Atmospheric corrosion of carbon steel in Colombia [J]. Corrosion Science, 2010, 52(1): 216-223.

[10] MA Y, LI Y, WANG F. Corrosion of low carbon steel in atmospheric environments of different chloride content [J]. Corrosion Science, 2009, 51(5): 997-1006.

6.2.4 参考文献

[1] 李钊，祁庆琚，马朝晖，等．湿热海洋大气环境下的钢铁腐蚀影响因素［J］．宝钢技术，2015（1）：77–80.

[2] 徐云泽．海洋管道内壁局部腐蚀的监测关键技术研究［D］．大连：大连理工大学，2019.

[3] 范静，李涛，孙明双．pH 值对汽车板在苛刻海洋大气环境中腐蚀行为的影响［J］．内蒙古科技大学学报，2019，38（1）：50–53.

[4] 周启票．海洋工程用碳钢的冲刷腐蚀耦合损伤机理研究［D］．大连：大连理工大学，2021.

[5] 李婷，张耀武，吕岩．船体材料腐蚀与防护措施的研究进展［J］．广州化工，2015，43（6）：33–35，71.

[6] 侯世忠．阴极保护技术的研究与应用［J］．全面腐蚀控制，2018，32（10）：39–44，65.

[7]（文献同英文）

[8]（文献同英文）

[9]（文献同英文）

[10]（文献同英文）

6.3 Soil corrosion

A large number of metal pipelines (oil, gas, water pipelines), communication cables, foundation steel columns, high-voltage transmission lines and TV towers and other metal bases are buried underground. Due to soil corrosion, the pipelines are damaged by perforation, causing oil, gas and water seepage. leakage or failure of telecommunications, and even cause fire and explosion accidents. These underground equipment are often difficult to overhaul, which brings great loss and harm to production. With the modernization of industry, especially the development of petroleum industry, it is more and more important to study soil corrosion(Figure 6-9) problems.

6.3 土壤腐蚀

大量的金属管道（油、气、水管线）、通信电缆、地基钢柱、高压输电线及电视塔等金属基座埋设在地下，由于土壤腐蚀造成管道穿孔损坏，引起油、气、水的渗漏或使电信发生故障，甚至造成火灾、爆炸事故。这些地下设备难于检修，给生产带来很大的损失和危害。随着工业现代化，尤其是石油工业的发展，研究土壤腐蚀（图6-9）问题越发重要。

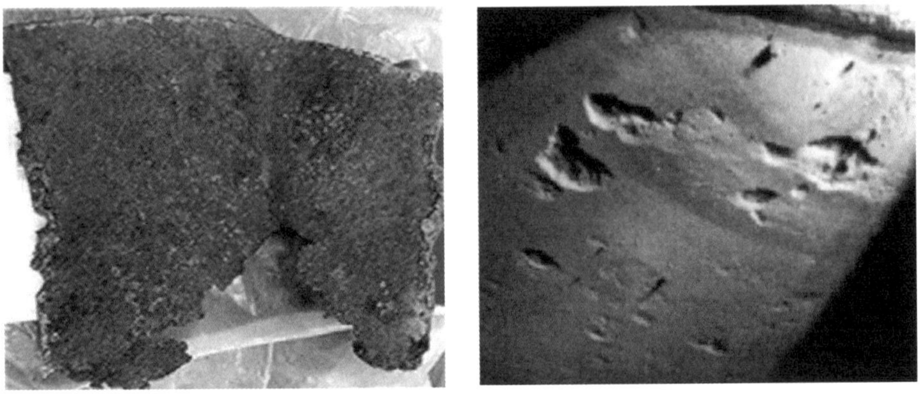

Figure 6-9 Soil corrosion

图6-9 土壤腐蚀

Soil corrosion is a type of electrochemical corrosion, and the soil contains moisture, salts and oxygen. Most soils are neutral, with some alkaline sandy clays and saline-alkali soils with a pH of 7.5 to 9.5, and acidic humus and swamp soils with a pH of 3 to 6. Soil contains solid particles of sand, stucco slag and humus after plant decay. Soil is a collection of inorganic and

土壤腐蚀是一种电化学腐蚀，土壤中含有水分、盐类和氧。大多数土壤是中性的，也有些pH值在7.5~9.5的碱性砂质黏土和盐碱土，以及pH值在3~6的酸性腐殖土和沼泽土。土壤含有固体颗粒砂子、灰泥

organic colloidal mixed particles. It is composed of soil particles, water and air, and is a complex multiphase structure. A large number of capillary pores or pores are formed between soil particles, and the pores are filled with air and water, often forming a colloid system, which is an ionic conductor. Soil water in which salts and other substances are dissolved is an electrolyte solution, and the conductivity of the soil is related to the dryness, wetness and salt content of the soil. The properties and structure of soil are uniform and changeable. The solid part of the soil is fixed to the metal surface buried in the soil, while the gas and liquid phases in the soil can make limited movements. These physical and chemical properties of soil, especially electrochemical properties, directly affect the characteristics of soil corrosion process. The complexity and variability of soil composition and properties make the corrosiveness of different soils vary widely.

渣和植物腐烂后的腐殖土，土壤是无机胶质和有机胶质混合颗粒的集合，由土粒、水、空气所组成，是复杂的多相结构。土壤颗粒间形成大量毛细管微孔或孔隙，孔隙中充满空气和水，常形成胶体体系，是一种离子导体。溶解有盐类和其他物质的土壤水则是电解质溶液，土壤的导电性与土壤的干湿程度及含盐量有关。土壤的性质和结构是均匀的、多变的，土壤的固体部分对埋设在土壤中的金属表面来说，是固定不动的，而土壤中的气、液相则可做有限运动。土壤的这些物理化学性质尤其是电化学特性，直接影响土壤腐蚀过程的特点。土壤组成和性质的复杂多变性，使不同的土壤腐蚀性相差很大。

6.3.1 Electrode process and control factors of soil corrosion

6.3.1 土壤腐蚀的电极过程及控制因素

Soil corrosion, like corrosion in electrolyte, is an electrochemical corrosion. The corrosion of most metals in soil is oxygen depolarization corrosion, and hydrogen depolarization corrosion occurs only in strongly acidic soil.

土壤腐蚀与在电解液中腐蚀相同，是一种电化学腐蚀。大多数金属在土壤中的腐蚀属于氧去极化腐蚀，只有在强酸性土壤中才发生氢去极化腐蚀。

The anodic polarization process of iron in moist soil has no obvious hindrance, which is similar to corrosion in solution. In dry and well-permeable

铁在潮湿土壤中阳极极化过程无明显阻碍，与溶液中腐蚀相似。在干燥且透气性良

soil, the anodic polarization process is passivated and ion hydration is difficult, which is similar to the anodic polarization behavior of iron corrosion in the atmosphere. Due to the secondary corrosion reaction, the insoluble corrosion products bond with the soil to form a tight layer, which plays a shielding role. As time increases, the anodic polarization increases and the corrosion rate decreases[1].

好的土壤中，阳极极化过程钝化并离子水化困难，这种情况与铁在大气中腐蚀的阳极极化行为相接近。由于腐蚀二次反应，不溶性腐蚀产物与土黏结形成紧密层，起屏蔽作用。随着时间增长，阳极极化增大，腐蚀速度减小[1]。

The cathodic polarization process is mainly the depolarization process of oxygen, which includes two basic steps: the oxygen transport to the cathode and the cathodic reaction of oxygen ionization. The process of oxygen transport to the cathode is more complicated. In the soil with multiphase structure, it is transported by gas phase and liquid phase, through the directional flow and diffusion of gas and liquid phase in the soil, and finally formed by capillary condensation to form a thin layer of electrolyte and a layer of corrosion products. In some cases the cathode has a hydrogen depolarization process or a cathode reduction process with the participation of microorganisms.

阴极极化过程主要是氧的去极化过程，其中包括两个基本步骤：即氧输向阴极和氧离子化的阴极反应。氧输向阴极过程比较复杂。在多相结构的土壤中由气相和液相两条途径输送，通过土壤中气、液相的定向流动和扩散两种方式，最后通过毛细凝聚形成的电解液薄层及腐蚀产物层。在某些情况下，阴极有氢的去极化过程或有微生物参与的阴极还原过程。

The conditions of soil corrosion are extremely complex, and the control factors for the corrosion process are also quite different. There are roughly the following control characteristics: For most soils, when the corrosion is determined by the corrosion of micro-cells or macro-corrosion cells with a not too long distance, the corrosion is mainly controlled by the cathodic process [Figure 6-10 (a)], similar to the case of total immersion in a static electrolyte. In loose, dry soil, as the oxygen permeability increases, the corrosion changes to anodic control [Figure 6-10(b)], and the control characteristics of the corrosion process are close to that of damp atmospheric

土壤腐蚀的条件极为复杂，对腐蚀过程的控制因素差别也较大，大致有以下几种控制特征：对于大多数土壤来说，当腐蚀决定于微观腐蚀电池或距离不太长的宏观腐蚀电池时，腐蚀主要由阴极过程控制［图6-10（a）］，与全浸在静止电解液中的情况相似。在疏松、干燥的土壤中，随着氧渗透率增加，腐蚀则转变为阳极控制［图 6-10（b）］，此时腐蚀过程的控制特征接近于

corrosion. For soil corrosion under the action of long-distance macro batteries, such as when underground pipelines pass through soils with different permeability to form oxygen concentration corrosion batteries, soil resistance becomes the main corrosion control factor or cathode-resistance mixed control [Figure 6-10(c)] .

潮的大气腐蚀。对于在长距离宏观腐蚀电池作用下的土壤腐蚀，如地下管道经过透气性不同的土壤形成氧浓差腐蚀电池时，土壤电阻成为主要的腐蚀控制因素或阴极–电阻混合控制［图6–10（c）］。

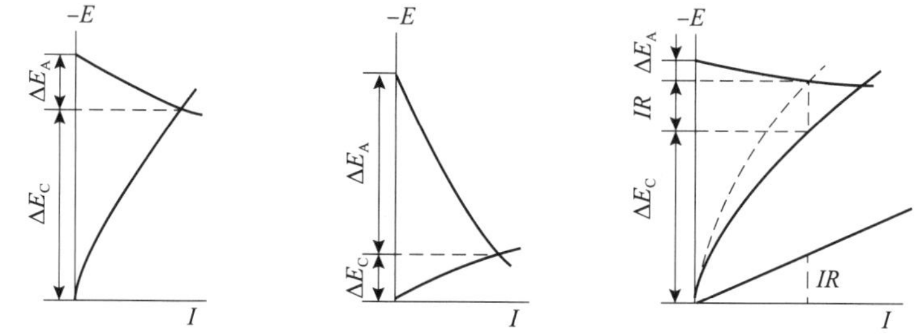

(a)潮湿土壤微观腐蚀电池，阴极控制　(b)疏松、干燥土壤微观腐蚀电池，阳极控制为主　(c)长距离宏观腐蚀电池阴极–电阻混合控制

Figure 6–10　Control characteristics of corrosion process under different soil conditions

图6–10　不同土壤条件下腐蚀过程控制特征

6.3.2　Types of soil erosion

1. Soil Corrosion Induced by Micro and Macro Batteries

In the case of soil corrosion, in addition to the micro-corrosion cells due to the inhomogeneity of the metal structure, there may also be macro-corrosion cells due to the inhomogeneity of the soil medium. Due to the different permeability of the soil, the penetration rate of oxygen is different. The inhomogeneity of the soil medium affects the potential of each part of the metal, which is the main factor to promote the formation of oxygen concentration cells.

For relatively short metal members, it can be considered that the surrounding soil structure, moisture, salinity, oxygen content, etc. are uniform. At this time, microbattery corrosion related to the non-uniformity of metal structure occurs. As shown in Figure 6-11.

6.3.2　土壤腐蚀的类型

1. 微观腐蚀电池和宏观腐蚀电池引起的土壤腐蚀

在土壤腐蚀的情况下，除因金属组织不均匀性引起的微观腐蚀电池外，还可能存在由于土壤介质的不均匀性引起的宏观腐蚀电池。由于土壤透气性不同，使氧的渗透速度不同。土壤介质的不均匀性影响金属各部分电位，这是促使形成氧浓差电池的主要因素。

对于比较短小的金属构件来说，可认为周围土壤结构、水分、盐分、氧量等是均匀的，这时发生和金属组织不均匀性有关的微观腐蚀电池，如图6–11所示。

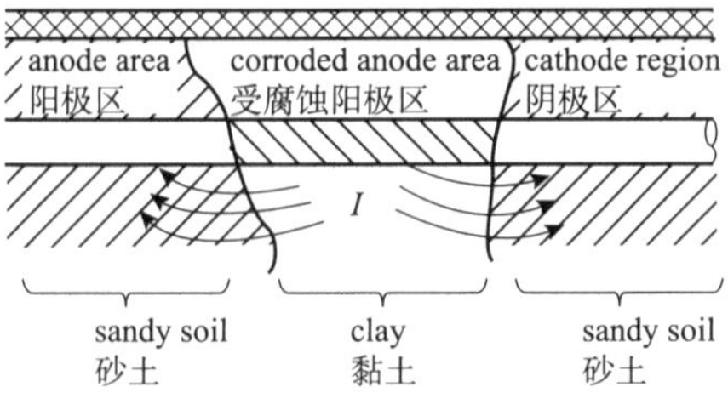

Figure 6-11 Oxygen concentration cells formed by pipes in soils with different structures

图6-11 管道在结构不同的土壤中所形成的氧浓差电池

As shown in Figure 6-12, for long metal components and pipes, due to the different oxygen permeability of each part, different structures such as clay and sand, and different burial depths, it can cause oxygen concentration cells and salt concentration cells. Such macroscopic cells cause localized corrosion, resulting in deep corrosion holes at the anode, causing serious damage to metal components. Diagram showing oxygen concentration cells formed by pipes in soils with different structures. Buried in dense, moist clay with poor oxygen permeability, the steel here acts as an anode and corrodes.

如图6-12所示，对于长的金属构件和管道，因各部分氧渗透率不同，黏土和砂土等结构不同，埋设深度不同，可引起氧浓差电池和盐浓差电池。这类宏观腐蚀电池造成局部腐蚀，在阳极部位产生较深的腐蚀孔，使金属构件遭到严重破坏。可以看到，管道在结构不同的土壤中所形成的氧浓差电池。埋在密实、潮湿的黏土中，氧渗透性差，这里的钢作为阳极而被腐蚀。

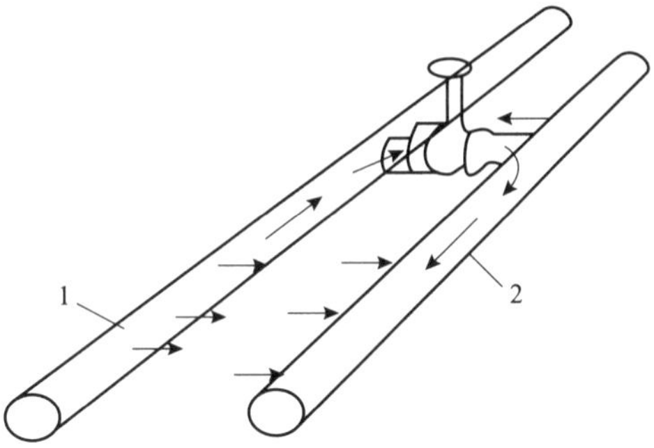

Figure 6-12

1—old tube (cathode); 2—new tube (anode)

图6-12

1—旧管（阴极）；2—新管（阳极）

2. Electrode Process of Soil Corrosion

(1) Anode process

A metal that is not significantly anodically polarized

2. 土壤腐蚀的电极过程

（1）阳极过程

阳极溶解时没有显著阳极

upon anodic dissolution. Mg, Zn, Al, Mn and Sn.

极化的金属，如Mg、Zn、Al、Mn和Sn。

The polarizability of anodic dissolution is low and depends on the overpotential of the metal ionization reaction. Such as Fe, carbon steel, Cu and Pb.

阳极溶解的极化率较低，并决定于金属离子化反应的过电位，如铁、碳钢、铜和铅。

Metals with high initial polarizability due to anode passivation. Such as Cr, Zr, chromium-containing or chromium-nickel stainless steel.

因阳极钝化而具有较高起始极化率的金属，如铬、锆、含铬或铬镍不锈钢。

Anodic dissolution of metals does not occur in soil conditions. Such as Ti and Ta.

在土壤条件不发生阳极溶解的金属，如钛、钽。

(2) Cathode Process

(2)阴极过程

The cathodic process is mainly the depolarization of oxygen.

阴极过程主要是氧的去极化作用。

3. Soil Corrosion Caused by Stray Currents

3.杂散电流引起的土壤腐蚀

The so-called stray current refers to the current flowing into other places from the original normal circuit leakage. Mainly from high-power DC electrical devices, such as electric locomotives, electric welding machines, electrolysis and electroplating tanks, electrochemical protection devices, etc. Metal structures, pipelines, storage tanks, cables, etc. buried underground are prone to corrosion due to this stray current. In addition, poor insulation of DC conductors in factories can also cause “self-stray currents” that can cause corrosion in pipes, tanks, equipment and other equipment.

杂散电流是指由原定的正常电路漏失而流入他处的电流。主要来自大功率直流电气装置，如电力机车、电焊机、电解和电镀槽、电化学保护装置等。地下埋设的金属构筑物、管道、贮槽、电缆等都容易因这种杂散电流引起腐蚀。此外，工厂中直流导线绝缘不良也可引起“自身杂散电流”，成为管道、贮槽、器械及其他设备腐蚀的原因。

Figure 6-13 is a schematic diagram of pipeline corrosion caused by stray current in soil. Under normal circumstances, the current flow is the positive pole of the power supply → the overhead line → the locomotive → the rail → the negative pole of the power supply. However, when the insulation between the rail and the soil is poor, a part of the current will leak from the rail

图6-13所示为土壤中因杂散电流而引起管道的腐蚀示意。正常情况下，电流流程为：电源正极→架空线→机车→铁轨→电源负极。但当路轨与土壤间绝缘不良时，就会有一部分电流从路轨漏到地下，进入

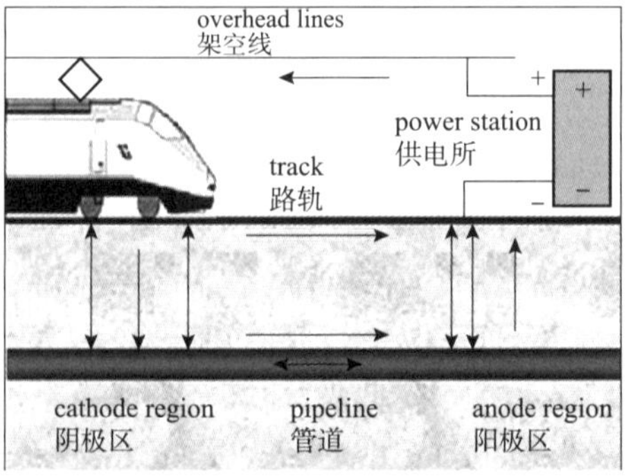

Figure 6-13 Schematic diagram of stray current corrosion in soil

图 6-13 土壤中杂散电流腐蚀示意

to the ground, enter somewhere in the underground pipeline, and then flow out from another part of the pipeline and return to the rail. When the current leaves the pipeline and enters the earth, it becomes the anode area of the corrosion battery, where the metal is corroded and damaged, and the degree of corrosion damage is proportional to the current intensity of the stray current. The higher the current intensity, the more severe the corrosion. Corrosion losses caused by stray currents are quite serious. Calculations show that 1 ampere of current for one year is equivalent to the amount of charge transported for electrochemical dissolution of 9 kg of iron. In areas with serious stray current interference, if the steel pipe is 8~9 thick, it will corrode and perforate in only 2~3 months. Stray currents can also cause intergranular corrosion of the lead sheath of the cable.

地下管道某处，再从管道另一处流出，回到路轨。电流离开管线进入大地处成为腐蚀电池的阳极区，该区金属遭到腐蚀破坏，腐蚀破坏程度与杂散电流的电流强度成正比。电流强度越大，腐蚀就越严重。杂散电流造成的腐蚀损失相当严重。计算表明：1 A 电流持续1年相当于9 kg的铁发生电化学溶解的电荷迁移量。杂散电流干扰比较严重的区域，若是8~9单位厚的钢管，只要2~3个月就会腐蚀穿孔。杂散电流还能引起电缆铅皮的晶间腐蚀。

4. Microbial Corrosion in Soil

4. 土壤中的微生物腐蚀

In oxygen-deficient soils, such as dense, moist clay, the metal corrosion process appears to be difficult, but such soil conditions favor the growth of certain microorganisms. It is often found that the strong corrosion of metals is caused by sulfate-reducing bacteria (anaerobic bacteria) and thiobacillus (there are two kinds of thiobacillus and thiooxidan, and the most

在缺氧土壤中，如密实、潮湿的黏土处，金属腐蚀过程似乎难以进行，但这种土壤条件却有利于某些微生物生长。研究发现，金属的强烈腐蚀是因硫酸盐还原菌（厌氧菌）和硫杆菌（有排硫杆菌和氧化硫

suitable temperature is 20~30℃). Moisture, Nutrients, temperature and pH are closely related to the growth of these microorganisms. For example, sulfate-reducing bacteria are easy to reproduce under neutral (pH=7.5) conditions, but when pH>9, it is difficult to reproduce and grow.

杆菌两种，最宜存在的温度为20~30℃）的温度而引起的。水分、养料、温度和pH值对这些微生物的生长密切相关，如硫酸盐还原菌易在中性（pH=7.5）条件下繁殖，在pH>9时，很难繁殖和生长。

These bacteria may cause inhomogeneity of soil physicochemical properties, thereby causing corrosion of oxygen concentration cells. Bacteria produce hydrogen sulfide, carbon dioxide, and acids during life activities, which corrode metals. Bacteria may also participate in the electrochemical process of corrosion. In anoxic neutral media, the reduction of cathode hydrogen ions is difficult due to the high hydrogen overpotential. There is only one layer of adsorbed hydrogen on the cathode, and the sulfate-reducing bacteria can consume the hydrogen atoms to make the depolarization reaction proceed smoothly.

这些细菌有可能引起土壤物理化学性质的不均匀性，从而造成氧浓差电池腐蚀。细菌在生命活动中产生硫化氢、二氧化碳和酸，对金属进行腐蚀。细菌还可能参与腐蚀的电化学过程，在缺氧的中性介质中，因氢过电位高，阴极氢离子还原困难。阴极上只有一层吸附氢，硫酸盐还原菌能消耗氢原子，使去极化反应顺利进行。

6.3.3 Influencing factors and preventive measures of soil corrosion

6.3.3 土壤腐蚀的影响因素及防止措施

The factors affecting soil corrosion include soil porosity (air permeability), water content, electrical conductivity, pH, salinity, and microorganisms, etc. These factors are related to each other, and several of the main influencing factors are[2].

影响土壤腐蚀的因素有土壤的孔隙度（透气性）、含水量、导电性、酸碱度、含盐量和微生物等，这些因素相互有联系，其中几种主要的影响因素如下[2]。

1. Porosity

Large porosity is beneficial to preserve moisture and oxygen infiltration. Good air permeability can accelerate the corrosion process, but too large air permeability hinders the anodic dissolution of metal, and is easy to form a protective corrosion product layer.

1.孔隙度

孔隙度大有利于保存水分和氧渗透。透气性好可加速腐蚀过程，但透气性太大又阻碍金属的阳极溶解，易生成具有保护能力的腐蚀产物层。

2. Water content

Moisture in soil can exist in a variety of ways, some adhering tightly around solid particles, and some flowing in micropores or bound to soil components. Electrolytes are formed when soluble salts in soil dissolve in it. The amount of water has a great influence on soil corrosion. When the water content is very low, the corrosion rate is not large. With the increase of water content, the dissolved salt in the soil increases, thus increasing the corrosion rate. When all the soluble salts are dissolved, the corrosion rate can reach the maximum value. If there is too much water, the pores of the soil will be blocked due to the expansion of the soil adhesive, the diffusion and penetration of oxygen will be blocked, and the corrosion will be reduced.

2. 含水量

土壤中的水分可以多种方式存在，有些紧密黏附在固体颗粒周围，有些在微孔中流动或与土壤组分结合在一起。当土壤中可溶性盐溶解在其中时，组成了电解液。水分的多少对土壤腐蚀影响很大，含水量很低时腐蚀速度不大，随着含水量增加，土壤中盐分的溶解量增大，因而加大腐蚀速度。当可溶性盐全部溶解时，腐蚀速度可达到最大值。若水分过多时，因土壤胶黏膨胀堵塞了土壤的孔隙，氧的扩散渗透受阻，腐蚀反而减小。

For the long-distance oxygen concentration macro cell, with the increase of water content, the resistivity of the soil decreases, and the effect of the oxygen concentration cell increases. However, when the water content increases to near saturation, the effect of oxygen concentration decreases instead.

对于长距离氧浓差电池来说，随着含水量增加，土壤的电阻率减少，氧浓差电池作用加大。但含水量增加到接近饱和时，氧浓差电池作用反而减低了。

3. Salt content

Soil generally contains inorganic salts such as sulfate, nitrate and sodium chloride. Usually the salt content per kilogram is about 80~500mg, and most of these salts are soluble. Except for Fe^{2+} ions (Fe^{2+} may enhance the destructive effect of anaerobic bacteria) on corrosion, generally cations have little effect on corrosion. Anions such as SO^{2+}, NO^{-} and Cl^{-} have a great influence on corrosion, and Cl^{-} ions have a promoting effect on soil corrosion. Soils in tidal areas near the sea or near salt pans are more corrosive. The higher the

3. 含盐量

土壤中一般含有硫酸盐、硝酸盐和氯化钠等无机盐类。通常每千克含盐量为80~500mg，这些盐类大多是可溶性的。除Fe^{2+}离子（Fe^{2+}可能增强厌氧菌的破坏作用）对腐蚀有影响外，一般阳离子对腐蚀影响不大。SO^{2+}、NO^{-}和Cl^{-}等阴离子对腐蚀影响较大，Cl^{-}离子对土壤腐蚀有促进作用。

salt content in the soil, the higher the conductivity of the soil and the higher the corrosiveness. In calcareous soils (non-acidic soils) with high content of calcium and magnesium ions, corrosion is slowed down by the formation of a protective layer of insoluble oxides or carbonates on the metal surface.

海边潮汐区或接近盐场的土壤，腐蚀性更强。土壤中含盐量大，土壤电导率增高，腐蚀性也增强。在钙、镁离子含量较高的石灰质土壤（非酸性土壤）中，因在金属表面形成难溶的氧化物或碳酸盐保护层而使腐蚀减缓。

4. Conductivity of soil

4.土壤的导电性

The conductivity of soil is affected by soil quality, water content and salinity. Soil with large porosity (such as sandy soil) is easy to infiltrate and lose water; while soil with small porosity (such as clay) is not easy to lose water. The water content is large, the soluble salts dissolve more, the conductivity is good, and the corrosiveness is strong. Especially for long-distance macro cell corrosion, the effect is more significant. General low-lying land and saline-alkali land are highly corrosive due to their good electrical conductivity.

土壤的导电性受土质、含水量及含盐量等影响，孔隙度大的土壤（如砂土），水分易渗透流失；而孔隙度小的土壤（如黏土），水分不易流失。含水量大，可溶性盐类溶解得多，导电性好，腐蚀性强。尤其是对长距离宏观腐蚀电池来说，影响更为显著。一般的低洼地和盐碱地因导电性好，所以有很强的腐蚀性。

5. Other factors

5.其他因素

Generally, the higher the acidity, the stronger the corrosiveness. This is because the hydrogen ion cathodic depolarization is prone to occur. When the soil contains a lot of organic acids, its pH value is close to neutral, but its corrosiveness is still very strong. Therefore, when measuring soil corrosivity, the total acidity of the soil should be measured.

通常酸度越大，腐蚀性越强。这是因为易发生氢离子阴极去极化作用的缘故。当土壤中含有大量有机酸时，其pH值虽然接近中性，但其腐蚀性仍然很强。因此，衡量土壤腐蚀性时，应测定土壤总酸度。

Elevated temperature can increase the conductivity of the soil electrolyte and accelerate the penetration and diffusion of oxygen, thus accelerating corrosion. Elevated temperatures, such as at 25℃, are optimal for microbial growth, which also accelerates corrosion.

温度升高能增加土壤电解液的导电性，加快氧的渗透扩散速度，因此加速腐蚀。温度升高，如处于25℃时，最适宜于微生物生长，从而也加速腐蚀。

6.3.4 Possible measures to prevent soil corrosion

The following measures can be adopted to prevent soil corrosion.

①Covering layer protection: Corrosion prevention is achieved by improving the insulation between the protected component and the soil. Anti-corrosion protective layers such as asphalt coating, epoxy powder coating and foam plastic can be used.

②Corrosion-resistant metal materials and metal coatings: use certain alloy steels and non-ferrous metals, or use zinc coatings to prevent soil corrosion. It is rarely used because it is not economical, and it is not suitable to be used in acidic soil.

③Treat soil to reduce its erosiveness: For example, use lime to treat acidic soil, or fill limestone fragments around underground components, transfer soil with less erosiveness, and strengthen drainage to improve soil environment and reduce corrosiveness.

④Cathodic protection: in addition to the above protection methods, cathodic protection measures can be added. This can not only make up for the deficiency of the protective layer, but also reduce the power consumption of cathodic protection. In general, the potential of the steel cathode is maintained at –0.85V (relative to the copper sulfate electrode) to obtain a complete protection effect. In the presence of sulfate-reducing bacteria, the potential should be maintained more negative, such as –0.95V (relative to the copper sulfate electrode) to inhibit bacterial growth. Cathodic protection is also used to protect underground lead

6.3.4 防止土壤腐蚀可采用的措施

防止土壤腐蚀可采用的措施如下。

①覆盖层保护：通过提高被保护构件与土壤间的绝缘性达到防蚀。可采用沥青涂层、环氧粉末涂层、泡沫塑料等防腐保护层。

②耐蚀金属材料和金属镀层：采用某些合金钢和有色金属，或采用锌镀层来防止土壤腐蚀。由于不经济而很少使用，且不宜用于酸性土壤。

③处理土壤减少其浸蚀性：如用石灰处理酸性土壤，或在地下构件周围填充石灰石碎块，移入浸蚀性小的土壤，加强排水，以改善土壤环境，降低腐蚀性。

④阴极保护：在上述保护方法的同时，可附加阴极保护措施。这样既可弥补保护层的不足，又可减少阴极保护的电能消耗。一般情况下，钢铁阴极的电势维持在–0.85V（相对于硫酸铜电极）可获得完全保护的效果。在有硫酸盐还原菌存在时，电势要维持得更负些，如–0.95V（相对于硫酸铜电极），以抑制细菌生长。阴极保护也用于保护地下铅皮电

sheathed cables, and its protective potential is about −0.7V (relative to copper sulfate electrodes).

缆，其保护电势约为−0.7V（相对于硫酸铜电极）。

6.3.5 References

[1] TAN Y, WANG H, WU W. Research progress on corrosion of outer floor of storage tanks [J]. Journal of Northwest University (Natural Science Edition), 2021, 51(4): 601-614.

[2] DING C J, YANG H X, WANG J D, et al. Corrosion and protection development of grounding grid materials [J]. Corrosion Science and Protection Technology, 2019, 31(1): 109-113.

6.3.5 参考文献

[1] 淡勇，王珅，武玮. 储罐外底板腐蚀的研究进展[J]. 西北大学学报（自然科学版），2021，51（4）：601−614.

[2] 丁长军，杨焕新，王建东，等. 接地网材料的腐蚀及其防护发展[J]. 腐蚀科学与防护技术，2019，31（1）：109−113.

Corrosion Protection Techniques

腐蚀防护技术

7

7.1 Design for corrosion and protection

7.1.1 Correct use of metal materials and machining processes

In industry, when designing product components, the correct selection of materials is a very important and complex task, the selection of materials directly affects the performance of the product or not. When selecting materials, in addition to attention to corrosion resistance, but also take into account mechanical properties, processing performance and the price of the material itself and other comprehensive factors. The following principles should be followed when selecting materials[1]:

①Various factors should be considered in a comprehensive and integrated manner according to the conditions of use. For example, in the presence of media,

7.1 腐蚀与防护设计

7.1.1 正确采用金属材料和加工工艺

工业上在进行产品构件设计时，正确选材是一项十分重要而又相当复杂的工作，选材的合理与否直接影响产品性能。在选材时，除注意耐蚀性外，还要考虑机械性能、加工性能及材料价格等综合因素。选材时应遵循以下原则[1]：

①应根据使用条件全面综合地考虑各种因素。如在有介质存在下，除考虑材料的断裂

in addition to considering the fracture toughness of the material K_{ic}, the threshold stress intensity factor K_{iscc} or stress corrosion fracture threshold stress σ_{th} value of the product should be considered.

韧性K_{ic}外，更应考虑产品的门槛应力强度因子K_{iscc}或应力腐蚀断裂门槛应力σ_{th}值。

②For the initial selection of materials it should be identified which types of corrosion they are sensitive to; which types of corrosion are likely to occur and the possibility of protection; whether they are compatible with the materials they come into contact with and whether contact corrosion can occur; and the state of the stresses they are subjected to.

②对初选材料应查明它们对哪些类型的腐蚀敏感；可能发生哪种腐蚀类型以及防护的可能性；与其接触的材料是否相容，能否发生接触腐蚀；以及承受应力的状态等。

③In parts prone to corrosion and not easily maintained, choose materials with high corrosion resistance.

③在容易产生腐蚀和不易维护的部位，应选择高耐蚀性材料。

④Choose materials and heat treatment states with low corrosion tendency. For example, when the tensile strength of 30CrMnSiA steel is below 1176 MPa, the sensitivity to stress corrosion and hydrogen embrittlement is high; however, when the tensile strength reaches 1373 MPa or more after heat treatment, the sensitivity of the material to stress corrosion and hydrogen embrittlement increases significantly. Therefore, a reasonable choice of material heat treatment state, control the upper limit of tensile strength of materials used is very necessary. Aluminum alloy, stainless steel in a certain heat treatment state or heating conditions, can produce intergranular corrosion, the selection of materials should also be considered.

④选择腐蚀倾向小的材料和热处理状态。例如，30CrMnSiA钢拉伸强度在1176 MPa以下时，对应力腐蚀和氢脆的敏感性高；但经热处理使其拉伸强度达到1373 MPa以上时，材料对应力腐蚀和氢脆的敏感性明显增高。因此，合理地选择材料的热处理状态，控制材料使用的拉伸强度上限是非常必要的。铝合金、不锈钢在一定的热处理状态或加热条件下，可产生晶间腐蚀，选材时也应予以考虑。

Choose materials with low impurity content to improve corrosion resistance. For high strength steel, aluminium alloy, magnesium alloy and other high strength materials, the presence of impurities will directly affect its ability to resist uniform corrosion and stress corrosion.

选用杂质含量低的材料，以提高耐蚀性。对高强度钢、铝合金、镁合金等强度高的材料，杂质的存在会直接影响其抗均匀腐蚀和应力腐蚀的能力。

7.1.2 Rational design of metal structures

To prevent corrosion, a reasonable design of the anti-corrosion structure should be carried out at the product design stage, with rigorous calculation and determination of the stresses to be used and correct selection of materials and protection systems. In mechanical structures, the design is reasonable and has a great influence on the sensitivity to stress corrosion, contact corrosion, uniform corrosion and crevice corrosion[2]. To reduce or prevent these corrosions, the following points should be noted:

①Strive for simplicity in shape to avoid accumulation of rainwater for protective construction, inspection of corrosion status and regular maintenance.

②Use airtight structures to prevent the intrusion of rainwater, fog and even seawater.

③Setting drainage holes where water can accumulate.

④Arrange suitable ventilation openings to prevent the collection of moisture and condensation.

⑤Avoid sharp corners, recesses and crevices as far as possible to prevent condensation from accumulating.

⑥The joint surface of rivets, screws or spot-welded joints should have isolation insulation to prevent contact corrosion or crevice corrosion.

⑦Materials with high water absorption should be used sparingly and, if unavoidable, the surrounding area should be sealed.

⑧In the grain orientation of the part, try to avoid tensile forces in the short transverse direction. Should

7.1.2 合理设计金属结构

为防止腐蚀，在产品设计阶段应当进行合理的防腐蚀结构设计，严格地计算和确定使用应力，正确选择材料和防护系统。在机械结构中，设计是否合理，对应力腐蚀、接触腐蚀、均匀腐蚀、缝隙腐蚀等敏感性影响很大[2]。为减少或防止这些腐蚀，应注意以下事项：

①外形力求简单，避免雨水积存，以便进行防护施工，对腐蚀状态进行检查并定期维修。

②采用密闭结构，以防雨水、雾及海水的侵入。

③能积水的地方设置排水孔。

④布置合适的通风口，以防湿气汇集和凝露。

⑤尽量避免尖角、凹槽和缝隙，以防冷凝水积聚。

⑥铆钉、螺钉或点焊连接头的结合面应当有隔离绝缘层，以防接触腐蚀或缝隙腐蚀。

⑦尽量少用吸水性强的材料，若不可避免时，周围应加以密封。

⑧在零件的晶粒取向上，尽量避免在短横向上受拉力。

avoid the use of stress, assembly stress and residual stress in the same-a direction superimposed to reduce and prevent stress corrosion fracture. Design forgings to ensure that the grain flow direction is adapted to the stress direction.

⑨ When contact between different metals cannot be avoided, the contact surface is treated with appropriate protection, for example, steel parts galvanised or cadmium plated can be in contact with anodised aluminium parts; or between two metals with corrosion inhibiting sealing paste. Insulating materials are used to separate the two metals.

7.2 Electrochemical protection

7.2.1 Definition

电化学防护

The use of the electrochemical principle of the primary battery to eliminate the primary battery reaction that causes electrochemical corrosion of metals, so that the metal is protected, this metal protection method is called electrochemical protection. Electrochemical protection is divided into two categories: anodic protection and cathodic protection[3].

Cathodic protection is the use of a metal component as a cathode to remove the electrochemical inhomogeneity of that metal surface by cathodic polarisation for protection purposes. Cathodic protection is an economical and effective protective measure. The cost of painting an offshore vessel accounts for 5% of the construction cost, while the cost of sacrificial

应避免使用应力、装配应力和残余应力在同一个方向上叠加，以减少和防止应力腐蚀断裂。设计锻件时要保证晶粒流向适应于应力方向。

⑨当无法避免不同金属间的接触时，对接触表面进行适当的防护处理，如钢零件镀锌或镀镉后可与阳极化的铝合金零件接触；或两种金属间用缓蚀密封膏、绝缘材料隔开。

7.2 电化学防护

7.2.1 定义

运用原电池的电化学原理，消除引起金属发生电化学腐蚀的原电池反应，使金属得到防护，这种金属防护法称为电化学防护。电化学防护分为阳极防护和阴极防护两大类[3]。

阴极防护是使用金属构件作为阴极，通过阴极极化来消除该金属表面的电化学不均匀性，以达到保护目的。阴极防护是一种经济有效的防护措施。一条海船在建造费中，涂装费占5%，而阴极防护用的牺牲阳极材料和施工费加起来不到1%。1座海上采油平台的建造费超过1亿元，而牺牲阳

anode materials and construction costs for cathodic protection add up to less than 1%. The construction cost of an offshore oil recovery platform is over 100 million RMB, while the cost of sacrificial anode materials and construction is only 1 to 2 million RMB. Without protection, the life of the platform is only 5 years, while it can last for more than 20 years under cathodic protection. The cost of cathodic protection for underground pipelines only accounts for 0.3%~0.6% of the total investment, which can significantly extend the service life. Some of the equipment required to be used in seawater and soil for decades, such as marine platforms, ships, docks, underground pipelines, cables, storage tanks, etc., must be cathodically protected to improve their corrosion resistance.

极材料和施工费只需100万~200万元。不采用保护，平台寿命只有5年，而阴极防护下可用20年以上。地下管线的阴极防护费只占总投资的0.3%~0.6%，就可显著延长使用寿命。一些要求在海水、土壤中使用几十年的设备，如海洋平台、轮船、码头，地下管线、电缆、贮槽等，都必须采用阴极防护，提高其抗蚀能力。

When no cathodic protection is applied, the anodic polarisation curve $\varphi_{0,A}A$ and the cathodic polarisation curve $\varphi_{0,C}C$ of the metal corrosion microcell intersect at the point S (ignoring the solution resistance), where the corresponding potential is the self-corrosion potential φ_{corr} of the metal and the corresponding current is the corrosion current I_{corr} of the metal. under the action of the wrrosion current the φ_{corr}, anode of the microcell is

阴极防护原理见图7-1。当未进行阴极防护时，金属腐蚀微电池的阳极极化曲线$\varphi_{0,A}A$和阴极极化曲线$\varphi_{0,C}C$相交于点S（忽略溶液电阻），此点对应的电势为金属的自腐蚀电势φ_{corr}，对应的电流为金属的腐蚀电流I_{corr}。在腐蚀电流I_{corr}作用

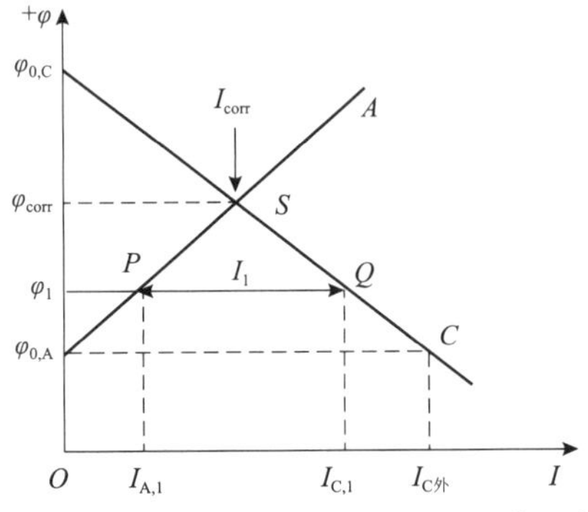

Figure 7-1 Schematic diagram of the principle of cathodic protection [4]

图7-1 阴极保护原理示意 [4]

dissolved continuously, resulting in corrosion damage.

下，微电池阳极不断溶解，导致腐蚀破坏。

In cathodic protection, under the polarization of the applied cathodic current I_1, the total metal potential becomes negative from φ_{corr} to φ_1, and part of the total cathodic current $I_{C,1}$ ($\varphi_1 Q$ section) is applied, i.e. $I_{A,1}$ (PQ section), and the other part of the current is still provided by the metal anode corrosion, i.e. $I_{A,1}$ ($\varphi_1 P$ section). Obviously, the anodic current of the metal microcell, $I_{A,1}$, is then reduced compared to the original corrosion current, I_{corr}, i.e. the corrosion rate is reduced and the metal is partially protected. The difference ($I_{corr}-I_{A,1}$) represents the reduced value of the corrosion microcell action on the metal after the applied cathodic polarisation, i.e. the reduced value of the corrosion current, known as the protection effect.

阴极防护时，在外加阴极电流I_1的极化下，金属总电势由φ_{corr}变负到φ_1，阴极总电流$I_{C,1}$（$\varphi_1 Q$段）中，一部分电流是外加的，即$I_{A,1}$（PQ段），另一部分电流仍然是由金属阳极腐蚀提供的，即$I_{A,1}$（$\varphi_1 P$段）。显然，这时金属微电池的阳极电流$I_{A,1}$，比原来的腐蚀电流I_{corr}减小了，即腐蚀速度降低了，金属得到部分保护。差值（$I_{corr}-I_{A,1}$）表示外加阴极极化后金属上微观腐蚀电池作用的减小值，即腐蚀电流的减小值，称为保护效应。

As the applied cathodic current continues to increase, the electrode potential of the metal system becomes even lower. A, the anode current on the metal is zero and all the current is outside the applied cathodic current I_C ($\varphi_{0,A} C$ section), at this time, only the cathodic reduction reaction occurs on the metal surface, and the metal dissolution reaction stops, so the metal is completely protected, and the potential of the metal at this time is called the minimum protective potential. The applied current density required for the metal to reach the minimum protective potential is called the minimum protective current density.

当外加阴极电流继续增大时，金属体系的电极电势变得更低。当金属总电势达到微电池阳极的起始电势$\varphi_{0,A}$时，金属上阳极电流为0，全部电流为外加阴极电流I_C外（$\varphi_{0,A} C$段），这时，金属表面只发生阴极还原反应，而金属溶解反应停止了，因此金属得到完全保护，这时金属的电势称为最小保护电势。金属达到最小保护电势所需的外加电流密度称为最小保护电流密度。

From this we can conclude that in order for the metal to be fully protected, it must be cathodically polarised to its equilibrium potential at the anode of the

由此可得出以下结论：要使金属得到完全保护，必须把它阴极极化到其微观腐蚀电池

corrosion microcell.

This is only the basic principle of cathodic protection, the actual situation is much more complex, for example, also take into account the influence of the time factor. Steel in seawater, for example, the original seawater without iron ions, to make the steel mixed potential down to the equilibrium potential of the anodic reaction, that is, iron dissolution reaction, it is necessary to have a corresponding concentration of iron ions in the seawater near the anode on the steel surface, for example 1×10^{-6}mol/L. This is difficult to do at the beginning of cathodic protection. In fact, in order to achieve a satisfactory protection effect, the chosen protection potential should always be lower than the corrosion microcell anode equilibrium potential.

阳极的平衡电势。

这只是阴极防护的基本原理，实际情况要复杂得多，如还要考虑时间因素的影响。以钢在海水中为例，原来海水中无铁离子，要使钢的混合电势降到阳极反应即铁溶解反应的平衡电势，需要在钢表面阳极附近的海水中有相应的铁离子浓度，如1×10^{-6}mol/L。这在阴极防护初期是很难做到的。实际上，为达到满意的保护效果，选用的保护电势总要低于腐蚀微电池阳极平衡电势。

Cathodic protection can be achieved by two methods: one is the sacrificial anode method, the second is the applied current method.

阴极防护可通过两种方法实现：一是牺牲阳极法，二是外加电流法。

(1) Sacrificial anode method

（1）牺牲阳极法

Sacrificial anode method is connected to the metal being protected electrode potential lower metal or alloy, as a sacrificial anode, rely on it to continuously dissolve the current generated by the cathodic polarization of the metal being protected, to achieve the purpose of protection. The sacrificial anode method includes the determination of the sacrificial anode material and the design and installation of two major parts.

牺牲阳极法是在被保护的金属上连接电极电势更低的金属或合金，作为牺牲阳极，靠它不断溶解所产生的电流对被保护金属进行阴极极化，达到保护的目的。牺牲阳极法包括牺牲阳极材料的确定和设计安装两大部分。

The sacrificial anode material must be able to create a sufficiently large potential difference (typically around 0.25V) with the metal component being protected. The sacrificial anode method for the protection of steel gates is shown in Figure 7-2 below. Therefore, the main requirements for sacrificial anode materials are:

牺牲阳极材料必须能与被保护的金属构件之间形成足够大的电势差（一般在0.25V左右）。图7-2所示为保护钢闸门的牺牲阳极法。对牺牲阳极材料的主要要求是：有足够低的

a sufficiently low electrode potential, and the anode polarisation rate should be small, the electric capacity should be large, that is, the consumption of a unit mass of metal to provide more electricity, the output current per unit area is large; self-corrosion is very small, the current efficiency is high; long-term use to maintain the anode activity, not easy to passivate, can maintain a stable potential and output current, anode dissolution uniform, corrosion products loose easy to fall off, not The anode dissolves evenly, the corrosion products are easy to fall off and do not adhere to the anode surface or form a high resistance crust.

电极电势，且阳极极化率小，电容量大，即消耗单位质量金属所提供的电量多，单位面积输出电流大；自腐蚀很小，电流效率高；长期使用时保持阳极活性，不易钝化，能维持稳定的电势和输出电流，阳极溶解均匀，腐蚀产物疏松易脱落、不黏附于阳极表面或形成高电阻硬壳；价格便宜，来源充分，制造工艺简单，无公害等。

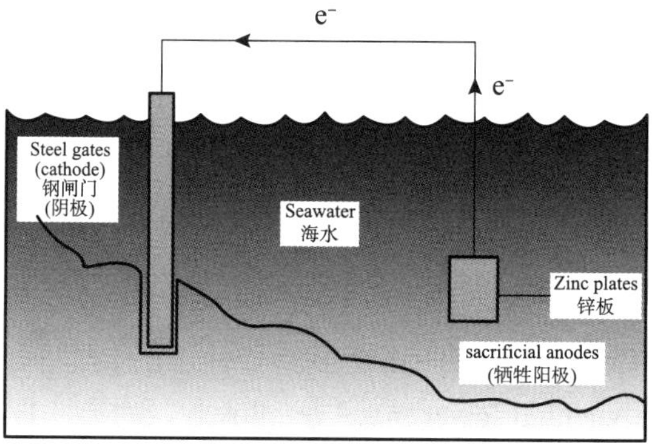

Figure 7-2　Protection of steel gates by the sacrificial anode method
图 7-2　牺牲阳极法保护钢闸门

Commonly used sacrificial anode materials are: Zn-0.6%Al-0.1%Cd, Al-2.5% Zn-0.02% In, Mg-6% Al-3% Zn-0.2% Mn, high purity zinc, etc., of which aluminium alloys are mostly used in seawater.

常用的牺牲阳极材料有：Zn-0.6% Al-0.1%Cd、Al-2.5% Zn-0.02%In、Mg-6% Al-3% Zn-0.2% Mn、高纯锌等，其中铝合金多用于海水中。

The design of the sacrificial anode protection system includes: calculation of the protection area, determination of the protection parameters, shape, size and number of sacrificial anodes, distribution and installation as well as evaluation of the cathodic protection effect.

牺牲阳极保护系统的设计包括：保护面积的计算，保护参数的确定，牺牲阳极的形状、大小和数量，分布和安装以及阴极保护效果的评定等。

(2) Applied current method of cathodic protection

The applied current method is to connect the protected metal to the negative terminal of the DC power supply and pass the cathodic current to polarise the metal to the protective potential range to achieve the purpose of corrosion prevention.

Although the sacrificial anode method does not require an external power supply and dedicated management, will not interfere with the adjacent metal facilities, and good current dispersion ability, convenient construction, but requires the consumption of a large amount of metal materials, the ability to automatically adjust the current is poor, and the installation of a large number of sacrificial anodes will increase the hull mass and resistance, or affect the stability and traction characteristics of the offshore oil platform. Therefore, after the 1950s, with the development of the electronics industry, the applied current cathodic protection technology has been greatly developed. Applied current cathodic protection systems have the advantages of small size and mass, automatic current and voltage regulation and a wide range of applications. If a reliable insoluble anode is used, its service life is long. A typical diagram of the applied current method is shown in Figure 7-3 below.

The external current method cathodic protection system consists of three main components: DC power supply, auxiliary anode and reference electrode. The DC power supply is usually a high-powered constant potential meter, which can automatically adjust the output current according to changes in external conditions, so that the potential of the protected body is always controlled within the protection potential range.

(2)外加电流法阴极保护

外加电流法是将被保护金属接到直流电源的负极，通以阴极电流，使金属极化到保护电势范围内，达到防蚀目的。

牺牲阳极法虽然不需要外加电源和专人管理，不会干扰邻近金属设施，电流分散能力好，施工方便，但需要消耗大量金属材料，自动调节电流的能力差，而且安装大量牺牲阳极会增加船体质量和阻力，或者影响海洋石油平台的稳定性和牵引特性。因此，20世纪50年代后随着电子工业的发展，外加电流阴极保护技术得到很大发展。外加电流阴极保护系统具有体积及质量小、能自动调节电流和电压、运用范围广等优点。若采用可靠的不溶性阳极，其使用寿命较长。图7-3所示为埋地管道外加电流阴极保护法示意。

外加电流法阴极保护系统主要由三部分组成：直流电源、辅助阳极和参比电极。直流电源通常用大功率的恒电位仪，可根据外界条件的变化，自动调节输出电流，使被保护体的电势始终控制在保护电势范围内。

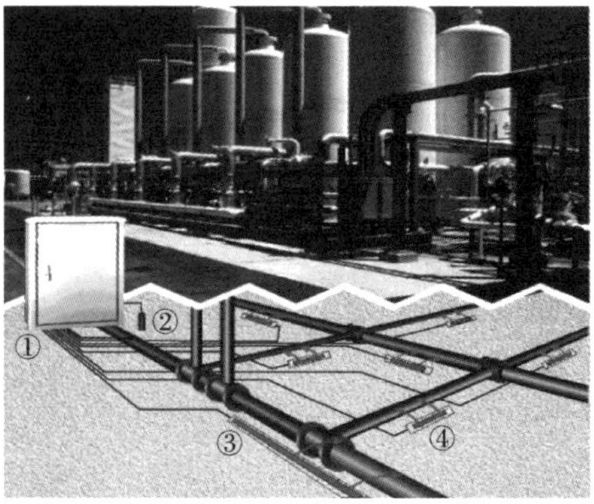

Figure 7-3 Schematic diagram of the applied current cathodic protection method for buried pipelines

① DC current/constant potential meter；② reference electrode；③、④ auxiliary anode

图 7-3 埋地管道外加电流阴极保护法示意

①直流电流 / 恒电位计；②参比电极；③、④辅助阳极

The auxiliary anode of the external current method is used to deliver current to the cathode (i.e. the metal being protected), which is very different from the anode material used in the sacrificial anode method. Auxiliary anode materials for the external current method should have good electrical conductivity, good corrosion resistance, long life, large discharge (i.e. large current per unit area at a certain voltage), small anodic polarisation, certain mechanical strength, easy processing, convenient sources, cheap and so on. Commonly used auxiliary anode materials are steel, graphite, high-silicon iron, magnetic iron oxide, lead-silver (2%) alloy, platinum-plated titanium, etc. These anode plates, with the exception of steel, are corrosion resistant and are available for long-term use. Titanium is coated with a layer of 2~5μm platinum as the anode, using an operating current density of 1000~2000A/m^2 and a consumption rate of only 4~10mg/(A·a) of platinum, which generally lasts 5~10 years.

外加电流法的辅助阳极用来把电流输送到阴极（被保护的金属），这与牺牲阳极法所用的阳极材料截然不同。外加电流法的辅助阳极材料应具有导电性好、耐蚀性好、寿命长、排流量大（一定电压下单位面积通过的电流大）、阳极极化小、有一定的机械强度、易于加工、来源方便、价格便宜等特点。常用的辅助阳极材料包括钢、石墨、高硅铁、磁性氧化铁、铅银（2%）合金、镀铂的钛等。这些阳极板除钢外，都是耐蚀的，可供长期使用。钛上镀一层2~5μm的铂作为阳极，使用工作电流密度为1000~2000 A/m^2，而铂消耗率只有4~10 mg/（A·a），一般可使用5~10年。

The reference electrode is used in conjunction with a constant potential meter to measure and control

参比电极用来与恒电位计配合使用，测量和控制保护

the protective potential, so it requires a good reversible reference electrode, which is not easily polarised and can maintain a stable, accurate and sensitive potential in long-term use, and is robust and durable. Commonly used reference electrodes are Ag/AgCl electrode, Cu/$CuSO_4$ electrode, Zn electrode, etc.

电势，因此要求参比电极可逆性好，不易极化，在长期使用中能保持电势稳定、准确、灵敏、坚固耐用等。常用的参比电极有Ag/AgCl电极、Cu/$CuSO_4$电极、Zn电极等。

The design of the external current protection system mainly includes: selecting the protection parameters, determining the material, quantity, size and installation position of the auxiliary anode, determining the material and size of the anode screen, calculating the capacity of the power supply, etc. As the auxiliary anode is insulated on the protected body, the current density near the anode is very high, which can easily cause “over-protection” and damage to the paint around the anode. Therefore, a special anode shield must be painted or installed in the vicinity of the anode. It should have a high bond with the steel, excellent insulation, good alkali resistance and seawater resistance. The anode shielding materials used for marine vessels are FRP anode screens, or thick slurry-type coatings coated with chlorinated rubber, or epoxy asphalt polyphthalamide coatings.

外加电流保护系统的设计主要包括：选择保护参数，确定辅助阳极材料、数量、尺寸和安装位置，确定阳极屏蔽材料和尺寸，计算供电电源的容量等。由于辅助阳极是绝缘地安装在被保护体上，故阳极附近的电流密度很高，易引起“过保护”，使阳极周围的涂料遭到破坏。因此，必须在阳极附近一定范围内涂刷或安装特殊的阳极屏蔽层。它应具有与钢结合力高、绝缘性优良，良好的耐碱、耐海水性能。对海船用的阳极屏蔽材料有玻璃钢阳极屏，或涂氯化橡胶厚浆型涂料，或环氧沥青聚酞胺涂料。

Cathodic protection is increasingly widely used, mainly for the protection of various metal components and equipment in the water and underground: such as ships, docks, bridges, sluices, floats, ocean platforms, submarine pipelines; cooling water systems in factories, heat exchangers, sewage treatment facilities, atomic power plants of all kinds of water supply systems; underground oil, gas, water pipelines, underground cables, etc., are available sacrificial anode method or applied current method cathodic protection, corrosion prevention effect is very good.

阴极防护的应用日益广泛，主要用于保护水中和地下的各种金属构件和设备，如舰船、码头、桥梁、水闸、浮筒、海洋平台、海底管线；工厂中的冷却水系统、热交换器、污水处理设施，原子能发电厂的各类给水系统；地下油、气、水管线、地下电缆等，都可用牺牲阳极法或外加电流法进行阴极防护，防蚀效果很好。

7.2.2 Anodic protection

The metal equipment to be protected is connected to the positive pole of the applied DC power supply, which is anodized in the corrosive medium to a stable passivation zone, and the metal equipment is protected. This method is called anodic protection method. This protection technique has been successfully used in industrial production to prevent corrosion in alkaline pulp cooking pots. It is now gradually being used in sulphuric acid, phosphoric acid, organic acid and liquid fertiliser production systems with good results.

This is shown in Figure 7-4. For metal equipment with passivation behaviour, it is anodically polarised with an external power source, so that its potential enters the passivation zone and the corrosion rate is minimal, i.e. anodic protection is obtained.

7.2.2 阳极防护

将被保护的金属设备与外加直流电源的正极相连，在腐蚀介质中使其阳极极化到稳定的钝化区，金属设备就得到保护，这种方法称为阳极防护法。这种防护技术已成功用于工业生产，以防止碱性纸浆蒸煮锅的腐蚀。现在逐渐用到硫酸、磷酸、有机酸和液体肥料生产系统中，取得了很好的效果。

如图7-4所示，对于具有钝化行为的金属设备，用外电源对它进行阳极极化，使其电势进入钝化区，腐蚀速度甚微，即得到阳极防护。

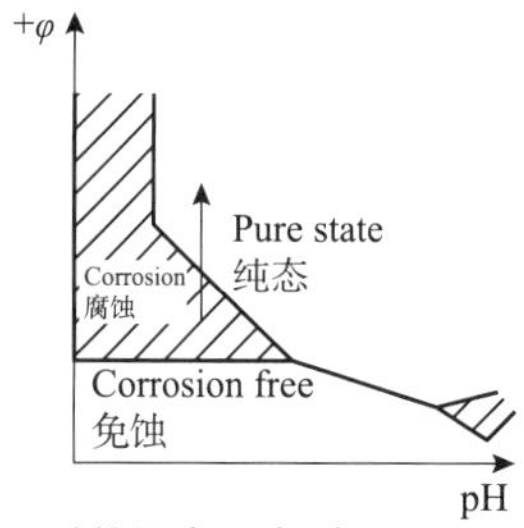

(a)Metal passivation process
金属钝化过程

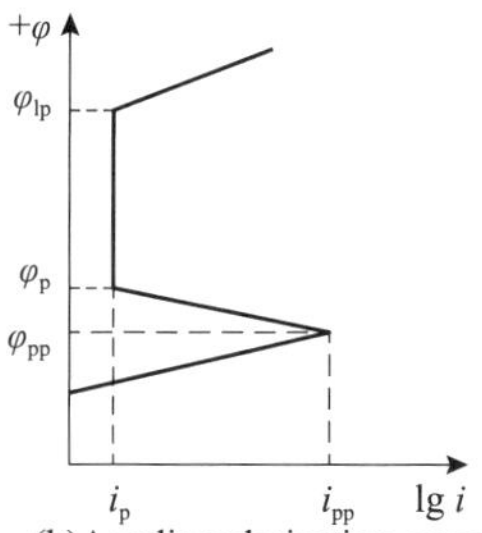

(b)Anodic polarization curve
阳级极化曲线

Figure 7-4 Anodic protection

图7-4 阳极防护原理示意

7.3 Metal protection layer

Metal cladding, sometimes called metal coating (plating), is manufactured by the following processes.

1. Electroplating

Electroplating is a method of depositing metal

7.3 金属保护层

金属覆盖层有时也称金属涂（镀）层，其制造方法包括下列工艺。

1.电镀

电镀是用电沉积的方法使

surfaces with a layer of metal or alloy. Plated metals include Ni, Cr, Cu, Sn, Zn, Cd, Fe, Pb, Co, Au, Ag, Pt and other single metal plating layers, as well as Zn-Ni, Cd-Ti, Cu-Zn, Cu-Sn and other alloy plating layers. In addition to protective, decorative, wear-resistant and heat-resistant effects, there are also various functional plating layers.

金属表面镀上一层金属或合金。镀层金属包括Ni、Cr、Cu、Sn、Zn、Cd、Fe、Pb、Co、Au、Ag、Pt等单金属电镀层，还有Zn–Ni、Cd–Ti、Cu–Zn、Cu–Sn等合金电镀层。除防护、装饰、耐磨、耐热等作用外，还有各种功能电镀层。

2. Hot plating

2.热镀

Hot plating, also known as hot dip plating, is the process of impregnating protected metal products in molten metal to form a protective metal layer on the surface. The liquid metal chosen is generally a low melting point, corrosion and heat resistant metal, such as Al, Zn, Sn, Pb, etc. Galvanised steel sheets (commonly known as white iron) and tin-plated steel sheets (commonly known as tinplate) are hot-dip galvanised. Usually the hot dip galvanising temperature is around 450℃. To improve the quality of the coating, a mass fraction of 0.2% Al and a small amount of MgO can be added to the zinc hot dip tin plating at a temperature of 310~330℃. Compared with the electroplating method, the metal hot plating layer is thicker and has a longer life in the same environment.

热镀亦称热浸镀，是将被保护金属制品浸渍在熔融金属中，使其表面形成一层保护性金属覆盖层。选用的液态金属一般是低熔点、耐蚀、耐热的金属，如Al、Zn、Sn、Pb等。镀锌钢板（俗称白铁）和镀锡钢板（俗称马口铁）就是采用热浸镀工艺。通常热镀锌温度在450℃左右，为改善镀层质量，可在锌中加入质量分数为0.2%的Al和少量MgO，热镀锡温度为310~330℃。与电镀法相比，金属热镀层较厚，在相同环境中，其寿命较长。

3. Thermal spraying

3.热喷涂

Thermal spraying technology is the use of gas combustion, explosive or electrical energy as a heat source, the filamentary or powdered metal heated to a molten or semi-molten state, and sprayed at high speed to the surface of the parts, so as to form a layer of special properties of the coating process method. Since the emergence of metal spray gun technology with an oxyacetylene flame in the 1930s, the thermal spraying

热喷涂技术是利用气体燃烧、爆炸或电能作热源，将丝状或粉状金属加热至熔化或半熔化状态，并以高速喷向零件表面，从而形成一层具有特殊性能涂层的工艺方法。自20世纪30年代出现氧乙炔焰的金属喷涂枪技术以来，热喷涂工艺从热源、介

process from the heat source . Media, spraying material form, etc. has been done in many ways to improve to meet the requirements of different applications. Figure 7-5 shows the various thermal spraying process methods. Commonly used spray material metals are Al, Zn, Sn, Pb, stainless steel, Ni-Al and Ni_3Al. Thick layers of spray metal can be used to repair worn shafts or other damaged components. Although the porosity of the spray metal layer is large, it can also provide good corrosion protection due to the mechanical isolation of the coating metal or the cathodic protection of the substrate.

质、喷涂材料形态等已做了多方面的改进，以满足不同应用的要求。图7–5所示为各种热喷涂的工艺方法。常用的喷料金属包括Al、Zn、Sn、Pb、不锈钢、Ni–Al和Ni_3Al等。厚的喷料金属层可用于修复已磨损的轴或其他损坏部件。虽然喷料金属层的孔隙度较大，但由于涂层金属的机械隔离或对基体的阴极保护作用，也能起到良好的防蚀效果。

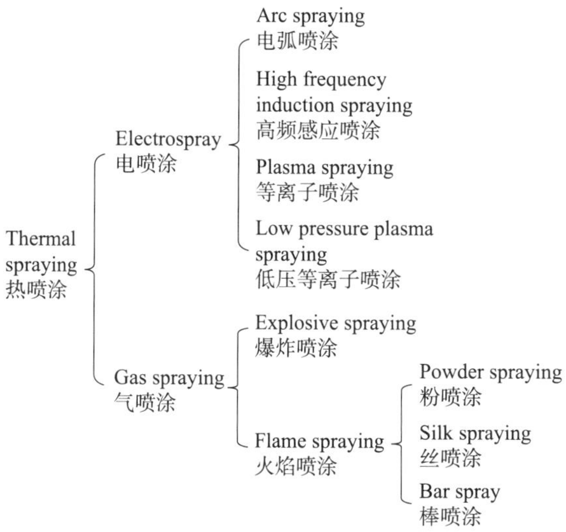

Figure 7–5　Types of thermal spraying

图 7–5　热喷涂种类

4. Infiltration plating

The diffusion of metal atoms at high temperatures is used to form an alloy diffusion layer on the surface of the metal being protected. The most common are Si, Cr, AI, Ti, B, W, Mo and other infiltrative plating layers. These layers are uniform in thickness, non-porous, thermally stable and firmly bonded to the substrate, and not only have good corrosion resistance, but also improve other physical and chemical properties of the material.

4. 渗镀

在高温下利用金属原子的扩散，在被保护金属表面形成合金扩散层。最常见的是Si、Cr、AI、Ti、B、W、Mo等渗镀层。这类镀层厚度均匀、无孔隙、热稳定性好、与基体结合牢，不但有良好的耐蚀性，还可改善材料的其他物理和化学性能。

5. Chemical plating

The use of redox reactions, so that the metal ions in the salt solution in the protected metal precipitation, the formation of a protective cover layer. Chemical plating has the advantages of uniform thickness, denseness and few pinholes, and it is easy to operate without a power supply and is suitable for the internal surfaces of parts and tubes with complex structural shapes. However, the chemical coating is thin (5~12μm) and the bath is difficult to maintain. Currently, the chemical plating of Ni and Ni-P amorphous alloys is becoming more and more widely studied and applied. This coating not only has good corrosion resistance, and the hardness value is also high, the hardness of the coating after heat treatment can be close to HV1000, is a very effective anti-vibration wear protection layer.

6. Overcoating

The metal with good corrosion resistance is wrapped on the protected metal or alloy by rolling, forming a cladding layer or bimetallic layer. For example, the surface of high-strength aluminium alloy is covered with a pure aluminium layer to form an aluminium alloy sheet with an aluminium cladding layer.

7. Mechanical plating

Mechanical plating is the impact material (such as glass balls) surface treatment agent, plating accelerator, metal powder and parts together into the plating with the roller, and through the roller rolling kinetic energy, the metal powder cold pressed to the surface of the parts to form a coating. If a metal powder is used, a single coating is obtained; if an alloy powder is used, an alloy coating is obtained; if two metal powders are added at the same time, a mixed coating is obtained; if a metal

5.化学镀

利用氧化还原反应，使盐溶液中的金属离子在被保护金属上析出，形成保护性覆盖层。化学镀层具有厚度均匀，致密、针孔少的优点，而且不用电源，操作简单，适于结构形状较复杂的零件和管子的内表面。但是化学镀层较薄（5~12μm），槽液维护较困难。目前，化学镀Ni和Ni-P非晶态合金研究和应用日臻广泛。这种镀层不仅具有良好的耐蚀性，且硬度值也较高，热处理后的镀层硬度可接近HV1000，是一种很有效的抗微振磨损防护层。

6.包镀

包镀是将耐蚀性良好的金属通过碾压的方法包覆在被保护的金属或合金上，形成包覆层或双金属层。如高强度铝合金表面包覆纯铝层，形成有包铝层的铝合金板材。

7.机械镀

机械镀是把冲击料（如玻璃球）表面处理剂、镀覆促进剂、金属粉和零件一起放入镀覆用的滚筒中，并通过滚筒滚动时产生的动能，把金属粉冷压到零件表面上形成镀层。若用一种金属粉，得到单一镀层；若用合金粉末，可得合金镀层；若同时加入两种金属粉末，可

powder is added first, after a certain time of coating, another metal powder is added, a multi-layer coating is obtained. Surface treatment agents and plating promoters keep the surface of the part clean and free of oxides, and control the plating speed.

得到混合镀层；若先加入一种金属粉，镀覆一定时间后，再加另一种金属粉，则可获得多层镀层。表面处理剂和镀覆促进剂可使零件表面保持无氧化物的清洁状态，并控制镀覆速度。

The advantages of mechanical plating are uniform layer thickness, no hydrogen embrittlement, room temperature operation, low energy consumption and low cost. Metals suitable for mechanical plating include Zn, Cd, Sn, Al, Cu and other soft metals. The lergth of the parts suitable for mechanical plating should not exceed 150mm and the mass should not exceed 0.5kg. such as screws, nuts, gaskets, iron needles, chains, reeds and other small parts. The mechanical plating process is particularly suitable for high strength steels and springs which are sensitive to hydrogen embrittlement. However, the holes in the parts must not be too small or too deep, and the shape of the parts must not be such that they jam each other in the drum.

机械镀的优点是镀层厚度均匀、无氢脆、室温操作、耗能少、成本低等。适于机械镀的金属包括Zn、Cd、Sn、Al、Cu等软金属。适于机械镀的零件长度一般不宜超过150mm，质量不超过0.5kg。例如螺钉、螺帽、垫片、铁针、铁链、簧片等小零件。机械镀工艺特别适于对氢脆敏感的高强度钢和弹簧。但零件上孔不能太小太深，零件外形不得使其在滚筒中互相卡死。

8. Vacuum plating

8.真空镀

Vacuum plating includes vacuum vapour plating, sputtering plating and ion plating, which are all methods of plating in a vacuum. They are non-polluting, non-hydrogen brittle, suitable for a wide range of substrates, both metallic and non-metallic, and simple. However, they have the disadvantages of thin plating, expensive equipment and limited size of the plated parts.

真空镀包括真空蒸镀、溅射镀和离子镀，它们都是在真空中镀覆的工艺方法。它们具有无污染、无氢脆、适于金属和非金属多种基材且工艺简单等特点。但有镀层薄、设备贵、镀件尺寸受限的缺点。

Vacuum vapour deposition is a process in which the plating material is heated in a vacuum (below 10^{-2} Pa), evaporated or sublimated and deposited on the plated part. Heating methods include resistance heating. Electron beam heating. High frequency induction heating, arc discharge or laser heating are commonly

真空蒸镀是在真空（10^{-2} Pa以下）中将镀料加热，使其蒸发或升华，并沉积在镀件上的工艺方法。加热方法包括电阻加热、电子束加热、高频感应加热，电弧放电或激光加热

used. Vacuum vapour deposition can be used to coat protective or decorative coatings such as Al, brass, Cd, Zn, etc; metal or metal compound coatings for electronic components such as resistors and capacitors; metal compound coatings for optical components such as lenses.

等，常用的是电阻加热。真空蒸镀可用来镀覆Al、黄铜、Cd、Zn等防护或装饰性镀层；电阻、电容等电子元件用的金属或金属化合物镀层；镜头等光学元件用的金属化合物镀层。

Sputtering is the process of bombarding a target with charge energy particles (usually positive gas ions) so that certain atoms escape from the surface of the target and are sputtered onto the parts near the target to form a coating. The vacuum level in the sputtering chamber (0.1~1 Pa) is lower than that of vacuum vapour deposition. Sputtering is divided into cathode sputtering, magnetron sputtering, plasma sputtering, high frequency sputtering, reactive sputtering, aspirant sputtering, bias sputtering and asymmetric AC sputtering.

溅射镀是用荷能粒子（通常为气体正离子）轰击靶材，使靶材表面某些原子逸出，溅射到靶材附近的零件上形成镀层。溅射室内的真空度（0.1~1 Pa）比真空蒸镀法低。溅射镀分为阴极溅射、磁控溅射、等离子溅射、高频溅射、反应溅射、吸气剂溅射、偏压溅射和非对称交流溅射等。

The most important feature of sputtering is the ability to coat the exact same composition as the target material, which makes it particularly suitable for the coating of high melting point metals, alloys, semiconductors and various compounds. The disadvantage is the high temperature rise of the plated parts (150~500℃). Currently sputter plating is mainly used to prepare the various thin films required for electronic components, but it can also be used to coat TiN imitations of gold and to coat hard coatings such as TiN and TiC on cutting tools to improve their service life.

溅射镀的最大特点是能镀覆与靶材成分完全相同的镀层，因此特别适用高熔点金属、合金、半导体和各类化合物的镀覆。缺点是镀件温升较高（150~500℃）。目前溅射镀主要用于制备电子元器件上所需的各种薄膜，也可用来镀覆TiN仿金镀层，以及在切削刀具上镀覆TiN、TiC等硬质镀层，以提高其使用寿命。

Ion plating requires first pumping the vacuum chamber to a vacuum of 10^{-3}Pa, and then introducing inert gas (usually argon) from the needle valve to maintain the vacuum at 0.1~1 Pa; Then connect the negative high pressure to generate glow discharge between the evaporation source (anode) and the plating

离子镀需要首先将真空室抽至10^{-3}Pa的真空度，再从针形阀通入惰性气体（通常为氩气），使真空度保持在0.1~1 Pa；然后接通负高压，使蒸发源（阳极）和镀件（阴极）之间产生辉光放

piece (cathode), and establish the plasma area and cathode area of low pressure gas discharge; Then the evaporation source is electrified to vaporize the plating metal and enter the plasma zone. Under the impact of high-speed electron bombardment, part of the metal gas is ionized, and under the action of electric field, it is accelerated to shoot on the surface of the plating piece to form the coating. Figure 7-6 shows the working principle of cathodic arc ion plating.

电，建立起低气压气体放电的等离子区和阴极区；最后将蒸发源通电，使镀料金属气化并进入等离子区，金属气体在高速电子轰击下，一部分被电离，并在电场作用下被加速射在镀件表面而形成镀层。图7-6所示为阴极电弧离子镀工作原理示意。

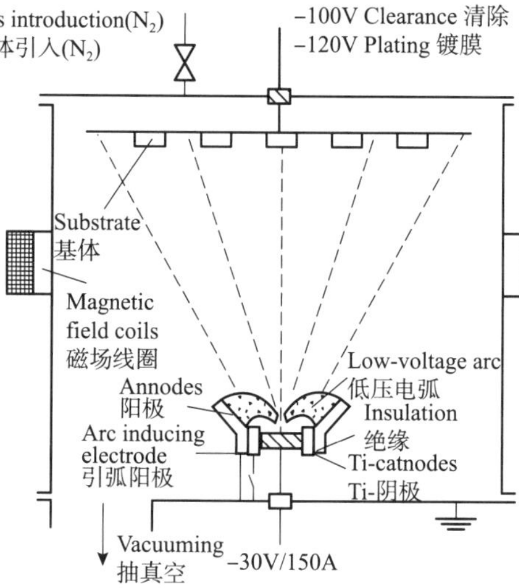

Figure 7-6　Schematic diagram of the working principle of cathodic arc ion plating

图7-6　阴极电弧离子镀工作原理示意

The main characteristics of ion plating are good coating performance and high coating adhesion. The good coating performance is due to the fact that the plating material is ionized and becomes positive ions, while the plating piece has negative charges, and the vaporized particles of the plating material collide with each other and are scattered in the space around the plating piece (cathode), so it can be coated on all surfaces of the part, while vacuum evaporation and sputtering can only be coated on the surface directly irradiated by the evaporation source or sputtering source. The

离子镀的主要特点是绕镀性好和镀层附着力高。绕镀性好是由于镀料被离子化而成为正离子，而镀件带负电荷，镀料的气化粒子相互碰撞，分散在镀件（阴极）周围空间，因此能镀在零件的所有表面上，而真空蒸镀和溅射镀则只能镀在蒸发源或溅射源直射的表面。附着力高的原因是已电离的惰性气体不断地对镀件进行

reason for the high adhesion is that the ionized inert gas continuously bombards the plated parts, so that the surface of the plated parts can be purified. In addition, the requirements of ion plating on the cleaning of parts before plating are not very strict. Ion plating can be used for decoration (such as TiN imitation gold coating), surface hardening, metal or compound coating for electronic components and optical coating.

轰击，使镀件表面得以净化。另外，离子镀对零件镀前清理的要求也不甚严格。离子镀可用于装饰（如TiN仿金镀层）、表面硬化，电子元器件用的金属或化合物镀层以及光学用镀层等方面。

9. Ion injection

9.离子注入

Ion implantation technology is used for hardening and corrosion prevention of metal surface, which can improve the wear resistance and corrosion resistance of materials. In the past, this technology was only used for semiconductor doping. Since J.R. Conrad first proposed plasma-based ion implantation technology in 1989, new surface modification technologies such as metal plasma-based ion implantation, plasma implantation, plasma-based ion implantation and mixing have emerged.

离子注入技术用于金属表面硬化和防蚀，可提高材料的耐磨、抗蚀性。以往，此项技术仅被用于半导体掺杂。自1989年J.R. Conrad首次提出等离子体基离子注入技术以来，相继出现了金属等离子体基离子注入、等离子体注渗、等离子体基离子注入混合等新兴表面改性技术。

The surface modification layer of any composition can be obtained on the metal surface by ion implantation, and the modified layer is amorphous. The thickness of single ion implantation layer is very thin, usually less than 300 nm. The plasma-based ion implantation hybrid technology can obtain several microns or even thicker surface modification layer.

利用离子注入可在金属表面获得任意成分的表面改性层，改性层为非晶结构。单一的离子注入层厚度极薄，通常小于300 nm。而等离子体基离子注入混合技术，可获得数微米甚至更厚的表面改性层。

7.4 Non-metallic protective layers

7.4 非金属保护层

Non-metallic coverings, also called non-metallic coatings, include inorganic and organic coatings.

非金属覆盖层也称非金属涂层，包括无机涂层和有机涂层。

1. Inorganic coatings

1.无机涂层

(1) Enamel coatings [5]

（1）搪瓷涂层 [5]

Enamel, also known as enamel, is a glass-like

搪瓷又称珐琅，是类似玻

substance. Enamel coating is made by adding silicate of potassium, sodium, calcium, aluminum and other metals into flux such as borax and spraying on the metal surface for sintering. The corrosion resistance of enamel can be improved by properly increasing the SiO_2 content (for example,>60%). Since the enamel coating has no micropores and cracks, it can completely separate the steel substrate from the medium and play a protective role.

璃的物质。搪瓷涂层是将钾、钠、钙、铝等金属的硅酸盐，加入硼砂等熔剂中，喷涂在金属表面上烧结而成。将其中的SiO_2成分适当增加（如>60%），可提高搪瓷的耐蚀性。由于搪瓷涂层没有微孔和裂缝，所以能将钢材基体与介质完全隔开，起到防护作用。

(2) Silicate cement coating

（2）硅酸盐水泥涂层

Silicate cement slurry is applied to the inner wall of large steel pipes and cured to form a coating. As it is inexpensive and easy to use, and the expansion coefficient is close to that of steel, it is not easy to crack due to temperature changes, so it is widely used for steel and cast iron pipelines in water and soil, with good corrosion protection. The coating thickness is 0.5 to 2.5 cm and has a service life of up to 60 years.

将硅酸盐水泥浆料涂覆在大型钢管内壁，固化后形成涂层。由于其价格低廉，使用方便，而且膨胀系数与钢接近，不易因温度变化而开裂，因此广泛用于水和土壤中的钢管和铸铁管线，防蚀效果良好。涂层厚度为0.5~2.5 cm，使用寿命最高可达到60年。

(3) Chemical conversion film

（3）化学转化膜

Chemical conversion film is the reaction between the atoms of the metal surface and the anions in the medium to produce a film with good adhesion and excellent corrosion resistance on the metal surface. The main chemical conversion films used for corrosion protection are the following:

化学转化膜是金属表层原子与介质中的阴离子反应，在金属表面生成附着性好，耐蚀性优良的薄膜。用于防蚀的化学转化膜主要有以下几种：

① Chromate passivation film. Zinc, cadmium, tin and other metals or coatings are passivated in solutions containing chromic acid, chromate or dichromate, and the metal surface forms a passivation film composed of trivalent chromium and hexavalent chromium compounds [such as $Cr(OH)_3 \cdot CrOH \cdot CrO_4$] . Thickness generally 0.01~0.15μm. The color of chromate can change from

①铬酸盐钝化膜。锌、镉、锡等金属或镀层在含有铬酸、铬酸盐或重铬酸盐溶液中进行钝化处理，其金属表面形成由三价铬和六价铬化合物[如$Cr(OH)_3 \cdot CrOH \cdot CrO_4$]组成的钝化膜，厚度为0.01~0.15μm。

colorless and transparent to golden yellow, green, brown or even black with different thickness. Figure 7-7 below shows the micro-morphology of trivalent chromium conversion film on AZ91D magnesium alloy.

由于厚度不同，铬酸盐的颜色可从无色透明转变为金黄色、绿色、褐色，甚至黑色。图7-7所示为AZ91D镁合金上的三价铬转化膜微观形貌。

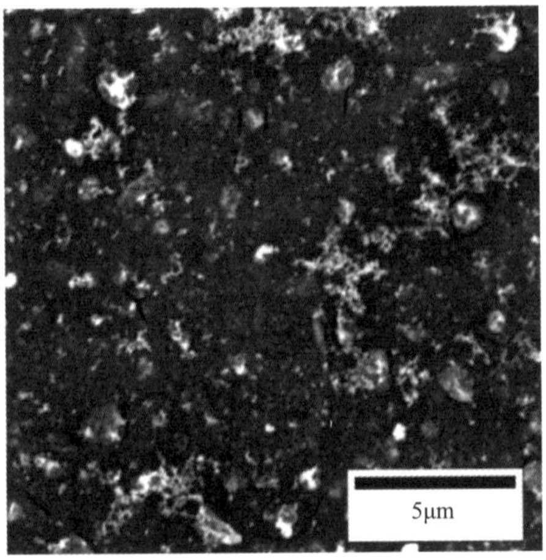

Figure 7-7 Below shows the microscopic morphology of the trivalent chromium conversion film on AZ91D magnesium alloy

图7-7 AZ91D镁合金上的三价铬转化膜微观形貌

In chromate passivation films, the insoluble trivalent chromium compounds form the skeleton of the film, giving it a certain thickness and mechanical strength; the hexavalent chromium compounds are dispersed within the film and act as a filler. When the membrane is lightly damaged, the hexavalent chromium dissolves from the membrane into the condensate, passivating the exposed metal surface and playing a role in repairing the passivated membrane. Therefore, the effective corrosion protection period of chromate films depends mainly on the rate of dissolution of hexavalent chromium in the film.

在铬酸盐钝化膜中，不溶性的三价铬化合物构成膜的骨架，使膜具有一定的厚度和机械强度；六价铬化合物则分散在膜内部，起填充作用。当膜受到轻度损伤时，六价铬会从膜中溶入凝结水中，使露出的金属表面钝化，起到修补钝化膜的作用。因此，铬酸盐膜的有效防蚀期主要取决于膜中六价铬的溶出速率。

② Phosphate film. Phosphate film, also known as phosphate film, is the steel parts in a solution containing phosphoric acid and soluble phosphate, with the help

②磷化膜。磷化膜又称磷酸盐膜，是钢铁零件在含磷酸和可溶性磷酸盐的溶液

of chemical reactions on the metal surface to generate insoluble, well-adhered to the protective film. This film-forming process is often referred to as phosphating or phosphate treatment. Phosphating process is divided into high temperature (90~98 ℃), medium temperature (50~70 ℃) and room temperature phosphating. Room temperature phosphating, also known as cold phosphating, that is, at room temperature (15~35 ℃). With different phosphating solution, the most widely used in industry there are three phosphate film: iron phosphate film, manganese phosphate film and zinc phosphate film. Phosphate film thickness is thin, generally only 5 to 6 pm. because of the larger pores, corrosion resistance is poor, so the phosphate must be used after the potassium dichromate solution passivation or oil immersion for sealing treatment. This treatment of the metal surface in the atmosphere has a very high corrosion resistance. In addition, the phosphate film is often used as a base layer of spray paint, that is, phosphate directly after painting, can greatly improve the adhesion of paint. The following Figure 7-8 for AZ91D magnesium alloy on the formation of phosphate film microscopic morphology.

中，借助化学反应在金属表面生成不可溶的、附着性良好的保护膜。这种成膜过程通常称为磷化或磷酸盐处理。磷化工艺分为高温（90~98℃）、中温（50 ~70℃）和常温磷化。常温磷化又称冷磷化，即在室温（15~35℃）下进行。由于磷化液不同，工业上最广泛应用的有三种磷化膜：磷酸铁膜、磷酸锰膜和磷酸锌膜。磷化膜厚度较薄，一般仅5~6pm。因孔隙较大，耐蚀性较差，因此磷化后必须用重铬酸钾溶液钝化或浸油进行封闭处理。这样处理的金属表面在大气中有很高的耐蚀性。另外，磷化膜经常作为喷漆底层，即磷化后直接涂漆，可大大提高油漆的附着力。图7-8所示为AZ91D镁合金上形成磷化膜微观形貌。

Figure 7-8　Microstructure of Phosphate Coating Formed on AZ91D Magnesium Alloy
图7-8　AZ91D镁合金上形成磷化膜微观形貌

③ Chemical oxide film of steel. A protective oxide film (Fe_3O_4) can be formed on the surface of steel by chemical method. The alkaline oxidation method can produce a blue-black protective film on the surface of steel, so it is also called bluing. Alkaline blueing is to immerse iron and steel products in a mixed solution containing NaOH, $NaNO_2$ or $NaNO_3$, and conduct oxidation treatment at about 140 ℃ to obtain 0.6~0.8 μm thick oxide film. In addition to alkaline bluing, there are also acid room temperature blackening and other iron and steel oxidation processes. The corrosion resistance of steel chemical oxide film is poor, and it is usually necessary to coat oil or wax to have good atmospheric corrosion resistance.

③钢铁的化学氧化膜。利用化学方法可在钢铁表面生成一层保护性氧化膜（Fe_3O_4）。碱性氧化法可使钢铁表面生成蓝黑色保护膜，故又称发蓝。碱性发蓝是将钢铁制品浸入含NaOH、$NaNO_2$或$NaNO_3$的混合溶液中，在约140℃下进行氧化处理，得到0.6~0.8 μm厚的氧化膜。除碱性发蓝外，还有酸性常温发黑等钢铁氧化处理法。钢铁化学氧化膜的耐蚀性较差，通常要涂油或涂蜡才有良好的耐大气腐蚀作用。

④ Anodic oxide film of aluminium and aluminium alloys. Aluminium and aluminium alloys are anodised in sulphuric, chromic or oxalic acid solutions, resulting in porous oxide films of several tens to hundreds of microns in thickness. After further sealing treatment or colouring, a protective film with good corrosion and wear resistance can be obtained.

④铝及铝合金的阳极氧化膜。铝及铝合金在硫酸、铬酸或草酸溶液中进行阳极氧化处理，可得到几十至几百微米厚的多孔氧化膜。经进一步封闭处理或着色后，可得到耐蚀和耐磨性很好的保护膜。

2. Organic coatings

2.有机涂层

(1) Paint coating [6]

（1）涂料涂层[6]

Paint coating is also called paint coating, because paint is commonly known as paint. In the early days, oil was used as the main raw material for paints, but now various organic synthetic resins are widely used. Paint coatings are therefore divided into two categories: oil-based coatings (film-forming substances are dry oils) and resin-based coatings. Commonly used organic paints are grease paints, alkyd resin paints, phenolic resin paints. perchloroethylene paint, nitrocellulose paint, asphalt paint, epoxy resin paint, polyurethane paint, silicone heat

涂料涂层也称油漆涂层，因为涂料俗称为油漆。早期油漆以油为主要原料，现在各种有机合成树脂得到广泛应用。油漆涂料分为油基涂料（成膜物质为干性油类）和树脂基涂料两大类。常用的有机涂料有油脂漆、醇酸树脂漆、酚醛树脂漆、过氯乙烯漆、硝基漆、沥青漆、环氧树脂漆、聚氨酯

resistant paint, etc. Figure 7-9 below shows epoxy resin coatings for sewage pipes.

漆、有机硅耐热漆等。图 7-9 所示为污水管道用环氧树脂涂料。

Figure 7-9 Epoxy resin coatings for sewage pipes [6]
图 7-9 污水管道用环氧树脂涂料 [6]

Paints of a certain viscosity are applied to clean metal surfaces by various methods and after drying and curing, paint films of different thicknesses are obtained. In addition to separating the metal from the corrosive medium, they may also passivate the metal with the help of certain pigments in the paint (e.g. lead dans, zinc chromate, etc.); or use the cathodic protection of steel by zinc powder in zinc-rich paint to improve the protective properties.

将一定黏度的涂料用各种方法涂覆在清洁的金属表面，干燥固化后，可得到不同厚度的漆膜。它们除了把金属与腐蚀介质隔开外，还可能借助于涂料中的某些颜料（如铅丹、铬酸锌等）使金属钝化；或者利用富锌涂料中锌粉对钢铁的阴极保护作用，提高防护性能。

(2) Plastic coating

（2）塑料涂层

In addition to the plastic powder sprayed on the metal surface, heated and cured to form a plastic coating (spraying method), with the lamination method of plastic film directly bonded to the metal surface, can also form a plastic coating. Organic coated metal sheets have developed rapidly in recent years, not only to improve corrosion resistance, but also to make a variety of colours and patterns (colour coated steel sheets), which are extremely versatile. Commonly used plastic films are:

除了将塑料粉末喷涂在金属表面，经加热固化形成塑料涂层（喷塑法）外，用层压法将塑料薄膜直接黏结在金属表面，也可形成塑料涂层。有机涂层金属板近年来发展很快，不仅可提高耐蚀性，而且可制成各种颜色，花纹的板材（彩色涂层钢板），用途极为广泛。

acrylic resin film, polyvinyl chloride film, polyethylene film and polyfluoroethylene film, etc.

常用的塑料薄膜包括：丙烯酸树脂薄膜、聚氯乙烯薄膜、聚乙烯薄膜和聚氟乙烯薄膜等。

(3) Hard rubber cover layer

（3）硬橡皮覆盖层

In the rubber mixed with a mass fraction of 30%~50% of the sulphur vulcanization, can be made of hard rubber. It has the characteristics of acid and alkali corrosion resistance, so it can be used to cover the surface of steel or other metals. Many chemical equipment using hard rubber for lining, its main disadvantage is that after heating will be aged brittle, can only be used at 50 ℃ or below.

在橡胶中混入质量分数为30%~50%的硫进行硫化，可制成硬橡皮。它具有耐酸、碱腐蚀的特性，故可用于覆盖钢铁或其他金属表面。许多化工设备采用硬橡皮做衬里，其主要缺点是加热后会老化变脆，只能在50℃以下使用。

With the continuous development of surface engineering technology, a variety of surface protection coverings and manufacturing methods have emerged one after another. Such as high-temperature fusion burning, laser coating and other processes have begun to be used for the surface protection of the combustion chamber of internal combustion engines. At present, it has been possible to prepare a surface protection layer of suitable composition according to the service environment and bearing conditions of the workpiece. However, how to ensure sufficient bonding between the protective layer and the substrate is often the key to evaluating the success or failure of the process. For example, diamond-like carbon films (DLC) are chemically stable and have good corrosion resistance in most corrosive media, but most preparation processes have difficulty in obtaining good film-substrate bonding. It can be seen that with the continuous improvement of the surface modification process, the corrosion phenomenon will certainly be more effectively controlled.

随着表面工程技术的不断发展，相继出现了多种表面保护覆盖层及制造方法。如高温熔烧、激光涂覆等工艺已开始用于内燃机燃烧室的表面防护。目前，已经可依据工件的服役环境及承载条件，制备出适宜成分的表面防护层。然而，如何保证防护层与基体间有足够的结合力是评定工艺方法成败的关键。如类金刚石碳膜（DLC）的化学稳定性极强，在绝大多数的腐蚀介质中都具有良好的耐蚀性，但多数制备工艺都难以获得良好的膜-基结合力。可见，随着表面改性工艺的不断完善，腐蚀现象必将会得到更有效的控制。

7.5 Corrosion inhibitors

7.5.1 Overview

缓蚀剂相关知识

Corrosion inhibitor is a number of small amounts added to the corrosive medium can significantly slow or stop the corrosion of metal substances, also known as corrosion inhibitors (Corrosion Inhibitor). Corrosion inhibitor protection of metals has the advantage of low dosage, quick results, low cost, easy to use, has been widely used in machinery, chemical industry, metallurgy, energy and many other sectors. For example, in the pickling process, sulfuric acid and hydrochloric acid to remove the surface oxidation of steel at the same time, but also the metal itself will quickly dissolve, if you add the appropriate corrosion inhibitor, it can inhibit the metal itself excessive corrosion. The disadvantage of corrosion inhibitor protection is that it is only applicable to the limited volume of corrosive media, such as electroplating and painting before the metal pickling rust removal, chemical cleaning of the inner wall of boilers, acidification of oil and gas wells, internal combustion engines and industrial cooling water anti-corrosion treatment and metal products between the process of rust prevention and product packaging; but for drilling platforms, wharf bridges and other open systems, it is not applicable.

The amount of corrosion inhibitor added to the corrosive medium is very small, usually 0.1%~1%. Corrosion inhibitor protection effect and the type of metal materials, the nature and nature of the corrosive medium, temperature, flow conditions, etc. have a close

7.5 缓蚀剂

7.5.1 概述

缓蚀剂是一些少量加入腐蚀介质中能显著减缓或阻止金属腐蚀的物质，也称腐蚀抑制剂。缓蚀剂防护金属的优点是用量少、见效快，成本较低，使用方便，目前已广泛用于机械、化工、冶金、能源等许多部门。如在酸洗过程中，硫酸和盐酸除去钢铁表面氧化皮的同时，也会使金属本身迅速溶解，若加入适当的缓蚀剂，则可抑制金属本身过分的腐蚀。缓蚀剂保护的缺点是，它只适用于腐蚀介质的体积量有限的情况，如电镀和喷漆前金属的酸洗除锈、锅炉内壁的化学清洗、油气井的酸化、内燃机及工业冷却水的防腐蚀处理和金属产品的工序间防锈与产品包装等；但对于钻井平台、码头桥梁等敞开体系，则不适用。

腐蚀介质中缓蚀剂的加入量很少，通常为0.1%~1%。缓蚀剂的保护效果与金属材料的种类、性质和腐蚀介质的性质、温度、流动情况等有密切

relationship, that is, corrosion inhibitor protection has a strong selectivity. Such as sodium nitrite or carbonic acid cyclohexylamine corrosion inhibition of steel, copper alloy is not only ineffective, but will accelerate its corrosion. There is no universal corrosion inhibitor for various metals in different media.

关系，即缓蚀剂保护有强烈的选择性。如亚硝酸钠或碳酸环己胺对钢铁有缓蚀作用，对铜合金不但无效，反而会加速其腐蚀。目前还没有对各种金属在不同介质中普遍适用的通用缓蚀剂。

Corrosion inhibitor corrosion inhibition efficiency ε is defined as:

缓蚀剂的缓蚀效率ε定义为：

$$\varepsilon / \% = \frac{v_{corr}^0 - v_{corr}}{v_{corr}^0} \times 100 \quad (7\text{–}1)$$

Where ε for corrosion inhibition efficiency, expressed in %; and respectively, before and after adding corrosion inhibitor corrosion rate of the metal, available in any common unit, but and the unit must be consistent. If the and said before and after adding corrosion inhibitor corrosion current density, the corrosion inhibition efficiency can be written as follows:

式中，ε为缓蚀效率，用%表示；v_{corr}^0 和 v_{corr} 分别为加入缓蚀剂前、后金属的腐蚀速度，可用任何通用单位，但 v_{corr}^0 和 v_{corr} 的单位必须一致。如果用 i_{corr}^0 和 i_{corr} 表示加入缓蚀剂前、后的腐蚀电流密度，则缓蚀效率可写为：

$$\varepsilon / \% = \frac{i_{corr}^0 - i_{corr}}{i_{corr}^0} \times 100 \quad (7\text{–}2)$$

Corrosion inhibition efficiency can reach 90% or more for good corrosion inhibitor, if ε reaches 100%, it means to achieve complete protection.

缓蚀效率能达到90%以上的为良好的缓蚀剂，若ε达到100%，则意味着达到完全保护。

It can be seen that the test method of corrosion inhibition efficiency is also the test method of metal corrosion rate. Therefore, the loss of weight method, various electrochemical test methods (such as Tafel linear extrapolation method, linear polarization method, AC impedance method, etc.) to determine the corrosion inhibitor corrosion inhibition efficiency. In order to maximise the simulation of corrosion in the field, and

可见，缓蚀效率的测试方法也是金属腐蚀速度的测试方法。因此可用失重法、各种电化学测试方法（如塔菲尔直线外推法、线性极化法、交流阻抗法等）来测定缓蚀剂的缓蚀效率。为最大限度地模拟现场的腐蚀情况，并使结果易于

to make the results easily reproducible and comparable between laboratory measurements, the test conditions must be strictly defined and standardised.

重现和各实验室的测量能够相互对比，测试条件必须严格规定，并予以标准化。

7.5.2 Classification of corrosion inhibitors

7.5.2 缓蚀剂分类

1. Classification by chemical composition

(1) inorganic corrosion inhibitors

Inorganic corrosion inhibitors are nitrite, chromate, silicate, polyphosphate, aluminate, borate and arsenite, etc., which often react with the metal surface, prompting the formation of passivation film or metal salt film to prevent the anodic dissolution process[7].

1.按化学成分分类

（1）无机缓蚀剂

无机缓蚀剂包括亚硝酸盐、铬酸盐、硅酸盐、聚磷酸盐、铝酸盐、硼酸盐和亚砷酸盐等，它们与金属表面反应，促使形成钝化膜或金属盐膜，以阻止阳极溶解过程[7]。

(2) Organic corrosion inhibitors

Organic corrosion inhibitors are far more than inorganic corrosion inhibitors, organic corrosion inhibitors often occur on the metal surface physical or chemical adsorption, thereby preventing corrosive substances close to the metal surface, or block the cathode, anodic process.

（2）有机缓蚀剂

有机缓蚀剂远比无机缓蚀剂多，有机缓蚀剂在金属表面发生物理吸附或化学吸附，从而阻止腐蚀性物质接近金属表面，或者阻滞阴、阳极过程。

2. Classification according to the mechanism of ac tion

According to the corrosion inhibitor on the corrosion of the main impact of the electrode process, the corrosion inhibitor can be divided into anodic, cathodic and mixed type three. From Figure 7-10 can be clearly seen, due to the addition of corrosion inhibitors, respectively, to increase the anodic polarization, cathodic polarization or both increase, so that the corrosion current from the original I_{corr}^{0} reduced to I_{corr}.

2.按作用机理分类

根据缓蚀剂对腐蚀电极过程发生的主要影响，可把缓蚀剂分为阳极型、阴极型和混合型三种。从图7-10中可明显看出，由于缓蚀剂的加入，分别增加阳极极化、阴极极化或二者同时增大，使腐蚀电流由原来的I_{corr}^{0}减小到I_{corr}。

(1) Anodic corrosion inhibitor

Anodic corrosion inhibitor can inhibit the anodic reaction, increase the anodic polarization, so that the corrosion current decreases, and the corrosion electrode

（1）阳极型缓蚀剂

阳极型缓蚀剂能抑制阳极反应，增大阳极极化，从而使腐蚀电流下降，且使腐蚀电极

potential becomes high [Figure 7-10 (a)] .

电势变高[图7-10(a)]。

(2) Cathodic corrosion inhibitor

(2)阴极型缓蚀剂

Cathodic corrosion inhibitor can inhibit the cathodic reaction, increase cathodic polarization, reduce the corrosion current, and make the self-corrosion electrode potential becomes lower [Figure 7-10 (b)] .

阴极型缓蚀剂能抑制阴极反应，增大阴极极化，降低腐蚀电流，且使自腐蚀电极电势变低[图7-10(b)]。

(3) Mixed corrosion inhibitor

(3)混合型缓蚀剂

Mixed corrosion inhibitors on the cathodic and anodic polarization processes are inhibited, corrosion potential may not change significantly, but the corrosion current is significantly reduced [Figure 7-10 (c)] .

混合型缓蚀剂对阴、阳极极化过程都起抑制作用，腐蚀电势可能变化不大，但腐蚀电流显著降低[图7-10(c)]。

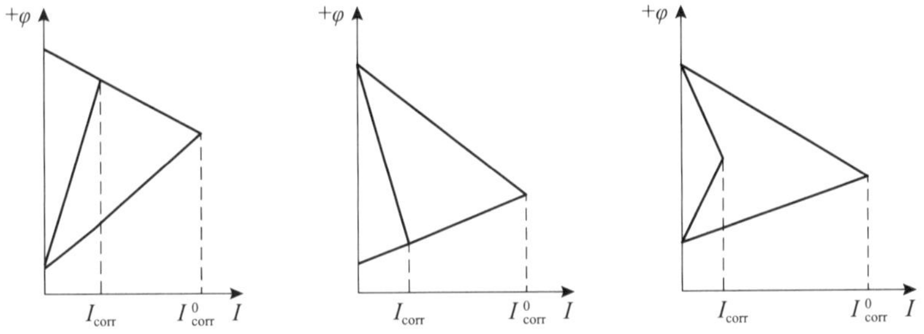

(a)Anodic corrosion inhibitor 阳极型缓蚀剂　(b)Cathodic corrosion inhibitor 阴极型缓蚀剂　(c)Mixed corrosion inhibitor 混合型缓蚀剂

Figure 7-10 Polarisation diagram for different types of corrosion inhibitors [8]

图7-10 不同类型缓蚀剂极化 [8]

3. Classification according to the characteristics of the protective film formed by corrosion inhibitors

3.按缓蚀剂形成的保护膜特征分类

(1) Oxidation (film) type corrosion inhibitor

(1)氧化(膜)型缓蚀剂

Oxidation (film) type corrosion inhibitor can make the metal surface to generate a dense and good adhesion of the oxide film, so as to inhibit the corrosion of metal. Because of its passivation, it is also known as passivation-type corrosion inhibitor, or directly known as passivator (Passivator). Steel in neutral media commonly used corrosion inhibitors such as Na, CrO_4, $NaNO_2$, NaMoOa, etc. belong to this category.

氧化(膜)型缓蚀剂能使金属表面生成致密而附着力好的氧化物膜，从而抑制金属腐蚀。因其有钝化作用，故又称钝化型缓蚀剂，或者直接称为钝化剂。钢在中性介质中常用的缓蚀剂如Na、CrO_4、$NaNO_2$、NaMoOa等都属于此类。

(2) Precipitation (film) type corrosion inhibitors

Precipitation (film) type corrosion inhibitor itself no oxidation, but they can be with the metal corrosion products (such as Fe^{2+}, Fe^{3+}) or with a common cathode reaction products (generally OH^-) to generate precipitation, can effectively repair the metal oxide film breakage, so as to play the role of corrosion inhibition, this material is called precipitation type corrosion inhibitor. For example, neutral aqueous solutions commonly used corrosion inhibitors sodium silicate (hydrolysis to produce SiO_2 gels) zinc salts [and OH^- reaction to produce Zn $(OH)_2$ precipitation film] phosphate type (and Fe^{2+} reaction to form $FePO_4$ film). Obviously they must have O_2, NO_2^- or CrO_4^{2-} and other depolarizing agents present to play a role.

（2）沉淀（膜）型缓蚀剂

沉淀（膜）型缓蚀剂本身无氧化性，但它们能与金属的腐蚀产物（如Fe^{2+}、Fe^{3+}）或与共轭阴极反应的产物（一般是OH^-）生成沉淀，能够有效地修补金属氧化膜的破损处，从而起到缓蚀作用，这种物质称为沉淀型缓蚀剂。例如，中性水溶液中常用的缓蚀剂硅酸钠（水解产生SiO_2胶凝物）锌盐［与OH^-反应产生Zn（OH）$_2$沉淀膜］磷酸盐类（与Fe^{2+}反应形成$FePO_4$膜）。显然它们必须有O_2、NO_2^-或CrO_4^{2-}等去极化剂存在时才起作用。

(3) Adsorption-type corrosion inhibitors

Adsorption-type corrosion inhibitors can form a dense adsorption layer on the metal/media interface, blocking moisture and aggressive substances close to the metal, or inhibit the metal corrosion process, to play a role in corrosion inhibition. Most of these corrosion inhibitors contain O, N, S, P polar groups or unsaturated bonds of organic compounds. Such as steel in the acid commonly used corrosion inhibitors thiourea, quinoline, acetyl alcohol and other types of derivatives, copper in neutral media commonly used corrosion inhibitors such as benzotriazole and its derivatives.

The above oxidation and precipitation type two types of corrosion inhibitors are also often together called film-type corrosion inhibitors. Because of the formation of the film, a three-dimensional new phase, so also known as three-dimensional corrosion inhibitors.

（3）吸附型缓蚀剂

吸附型缓蚀剂能在金属/介质界面上形成致密的吸附层，阻挡水分和侵蚀性物质接近金属，或者抑制金属腐蚀过程，起到缓蚀作用。这类缓蚀剂大多含有O、N、S、P的极性基团或不饱和键的有机化合物。如钢在酸中常用的缓蚀剂硫脲、喹啉、炔醇等类的衍生物，铜在中性介质中常用的缓蚀剂苯丙三氮唑及其衍生物等。

上述氧化型和沉淀型两类缓蚀剂也常合称为被膜型缓蚀剂。因为被膜的形成，产生了三维新相，故也称为三维缓蚀剂。而单纯吸附形成的缓蚀

The formation of a single molecular layer of corrosion inhibitor by adsorption alone is two-dimensional, also known as two-dimensional corrosion inhibitor. In fact, for specific high-efficiency corrosion inhibitors, the mechanism of action is quite complex, it is difficult to simply attribute to a certain type. Even the typical oxidation-type corrosion inhibitor NO_2^-, CrO_4^{2-}, etc., due to the addition of a small amount, but also must first adsorption (enrichment) in the metal surface oxide film weak points or breakage, in order to make the place passivated, play a protective role. And organic corrosion inhibitors, in addition to the initial adsorption effect, may be produced through the secondary chemical effect of three-dimensional film, to play a better role in corrosion inhibition. In addition, the above classifications and names are often crossed. For example, oxidation-type corrosion inhibitors are generally anodic; precipitation-type corrosion inhibitors are often cathodic; and adsorption-type corrosion inhibitors are likely to be mixed. But these are not absolute.

剂单分子层，是二维的，也称二维缓蚀剂。实际上，对于具体的高效缓蚀剂，其作用机理相当复杂，很难简单地归于某一类型。即使是典型的氧化型缓蚀剂NO_2^-、CrO_4^{2-}等，由于加入量很少，也必须首先吸附（富集）于金属表面氧化膜的薄弱点或破损处，才能使该处钝化，起到保护作用。而且有机缓蚀剂，除初始的吸附作用外，可能通过二次化学作用产生三维被膜，起到更好的缓蚀作用。另外，上述分类和名称是交叉的。例如，氧化型缓蚀剂一般是阳极型的；沉淀型缓蚀剂是阴极型的；而吸附型缓蚀剂很可能是混合型的。但这些并不是绝对的。

4. Classification by physical properties

4.按物理性质分类

(1) Water-soluble corrosion inhibitor

（1）水溶性缓蚀剂

Water-soluble corrosion inhibitors can be dissolved in aqueous solutions, usually as acid, brine solution and cooling water corrosion inhibitors, also used as inter-process rust prevention water, rust prevention lubrication cutting fluid.

水溶性缓蚀剂可溶于水溶液中，通常作为酸、盐水溶液及冷却水中的缓蚀剂，也用作工序间防锈水、防锈润滑切削液中。

(2) Oil-soluble corrosion inhibitor

（2）油溶性缓蚀剂

Oil-soluble corrosion inhibitors can be dissolved in mineral oil, as the main additive of rust prevention oil (grease). They are mostly organic corrosion inhibitors, and there are polar groups (pro-metal and water) and non-polar groups (pro-oil hydrocarbon chains) in the

油溶性缓蚀剂可溶于矿物油中，作为防锈油（脂）的主要添加剂。它们大多是有机缓蚀剂，分子中存在极性基团（亲金属和水）和非极性基团

molecule. Therefore, such corrosion inhibitors can occur at the metal/oil interface in the direction of adsorption, forming a tight adsorption film, blocking moisture and corrosive substances close to the metal.

（亲油的碳氢链）。因此，这类缓蚀剂可在金属/油界面发生定向吸附，构成紧密的吸附膜，阻挡水分和腐蚀性物质接近金属。

(3) Vapour phase corrosion inhibitors

Vapour phase corrosion inhibitor is a metal corrosion inhibitor that can evaporate into a gas at room temperature. If it is a solid, it must be able to sublimate, and if it is a liquid, it must have a large enough vapour pressure. This type of corrosion inhibitor must be used in a limited space, such as putting the vapour phase corrosion inhibitor into a sealed bag.

（3）气相缓蚀剂

气相缓蚀剂是在常温下能挥发成气体的金属缓蚀剂。若是固体，必须能升华；若是液体，必须具有足够大的蒸气压。这类缓蚀剂必须在有限的空间内使用，如将气相缓蚀放入密封袋中。

5. Classification according to use

According to the use of corrosion inhibitors can be divided into cooling water corrosion inhibitors, boiler corrosion inhibitors, pickling corrosion inhibitors, oil and gas well corrosion inhibitors, petrochemical corrosion inhibitors, rust inhibitors between processes, etc.

5.按用途分类

根据缓蚀剂的用途可分为冷却水缓蚀剂、锅炉缓蚀剂、酸洗缓蚀剂、油气井缓蚀剂、石油化工缓蚀剂、工序间防锈缓蚀剂等。

7.5.3 Selection and applications of corrosion inhibitors

7.5.3 缓蚀剂的选择和应用

Corrosion inhibitor is widely used in acid cleaning, rust prevention between processes, oil seal packaging, cooling water treatment, petrochemical and other aspects due to its simple equipment, convenient use, low investment and high efficiency.

采用缓蚀剂防腐蚀，由于设备简单，使用方便，投资少，收效大，因而广泛应用于酸洗、工序间防锈、油封包装、冷却水处理，石油化工等许多方面。

Corrosion inhibitors have obvious selectivity, so the appropriate corrosion inhibitor should be selected according to the metal and the medium. Metals are different, the applicable corrosion inhibitors are different, such as Fe is a transition group metal, there is a vacant

缓蚀剂有明显的选择性，因此应根据金属和介质选用合适的缓蚀剂。金属不同，适用的缓蚀剂也不同，如Fe是过渡族金属，有空位的d轨道，易

d orbit, easy to accept electrons, many with a lone pair of electron groups to produce adsorption. And copper d orbitals have been filled with electrons, so the corrosion inhibitor for steel efficient, the effect on copper is not good, or even harmful (such as amines); and copper has a special effect on the corrosion inhibitor BTA, the effect on steel is also very poor. Therefore, for a variety of metal composition of the system, such as cars, train engines, cooling water systems, including a variety of metals, so we must choose a multi-effect corrosion inhibitor or use a variety of corrosion inhibitors with the use.

接受电子，对许多带孤对电子的基团产生吸附。而铜的d轨道已填满电子，因此对钢铁高效的缓蚀剂，对铜效果不好，甚至有害（如胺类）；而对铜有特效的缓蚀剂BTA，对钢铁的效果也很差。对于多种金属组成的系统，如汽车、火车发动机的冷却水系统，包含多种金属，因此要选用多效缓蚀剂或用多种缓蚀剂配合使用。

Different media, but also to choose a different corrosion inhibitor. -General neutral water medium with inorganic corrosion inhibitors, mainly oxidation type and precipitation type. Acidic media in more organic corrosion inhibitors, mainly adsorption type. Oil media to use oil-soluble adsorption-type corrosion inhibitors. Vapour phase corrosion inhibitors must have a certain vapour pressure and sealed environment.

介质不同，也要选不同的缓蚀剂。一般中性水介质中多用无机缓蚀剂，以氧化型和沉淀型为主。酸性介质中有机缓蚀剂较多，以吸附型为主。油类介质中要选用油溶性吸附型缓蚀剂。气相缓蚀剂必须有一定的蒸汽压和密封的环境。

Corrosion inhibitors must not only choose their species, but also to determine the appropriate amount. Corrosion inhibitor dosage is too much, may change the nature of the medium (such as pH), higher costs, corrosion inhibition effect may not be good; too little to achieve the corrosion inhibition effect, for the anodic corrosion inhibitor, but also to accelerate corrosion or local corrosion. Therefore, there is usually a critical corrosion inhibitor concentration. Critical concentration varies with the corrosion system, in the selection of corrosion inhibitors must be tested to determine the appropriate amount. For the film-type corrosion inhibitor, the initial use of the dosage is often increased, sometimes

缓蚀剂不但要选择其品种，还必须确定其合适的用量。缓蚀剂用量过多，可能改变介质的性质（如pH值），成本增高，缓蚀效果也未必好；过少则达不到缓蚀作用。对于阳极型缓蚀剂，还会加速腐蚀或产生局部腐蚀。因此，通常存在临界缓蚀剂浓度。临界浓度因腐蚀体系而不同，在选用缓蚀剂时必须进行试验，以确定合适的用量。对于被膜型缓蚀剂，初始使用时要加大用量，有时比

more than ten times the normal dosage, in order to quickly generate a good protective film, which is called the "pre-film" treatment.

正常用量高出十几倍，以便快速生成完好的保护膜，这就是"预膜"处理。

The use of a corrosion inhibitor alone often does not achieve good results. A variety of corrosion inhibiting substances used in combination, often much better than the total effect when used alone, this phenomenon is called synergistic effect. The mechanism of the synergistic effect varies with the system, many are still not very clear, generally consider the cathodic and anodic type compounding. Different adsorption groups of compounding, corrosion inhibitors and solubilizing dispersant compounding, taking care of the compounding of different metals and the development of efficient multi-functional corrosion inhibitors, these are the focus of the current corrosion inhibitor research.

单独使用一种缓蚀剂达不到良好的效果。多种缓蚀物质复配使用时，常常比单独使用时的总效果好得多，这种现象称为协同效应。产生协同效应的机理因体系而不同，一般考虑阴极型和阳极型复配。不同吸附基团的复配，缓蚀剂与增溶分散剂复配、照顾不同金属复配以及高效多功能缓蚀剂的研制，这些都是目前缓蚀剂研究的重点。

The choice of corrosion inhibitors, in addition to corrosion prevention purposes, should also take into account the total effect of industrial system operation (such as cooling water systems to consider corrosion prevention. Scale prevention, sterilization, cooling efficiency and smooth operation, etc.) and environmental protection issues.

缓蚀剂的选用，除防蚀目的外，还应考虑工业系统运行的总效果（如冷却水系统要考虑防蚀、防垢、杀菌、冷却效率及运行通畅等）和环境保护等问题。

7.6 References

[1] BAI C C, ZHANG H Y. Analysis of chemical pressure vessel design and material selection [J]. Petrochemical Technology, 2020, 27(5):91~97.

[2] FANG Y. Corrosion types and influencing factors of stainless steel [J]. Contemporary chemical research, 2016, 12:11-12.

[3] LI X B. The nature of electrochemical corrosion of metals and reasonable protection [J]. Secondary School Chemistry,

7.6 参考文献

[1] 白翠翠，张红岩. 化工压力容器设计选材问题解析 [J]. 石化技术，2020，27（5）：91~97.

[2] 方毅. 不锈钢的腐蚀种类及影响因素 [J]. 当代化工研究，2016（12）：11–12.

[3] 黎小兵. 金属电化学腐蚀的本质与合理防护 [J]. 中学化学，

2021, 10:27-28.

[4] LIU J, GUO C, ZHANG X, et al. Application of cathodic protection in engineering [J]. China New Technology and New Products, 2015, 8: 44.

[5] WANG X R.Properties of chromium-free chemical conversion films on AZ91D magnesium alloy [J]. Electroplating and Coating, 2020, 39(23):1643-1647.

[6] FAN C H. Research on graphene oxide modified epoxy resin coating to enhance the anti-carbonation performance of concrete [D]. China University of Mining and Technology, 2021.

[7] WANG C, SONG X, WANG C Q, et al. Current status and outlook of environment-friendly corrosion inhibitors [J]. Synthetic Lubricating Materials, 2021, 48(4):44-48.

[8] ZHANG M, CHENG G, FANG Y, et al. Research status and development trend of corrosion inhibitors [J]. Chemical Technology and Development, 2020, 49(4):43-45.

2021(10):27–28.

[4] 刘佳，郭超，张欣，等.阴极保护在工程中的应用[J].中国新技术新产品，2015(8):44.

[5] 王向荣.AZ91D镁合金无铬化学转化膜的性能[J].电镀与涂饰，2020，39(23):1643–1647.

[6] 范春华.氧化石墨烯改性环氧树脂涂层提升混凝土抗碳化性能的研究[D].徐州：中国矿业大学，2021.

[7] 王川，宋星，王超群，等.环境友好型缓蚀剂的现状及展望[J].合成润滑材料，2021，48(4):44–48.

[8] 张明，程刚，方勇，等.缓蚀剂的研究现状及发展趋势[J].化工技术与开发，2020，49(4):43–45.

8 Application of Corrosion and Protection in Refinery Equipment

炼油装备腐蚀防护案例

8.1 Principles and significance of formulating anticorrosion strategies

8.1 防腐策略制定原则与意义

8.1.1 Significance of the development of an anti-corrosion strategy

8.1.1 防腐策略制定意义

1.Corrosive media in crude oil [1]

In addition to carbon and hydrogen, there are also sulfur, nitrogen, oxygen, chlorine, heavy metals and impurities in the crude oil. It is the non-hydrocarbon elements in the crude oil that are converted into various corrosive media under the action of high temperature, high pressure and catalyst in the oil processing process, and form a complex and volatile corrosive environment together with the chemicals added in the oil processing process.

1.原油中的腐蚀介质 [1]

原油中除存在碳、氢元素外，还存在硫、氮、氧、氯及重金属和杂质等，正是原油中存在的非碳氢元素在石油加工过程中的高温、高压、催化剂作用下转化为各种各样的腐蚀性介质，并与石油加工过程中加入的化学物质一起形成复杂多变的腐蚀环境。

(1) Inorganic salts

Inorganic salts includ: NaCl, $MgCl_2$, $CaCl_2$, etc., the content is generally $(5\sim130)\times10^{-6}$.

(1)无机盐类

无机盐类包括NaCl、$MgCl_2$、$CaCl_2$等，含量一般为（5~130）$\times10^{-6}$。

NaCl: 75%、$MgCl_2$: 15%、$CaCl_2$: 10%

$$MgCl_2+2H_2O \longrightarrow Mg(OH)_2+2HCl \quad (8\text{-}1)$$

$$CaCl_2+2H_2O \longrightarrow Ca(OH)_2+2HCl \quad (8\text{-}2)$$

The low temperature parts of the primary and secondary processing units form a composite corrosive environment of hydrochloric acid, resulting in severe corrosion of the low temperature parts. Types of corrosion include uniform corrosion and chloride ion stress corrosion cracking of stainless steel materials.

一次和二次加工装置的低温部位形成了盐酸复合腐蚀环境，造成低温部位的严重腐蚀。腐蚀类型包括均匀腐蚀和不锈钢材料的氯离子应力腐蚀开裂。

(2) Sulphides

Sulphides include Hydrogen sulfide, mercaptans, sulfur ethers, disulfides, and cyclic sulfides, etc., sulfur corrosion of refinery units throughout the primary and secondary processing units, resulting in serious corrosion of the unit, corrosion types include low temperature wet hydrogen sulfide corrosion, high temperature sulfur corrosion, even more sulfuric acid corrosion, flue gas sulfuric acid dew point corrosion.

(2)硫化物

硫化物包括硫化氢、硫醇、硫醚、二硫化物及环状硫化物等，炼油装置的硫腐蚀贯穿一次和二次加工装置，对装置产生严重腐蚀，腐蚀类型包括低温湿硫化氢腐蚀、高温硫腐蚀、连多硫酸腐蚀、烟气硫酸露点腐蚀等。

(3) Naphthenic acid

Naphthenic acid is an organic acid with a saturated ring structure presenting in petroleum, and its general formula is RCH_2COOH. Acidic compounds in petroleum include naphthenic acid, fatty acids, and phenols, and the content of naphthenic acid is the most, so it is generally referred to as naphthenic acid in petroleum, so the acid in petroleum is a very complex mixture. During the processing of crude oil, serious corrosion is caused to

(3)环烷酸

环烷酸是一种存在于石油中的含饱和环状结构的有机酸，其通式为RCH_2COOH，石油中的酸性化合物包括环烷酸、脂肪酸及酚类，而以环烷酸含量最多，故一般称石油中的酸为环烷酸，因此石油中的酸是一种非常复杂的混合物。在原油

high temperature parts of the unit such as atmospheric pressure reduction, and therefore atmospheric pressure reduction units processing highly sour crude oil should be upgraded with comprehensive materials to cope with the corrosion of naphthenic acids.

(4) Nitrides

The main ones are pyridine, pyrrole and their derivatives. Nitrogenous compounds in crude oil can be converted to ammonia after high temperature, pressure and catalyst action in secondary processing units, forming ammonia salt crystals in catalytic cracking, coking and hydrocracking effluent systems, which can severely block equipment and pipelines and cause subscale corrosion. Cyanide can also cause uniform corrosion, hydrogen bulging and stress corrosion cracking in catalytic cracking absorption, stabilisation, desorption tower tops and their condensing cooling systems.

(5) Heavy metal compounds

Heavy metal compounds in crude oil remain in heavy oil components during crude oil processing and enter the secondary processing unit, causing catalyst failure and seriously affecting the normal operation of the unit. Heavy metal V in crude oil will form low melting point compounds on the outer wall of heating furnace tubes during crude oil processing, causing molten ash corrosion of alloy components.

2. Serious trend of crude oil degradation

With the deep exploitation of oil resources and the increasing import of high-sulphur, high-acid crude oil, the trend of crude oil degradation is becoming more and more obvious. On the one hand, the deep exploitation of

加工过程中，对常减压等装置高温部位产生严重腐蚀，因而加工高酸原油的常减压装置应进行全面材料升级以应对环烷酸的腐蚀问题。

（4）氮化物

氮化物主要包括吡啶、吡咯及其衍生物。原油中的含氮化合物经过二次加工装置高温、高压和催化剂的作用后可转化为氨，在催化裂化、焦化、加氢裂化流出物系统形成氨盐结晶，严重时可堵塞设备和管线，而且会引起垢下腐蚀。氰化物还会造成催化裂化吸收、稳定、解吸塔顶及其冷凝冷却系统的均匀腐蚀、氢鼓泡和应力腐蚀开裂。

（5）重金属化合物

原油中的重金属化合物在原油加工过程中残存于重油组分中，进入二次加工装置，引起催化剂失效，严重影响装置的正常运转，原油中的重金属V在原油加工过程中会在加热炉炉管外壁形成低熔点化合物，造成合金构件的熔灰腐蚀。

2.原油劣质化趋势严重

随着石油资源的深度开采，以及进口高硫、高酸原油不断增加，原油劣质化趋势日趋明显。一方面，国内原油深

domestic crude oil, the density and acid value of crude oil are increasing, and many additives are added in the process of tertiary oil extraction, making the degree of corrosion of refining devices more serious. On the other hand, changes in the world crude oil supply market, in order to obtain better economic benefits, China's imports of high sulphur, high acid poor quality crude oil have been increasing year by year. Both aspects lead to serious corrosion of refinery units and great pressure on long term safety production.

度开采，原油的密度和酸值不断提高，而且在三次采油过程中加入许多助剂，使得炼油装置腐蚀程度更为严重。另一方面，世界原油供应市场的变化，为获取更好的经济效益，中国进口高硫、高酸劣质原油量逐年增加。这两方面都导致炼油装置腐蚀严重，长时间安全生产也面临着很大压力。

3. Some device designs cannot meet existing requirements

3.部分装置设计不能满足现有要求

Some enterprises have the problems of low design material standards and poor adaptability to raw materials. Under the circumstances of frequent changes in the quality and nature of crude oil, the acid value and sulfur content of the crude oil actually processed have exceeded the design standards, which is bound to cause serious corrosion to the unit.

部分企业存在装置设计材质标准低，对原材料适应性差的问题。在原油质量和性质变化频繁的情况下，实际加工的原油酸值和硫含量已超过设计标准，对装置造成严重腐蚀。

There are also enterprises that adapt to the changes in the nature of crude oil, but due to the limitations of technology and cost, the equipment material may not be upgraded completely, and there are still weak links. The poor adaptability to the processing of inferior crude oil causes corrosion problems in the refining unit.

也有企业为适应原油性质变化对炼油进行适应性改造，但由于技术、费用等方面的限制，设备材质可能升级并不彻底，仍存在薄弱环节，对加工劣质原油的适应性差，造成炼油装置的腐蚀问题。

4. Strengthen the process of corrosion prevention

4.炼油装置长周期安全运转的要求

With the extension of the operation period of production units, the requirements for corrosion control and corrosion management of units are becoming higher and higher. At present, some enterprises have been

随着生产装置运行周期延长，对装置腐蚀控制和腐蚀管理的要求越来越高。目前，有的企业长期加工低硫低酸原

processing low-sulfur and low-acid crude oil for a long time, and the corrosion control measures of the device are not perfect and the corrosion management level is low. After the adaptive transformation of processing high-sulfur and high-acid crude oil, although the device hardware meets the requirements of processing high-sulfur and high-acid crude oil, there are still great defects in the corrosion control technology, corrosion management and other software of the device, resulting in frequent corrosion accidents of the device.

油，装置的腐蚀控制措施不完善和腐蚀管理水平低下，当进行加工高硫、高酸原油适应性改造后，虽然装置硬件满足了加工高硫、高酸原油的要求，但装置的腐蚀控制技术、腐蚀管理等软件方面仍存在很大缺陷，使得装置腐蚀事故频发。

8.1.2 Principles of anti-corrosion strategy development

1. Strengthening raw material control

From the perspective of corrosion and protection of refinery equipment, the main corrosive media in the incoming crude oil and feed oil of the unit should be strictly controlled to ensure long-term safe operation of the unit:

(1) The crude oil is purchased uniformly through the headquarters to ensure the stability of crude oil varieties.

(2) The incoming crude oil shall be stored and refined separately. If the acid value and sulfur content of the incoming crude oil cannot meet the standard value of the equipment, the blending of crude oil can be considered, and effective measures can be taken to make it mixed evenly to avoid impact and equipment corrosion caused by uneven crude oil.

(3) The sulfur content, acid value, salt content, water content and density of crude oil need to be controlled during primary and secondary processing.

8.1.2 防腐策略制定原则

1.加强原料控制

出于对炼油厂设备腐蚀与防护角度考虑，进厂原油和进装置原料油中主要的腐蚀介质应严格控制，以确保装置长周期安全运转：

（1）通过总部统一进行原油收购，保证原油品种稳定。

（2）进厂原油要做到“分贮分炼”，如果再次进厂原油酸值和硫含量不能满足设备标准值时，可以考虑原油掺混，采取有效措施使其混合均匀，避免原油不均匀造成冲击与设备腐蚀。

（3）在进行一次加工和二次加工时都需要对原油的硫含量和酸值、原油含盐、含水、密度等进行控制。

2. Strengthen the process of corrosion prevention

Process corrosion prevention refers to the process corrosion prevention measures mainly including electric desalting, neutralizer injection, water injection, corrosion inhibitor injection, etc., which are taken to solve the corrosion of equipment and pipeline at the low temperature light oil part of the fractionation system of the atmospheric and vacuum distillation unit (primary distillation tower, atmospheric tower, vacuum tower top), and catalytic cracking, coking, reforming, hydrotreating, hydrocracking and other units. Process corrosion prevention shall follow the following principles:

(1) when the nature of the crude oil changes, timely electric desalination process evaluation and chemical screening.

(2) When monitoring the pH value of condensate less than 7, should consider the dew point of the aqueous solution before the injection of neutralising agents.

(3) When the inorganic salts in the oil and gas at the top of the tower are likely to scale after condensation, consideration should be given to filling water before the dew point temperature of the aqueous solution.

(4) When the tower top cold exchange equipment for carbon steel, and the media in the corrosive media content is high, should be considered in the aqueous solution before the dew point temperature of the injection of corrosion inhibitors.

(5) In the actual production process, should consider the crude oil processing scheme and process flow changes, operating conditions fluctuations brought about by corrosion problems.

2. 强化工艺防腐

工艺防腐是指为解决常减压装置“三顶”（初馏塔顶、常压塔顶、减压塔顶）系统，以及催化裂化、焦化、重整、加氢精制、加氢裂化等装置分馏系统中低温轻油部位设备、管道腐蚀所采取的以电脱盐、注中和剂、注水、注缓蚀剂等为主要内容的工艺防腐蚀措施。工艺防腐应遵循以下原则：

（1）原油性质发生改变时，及时进行电脱盐工艺评定及药剂筛选。

（2）当监控冷凝水的pH值小于7时，应考虑在水溶液露点前加注中和剂。

（3）当塔顶油气中的无机盐冷凝后有可能结垢时，应考虑在水溶液露点温度前加注水。

（4）当塔顶冷换设备为碳钢，且介质中腐蚀性介质含量较高时，应考虑在水溶液露点温度前加注缓蚀剂。

（5）在实际生产过程中，应考虑原油加工方案和工艺流程的变化、操作条件波动带来的腐蚀问题。

3. Reasonable selection of materials

Equipment and piping materials should be selected according to the normal operating conditions of the device corrosive media (such as sulfur, acid, chlorine, etc.)content, and fully consider the most demanding conditions may reach the maximum corrosive media content, combined with the site operating conditions (temperature, flow rate, etc.). The following principles should be followed in the selection of materials for oil refineries:

(1) Equipment selection should be based on a study of the sulphur distribution, acid distribution, flow rate and chlorine distribution of the main crude oil processed by the enterprise before selection, as the basis data for material selection.

(2) Equipment selection should refer to the device corrosion suitability assessment and RBI risk assessment, if the operating medium leakage will cause poisoning, fire and other malignant consequences, the selection of materials should be considered to raise a grade.

(3) The selection of materials for low-temperature corrosion parts is mainly carbon steel, which is mainly solved by process corrosion. If the corrosion is serious, higher materials can be considered, such as 0Cr13, 304, 316, duplex steel, Monel, etc., but the influence of anions (such as chloride ions, etc.) should be paid attention to when selecting austenitic stainless steel. Monel should consider ammonia corrosion to prevent pitting corrosion or stress corrosion cracking.

(4) For high sulphur and low acid crude oil, at high temperature mainly high temperature sulphur corrosion, the selection of materials should be based on field experience and API 581 standard, refer to McConomy

3. 合理选材

设备、管道选材要根据装置正常操作条件下腐蚀性介质（如硫、酸、氯等）含量，并充分考虑最苛刻条件下可能达到的最大腐蚀性介质含量，结合现场操作条件（温度、流速等）进行。炼油厂选材应遵循以下几个原则：

（1）设备选材前应对企业主要加工的原油品种进行硫分布、酸分布、流速和氯分布研究，以此作为选材基础数据。

（2）设备选材应参考装置腐蚀适应性评估和RBI风险评估，如果操作介质泄漏会造成人员中毒、着火等恶性后果，选材时应考虑升高一个等级。

（3）低温腐蚀部位选材以碳钢为主，主要靠工艺防腐来解决，如果腐蚀严重，可以考虑选用更高的材质，如0Cr13、304、316、双相钢、Monel等，但选用奥氏体不锈钢时要注意阴离子（如氯离子等）的影响，Monel应考虑氨腐蚀，防止出现点蚀或应力腐蚀开裂。

（4）对于高硫低酸原油，在高温下主要是高温硫腐蚀，选材要根据现场经验和API 581标准，参考McConomy

curve. For the material transport pipeline, the medium temperature <240 ℃ parts of the selection of materials to carbon steel; medium temperature between 240~288 ℃, in principle, the choice of 1Cr5Mo steel; medium temperature≥288 ℃ parts should be selected chromium content of 5% and above the alloy steel. If the pipeline is now made of carbon steel, it should be combined with the sulphur content of crude oil, in the presence of reliable data to prove that the corrosion rate <0.25mm/a, and the corrosion margin is sufficient, you can continue to use carbon steel, but must strengthen the monitoring.

曲线进行。对于物料输送管道，介质温度<240℃的部位选材以碳钢为主；介质温度在240~288℃时，原则上选用1Cr5Mo钢；介质温度≥288℃的部位应选用铬含量在5%及以上的合金钢。如现用为碳钢材质的管道，应结合原油含硫情况，在有可靠数据证明腐蚀速率<0.25mm/a，且腐蚀裕量足够时，可以继续使用碳钢，但必须加强监测。

(5) For high sulphur and high acid crude oil, the corrosion is mainly by naphthenic acid at high temperatures. Usually 316 stainless steel containing M72.2.5 has good resistance to high temperature naphthenic acid corrosion.

（5）对于高硫高酸原油，在高温下主要是环烷酸腐蚀。通常含M72.2.5的316不锈钢耐高温环烷酸腐蚀性能良好。

(6) For pro-complex equipment pipelines, if the system contains S, the selection of materials should be based on the Couper-Gorman curve. If it does not contain H_2S, then the material selection is based on the Nelson curve.

（6）对于临复设备管线，如果系统中含有S，则选材要依据Couper-Gorman曲线进行。如果不含H_2S，则依据Nelson曲线选材。

(7) In addition to the transport medium, the material of the heating protection pipe shall also be selected according to the furnace chamber temperature and fuel properties, which is resistant to high temperature oxidation, carbonization and decarburization.

（7）加热护管的选材除输送介质外，还应根据炉腔温度和燃料性质，选用耐高温氧化、碳化、脱碳的材质。

(8) For the material selection of pipes with the same operating conditions, attention should be paid to the coordination of materials on each pipe and the bones, pipe fittings, four-door instrument pipe spray and its root valve on the street root pipe. In principle, the same material or material with equivalent performance shall be

（8）对于具有同样操作条件的管道选材应注意每条管道上各材料的协调和街道根管上的骨子、管件、四门仪表管喷及其根部阀等。原则上应选取同种材料或性能相当的材料；

selected; The first valve of the branch pipe connected to the main pipe, the purging steam pipe, etc., and the front pipe shall be made of the same material or material with the same performance as the main pipe.

(9) When selecting materials, the operating conditions (such as temperature, pressure), flow rate and whether it is in the phase change part should also be considered, and special treatment should be carried out from material selection to structural design, such as increasing the flow area, reducing the flow rate, appropriately increasing the wall thickness, adding a support plate and upgrading local materials, to prevent serious local corrosion and improve the economic rationality of material selection.

(10) When the process technology is changed, it is necessary to consider whether the material of the equipment pipeline can meet the requirements of the changed process conditions (corrosive medium, temperature, flow rate, etc.). If not, the material shall be changed or upgraded.

(11) The application of material anti-corrosion technology should be strengthened, including coating, surface treatment, cathodic protection, etc. Before adopting the new technology, the material corrosion test should be carried out and its application effect in other enterprises should be investigated.

(12) If the residual wall thickness of pressure vessels and pressure pipes is less than the minimum allowable wall thickness, they shall be replaced immediately; If the annual corrosion rate of pressure vessels is greater than 0.3mm/a and that of pressure pipes is greater than 0.275mm/a, better corrosion resistant materials should be selected.

与主管相接的分管道、吹扫蒸汽管道等的第一道阀门及间前管道均应选取与主管道同种材料或性能相当的材料。

（9）在选用材料时，还应考虑介质的操作条件（如温度、压力）流速，以及是否处于相变部位等因素，从材料选择到结构设计进行特殊处理，如加大流通面积、降低流速，适当增加壁厚、增设持板及局部材料升级等，以防止局部产生严重腐蚀，同时提高选材的经济合理性。

（10）当工艺技术改动时要考虑设备管线材质能否满足改动后的工艺条件（腐蚀性介质、温度、流速等）要求，如果不能则需变进行材质更换或升级。

（11）要加强材料防腐技术的应用，包括涂料、表面处理、阴极防护等。在采用新技术前应进行材料腐蚀试验，并考察其在其他企业的应用效果。

（12）压力容器、压力管道若经检测剩余壁厚小于最小允许壁厚，应立即更换；若经检测年腐蚀速率，压力容器大于0.3mm/a，压力管道大于0.275mm/a，则应选用更好的耐腐蚀材料。

4. Strerythening the corrosion monitoring and detection [2]

Corrosion monitoring detection is the key to prevent corrosion in oil refineries, each enterprise should, according to its own situation, carry out as much corrosion monitoring and detection work as much as possible, timely discovery of equipment and pipeline corrosion problems, prevention before it happens. Equipment corrosion monitoring and detection should follow the following principles:

(1) Fixed-point thickness measurement requires thickness measurement specifications, accurate, timely data processing, and the realization of thickness measurement data computer management.

(2) It is recommended to use the corrosion online monitoring system at the low temperature corrosion site, and the specific monitoring site shall be jointly determined by the workshop, machinery, technology and other departments.

(3) The corrosion probe hanging piece can be used for corrosion monitoring at high temperature and high pressure. If conditions permit, a corrosion monitoring bypass can be added to ensure the safety of monitoring.

(4) Corrosion inspection of the shutdown device shall be carried out when the device is shut down. Before the corrosion inspection, the inspection plan shall be formulated according to the operation and characteristics of the device, and the inspection shall be carried out by professional anti-corrosion technicians.After the inspection, the inspection report and anti-corrosion suggestions shall be submitted.

(5) Before the commencement of the unit, the on-site corrosion coupon shall be carried out. The important and severely corroded tower, vessel, cold exchanger

4. 加强腐蚀监/检测[2]

腐蚀监/检测是炼油厂防腐蚀的关键，各企业要根据自身情况，尽可能多地开展各项腐蚀监/检测工作，及时发现装置设备管道存在的腐蚀问题，防患于未然。设备腐蚀监/检测应遵循以下原则：

（1）定点测厚时要求测厚规范、准确、数据处理及时，并实现测厚数据计算机管理。

（2）推荐在低温腐蚀部位采用腐蚀在线监测系统，具体监测部位应由车间、机动、技术等部门联合确定。

（3）在高温高压部位可采取腐蚀探针挂片进行腐蚀监测，有条件的可以增设腐蚀监测旁路，以保证监测的安全性。

（4）装置停工时应做好停工装置腐蚀检查工作。腐蚀检查前要根据装置运行情况和特点制定检查方案，并由专业防腐技术人员进行检查，检查完成后要提交检查报告和防腐建议。

（5）装置开工前应进行现场腐蚀挂片，各装置重要及腐蚀严重的塔、容器、冷换等部

and other parts of the unit can be put into the on-site corrosion coupon, and the corrosion rate can be measured after the unit is operated for one cycle.

位都可以挂入现场腐蚀挂片，随装置运行一个周期后再取出测量腐蚀速度。

(6) Condensate analysis after “three injections” is required to be carried out by allenterprises, and the monitoring frequency is three times a week. Enterprises with conditions can add pH online monitoring system and gradually realize automatic control dosing.

（6）“三注”后的冷凝水分析要求各企业必须进行，监测频率为每周3次。有条件的企业可增设pH在线监测系统，并逐步实现自动控制加药。

(7) Condensate or acid water analysis can also be considered for the fractionator overhead condensate cooling system of catalytic and coking units, the desorber overhead system of stable absorption, the high-pressure condensate cooling system of reaction units of hydrogenation units, and the fractionator overhead condensate cooling system of fractionation units.

（7）催化、焦化等装置分馏塔顶冷凝冷却系统，稳定吸收的解吸塔顶系统，各加氢装置反应单元的高压冷凝冷却系统，以及分馏单元分馏塔顶冷凝冷却系统等部位也可以考虑进行冷凝水或酸性水分析。

(8) In order to monitor the corrosion caused by high temperature naphthenic acid, the acid value, iron ion or iron-nickel ratio of feed oil and side oil can be tracked and monitored, and the equipment corrosion can be judged according to its change.

（8）为监测高温环烷酸造成的腐蚀，可进行原料油及侧线油酸值、铁离子或铁镍比的跟踪监测，根据其变化情况判断设备腐蚀情况。

(9) Solvents shall be analyzed for devices with solvents. For example, for amine desulfurization, the amount of acid gas absorption of amine liquid, the amount of amine aging substances, iron ions and mechanical impurities shall be analyzed; If Benfield method is used for hydrogen production unit and sulfolane solvent is used, the concentration of V_2O_5 and iron ion shall be analyzed.

（9）有溶剂的装置要分析溶剂，如胺脱硫要分析胺液的酸性气吸收量，胺老化物质的量，铁离子和机械杂质；如制氢装置采用本菲尔法用环丁砜溶剂分析缓蚀剂V_2O_5浓度和铁离子。

(10) Trace the corrosive media of the materials entering the unit, such as sulfur, acid, nitrogen, etc. of oil products, chlorine in system hydrogen, chemical water and circulating water.

（10）对进入装置的物料进行腐蚀介质的跟踪，如油品的硫、酸、氮等，系统氢气含氯、化学水和循环水等。

(11) Other corrosion detection technologies such as hydrogen flux detection, infrared imaging, dew point corrosion monitoring, nondestructive testing and so on can be adopted to improve the accuracy and efficiency of corrosion detection.

(12) It is necessary to timely track the latest corrosion monitoring and detection technology at home and abroad, and continue to promote its application in oil refining enterprises.

（11）可采取氢通量检测、红外成像、露点腐蚀监测、无损检测等其他腐蚀检测技术，提高腐蚀检测的准确性和工作效率。

（12）要及时跟踪国内外最新腐蚀监/检测新技术，并不断在炼油企业推广应用。

8.2 Corrosion and protection of atmospheric-vacuum distillation units

8.2 常减压装置的腐蚀与防护

8.2.1 Corrosion of atmospheric–vacuum distillation units

8.2.1 常减压装置的腐蚀

1. Corrosion at low temperature sites

(1) Corrosion of the HCl-H_2S-H_2O system

常减压装置相关知识

The initial distillation tower of the normal depressurization unit and the condensing cooling system at the top of the normal depressurization tower and the top of the tower, the temperature is generally around 100℃, which is low temperature corrosion, mainly caused by inorganic salts in crude oil, and belongs to the corrosion of HCl-H_2S-H_2O environmental medium. The corrosion form is manifested as a general thinning of carbon steel; pitting corrosion of Cr13; chloride stress corrosion cracking of 1Cr18Ni9Ti.

1. 低温部位的腐蚀

（1）HCl–H_2S–H_2O系统的腐蚀

常减压装置的初馏塔和常减压塔顶部及塔顶的冷凝冷却系统（图8–1），温度一般在100℃左右，为低温腐蚀，主要是由于原油中的无机盐引起的，属于HCl–H_2S–H_2O环境介质的腐蚀。腐蚀形态表现为对碳钢为普遍减薄；对Cr13为点蚀；对1Cr18Ni9Ti为氯化物应力腐蚀开裂。

$$Fe+2HCl \longrightarrow FeCl_2+H_2\uparrow \quad (8\text{-}3)$$

$$Fe+H_2S \longrightarrow FeS+H_2\uparrow \quad (8\text{-}4)$$

$$FeS+2HCl \longrightarrow FeCl_2+H_2S \quad (8\text{-}5)$$

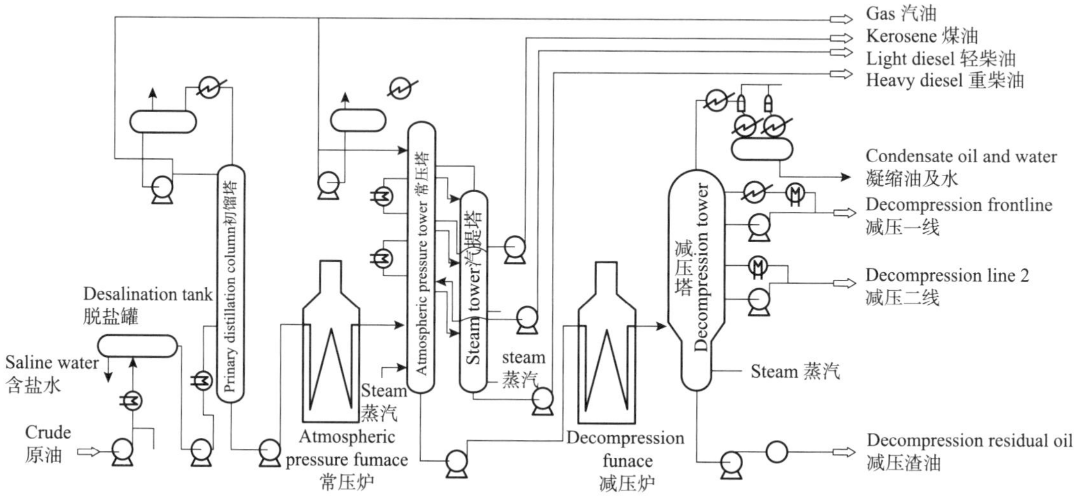

Figure 8-1 Atmospheric and reduced-pressure units: atmospheric and reduced-pressure distillation units, combined with electrodesalination and dehydration units [5]

图 8-1 常压和减压蒸馏装置，配合电脱盐和脱水装置[5]

In the absence of water, hydrogen sulfide and hydrogen chloride cause almost no corrosion to equipment. In the gas phase into the liquid phase of the site, the presence of dew water, then there will be HCl-H_2S-H_2O type of corrosion media.

(2) Low temperature flue gas dew point corrosion

Mainly occurs in the heating furnace, low temperature parts of boiler air preheater. Heating furnace, boilers with fuel containing sulfide, the common content of 1%~2.5%, SO_2 is generated completely after the sulfur is burnt, due to the presence of excess oxygen in the combustion chamber, so there is a small amount of SO_2 and oxidation further combined to form SO_3. In the usual excess air coefficient conditions, all SO_2 in about 1%~3% into SO_3. In the high temperature flue gas SO_3 does not corrode the metal, but when the flue gas temperature drops below 400℃, it will be combined with water vapour to form dilute sulphuric acid.

在没有水存在时，硫化氢和氯化氢对设备几乎没有腐蚀。在气相变液相的部位，出现露水后，则会出现HCl-H_2S-H_2O型的腐蚀介质。

（2）低温烟气的露点腐蚀

低温烟气的露点腐蚀主要发生在加热炉、锅炉空气预热器的低温部位。加热炉、锅炉用的燃料中含有硫化物，一般含量在1%~2.5%，硫燃烧后全部生成SO_2，由于燃烧室中有过量的氧气存在，所以又有少量的SO_2进一步再氧化形成SO_3。在通常的过剩空气系数条件下，全部SO_2中有1%~3%转化成SO_3。在高温烟气中的SO_3不腐蚀金属，但当烟气温度降到400℃以下时，将与水蒸气化合生成稀硫酸。

$$SO_3\uparrow + H_2O\uparrow \xrightarrow{400℃} H_2SO_4\uparrow \quad (8\text{–}6)$$

The temperature of the flue gas continues to fall. When the temperature gets down to 150~170 ℃, it has reached the dew temperature of sulfuric acid, then dilute sulfuric acid will condense to the heating surface of the furnace so that low-temperature sulfuric acid corrosion occurs. As this corrosion occurs in the dew temperature of sulfuric acid below, so also known as dew point corrosion.

烟气温度继续下降，当降至150~170℃时，已达到硫酸的结露温度，这时稀硫酸就会凝结到加热炉的受热面，从而发生低温硫酸腐蚀。由于这种腐蚀发生在硫酸的结露温度以下，所以又称露点腐蚀。

2. Corrosion in high-temperature parts

2. 高温部位的腐蚀

(1) Corrosion of high-temperature sulfide

（1）高温硫化物的腐蚀

When the wall temperature of oil refining equipment is higher than 250 ℃ and under the H_2S environment, it will be corroded by H_2S, mainly concentrated in the atmospheric furnace and outlet oil transfer line, atmospheric tower, vacuum furnace, vacuum tower, vacuum oil transfer line and other parts. In recent years, the sulfur content of crude oil has gradually increased. This kind of corrosion is characterized by the thinning of the equipment surface, which belongs to uniform corrosion.

当炼油设备壁温高于250℃且又处于H_2S环境下时，就会受到H_2S腐蚀，主要集中在常压炉及出口转油线、常压塔、减压炉、减压塔、减压转油线等部位，近年来原油的硫含量有逐步增大的趋势。这类腐蚀表现为设备表面减薄，属均匀腐蚀。

(2) Naphthenic acid corrosion

（2）环烷酸腐蚀

The corrosion of the second and third vacuum lines of atmospheric and vacuum distillation is serious. When the temperature is below 220 ℃, the corrosion of naphthenic acid is not severe, but it tends to increase gradually with the increase of temperature. When the temperature is above 280 ℃, the corrosion rate of naphthenic acid to carbon steel and low alloy steel will increase three times for every 55 ℃ rise in temperature, until 385 ℃. Since the boiling point of naphthenic acid is about 280 ℃, the corrosion is the most severe here. When it is higher than 350 ℃, it is aggravated by the

在常减压装置的减压二线、减压三线腐蚀严重，在220℃以下时，环烷酸腐蚀并不剧烈，但随着温度升高有逐步增大趋势。在280℃以上时，温度每升高55℃，环烷酸对碳钢和低合金钢的腐蚀速度就增加3倍，直到385℃时为止。由于环烷酸沸点在280℃左右，故在此时腐蚀最厉害，而当高于350℃时，又由于H_2S的影响

influence of H_2S. Later, the corrosion rate decreases with the increase of temperature.

而加剧，以后随着温度升高，腐蚀速度就下降了。

$$Fe+2RCOOH \longrightarrow Fe(RCOOH)_2+H_2\uparrow \quad (8-7)$$

$$FeS+2RCOOH \longrightarrow Fe(RCOOH)_2+H_2S \quad (8-8)$$

The corrosion is characterised by clean, smooth and scale free metal surfaces corroded by naphthenic acid. When the flow rate is high, the grooves with the same direction as the liquid flow can be generated; and sharp holes can be formed when the flow rate is low.

腐蚀特征为：环烷酸腐蚀的金属表面清洁、光滑无垢。流速高时能产生与液流同向的沟槽；流速低时能形成尖锐的孔洞。

8.2.2 Protection of atmospheric pressure-reducing devices

8.2.2 常减压装置的防护

1. One desalting and four injections [3]

1. 一脱四注[3]

(1) Desalination

（1）脱盐

Desalination is the most important link in process protection, which aims to remove the corrosive salts in crude oil. The removal of chlorides in crude oil can reduce the content of Cl^- in the tower top and reduce the corrosion. At present, deep desalting of crude oil is required. If the desalting depth is not enough, Ca and Mg salts cannot be effectively removed.

脱盐是工艺防护中最重要的一个环节，目的是去除原油中引起腐蚀的盐类。脱除原油中的氯化物减少塔顶Cl^-含量，可以减轻腐蚀。目前要求原油深度脱盐，如脱盐深度不够，则不能有效去除Ca、Mg盐类。

If the desalting is stabilized below 3 mg/L, the corrosive medium can be controlled in a lower range. The desalting effect is related to many factors such as crude oil properties (emulsion stability, specific gravity, viscosity), demulsifier, temperature, water injection and electric field strength. Generally, the desalting temperature is 100~120℃, the dosage of demulsifier is $(20\sim50)\times10^{-6}$, and water injection is 4%~10%.

如果将脱盐稳定在3mg/L以下，就能把腐蚀介质控制在一个较低范围。脱盐效果与原油性质（乳化液稳定性、比重、黏度）、破乳剂、温度、注水及电场强度等多种因素有关，一般脱盐温度为100~120℃，破乳剂用量为$(20\sim50)\times10^{-6}$，注水为4%~10%。

电脱盐罐按照布局分为水平、垂直和环形极板式，如8-2所示。

Figure 8-2 Electric desalting tank is divided into horizontal, vertical and annular plate type according to electrode layout[3]

图 8-2 水平、垂直和环形极板式电脱盐罐[3]

(2) Injection of alkali

NaCl is generally not hydrolysed and is easier to remove. The most easily hydrolysed $MgCl_2$ is the most difficult to remove. Inorganic salts will hydrolyse to produce HCl, while at the top of the atmospheric tower with water to produce hydrochloric acid, strong corrosion occurs, after the desalination also to inject alkali. The role of alkali injection (NaOH, Na_2CO_3) after the desalination of crude oil is manifested in three main ways:

① Partially control residual magnesium chloride, calcium chloride hydrolysis, so that the amount of hydrogen chloride occurrence is reduced.

（2）注碱

NaCl一般不水解，较容易脱去。最容易水解的$MgCl_2$则最难脱掉。无机盐会水解生成HCl，而在常压塔顶部与水生成盐酸，发生强烈的腐蚀，在脱盐后还要注碱。原油脱盐后注碱（NaOH、Na_2CO_3）的作用主要表现在以下三个方面：

①部分地控制残留$MgCl_2$、$CaCl_2$水解，使HCl发生量减小。

$$MgCl_2+2NaOH \longrightarrow 2NaCl+Mg(OH)_2 \quad (8\text{-}9)$$

$$MgCl_2+Na_2CO_3 \longrightarrow 2NaCl+MgCO_3 \quad (8\text{-}10)$$

② Once hydrolyzed, it can neutralize some of the generated hydrogen chloride.

②一旦水解，能中和一部分生成的HCl。

$$MgCl_2+2H_2O \longrightarrow 2HCl+Mg(OH)_2 \quad (8\text{-}11)$$

$$CaCl_2+2H_2O \longrightarrow 2HCl+Ca(OH)_2 \quad (8\text{-}12)$$

$$HCl+NaOH \longrightarrow NaCl+H_2O \quad (8\text{-}13)$$

$$2HCl+Na_2CO_3 \longrightarrow 2NaCl+H_2O+CO_2\uparrow \quad (8\text{-}14)$$

③ Alkali injection can also neutralize naphthenic acid and partial hydrogen sulfide in crude oil.

③注碱也可以中和原油中的环烷酸和部分HCl。

$$NaOH + RCOOH \longrightarrow RCOONa + H_2O \tag{8-15}$$

$$Na_2CO_3 + 2RCOOH \longrightarrow RCOONa + H_2O + CO_2\uparrow \tag{8-16}$$

$$H_2S + 2NaOH \longrightarrow Na_2S + 2H_2O \tag{8-17}$$

According to the test results of Shengli Refinery, when 18~27 g Na_2CO_3 is added to each ton of crude oil, the content of Cl^- in the overhead condensate can be reduced by 80%~85%, and the iron ion can be reduced by 60%~90%, that is, the corrosion rate can be reduced. Alkali injection is effective for neutralizing naphthenic acid, but it is harmful due to high energy consumption. In refineries with catalytic cracking units, the content of Na^+ is required to be less than 1×10^{-6}. Therefore, China Petrochemical Corporation requires to stop alkali injection.

根据胜利炼油厂的试验结果，每吨原油加入18~27g Na_2CO_3时，塔顶冷凝水中Cl^-含量可降低80%~85%，铁离子可降低60%~90%，即腐蚀速度降低。注碱中和环烷酸虽然有效，但耗能大。在有催化裂化装置的炼油厂要求Na^+含量小于1×10^{-6}，因此，石化总公司要求停止注碱。

(3) Ammonia injection

(3)注氨

Neutralize the HCl and H_2S in the overhead distillate system, and adjust the pH value of the condensate in the overhead distillate system.

中和塔顶馏出系统中的HCl和H_2S，调节塔顶馏出系统冷凝水的pH值。

$$NH_3 + HCl \longrightarrow NH_4Cl \tag{8-18}$$

$$NH_3 + H_2S \longrightarrow NH_4HS \tag{8-19}$$

When the concentration of ammonia chloride generated is high, it will precipitate in the form of solid, causing corrosion under the scale. Ammonia injection is an important measure to adjust pH value and slow down corrosion. At present, the petrochemical corporation are all injected with ammonia water, and the use of organic amines to replace ammonia water in foreign countries has better effects, because the dew point of organic amines is high, which helps to prevent dew point corrosion in the water-cooled condensation zone, and can condense with

生成的NH_4Cl在浓度较高时会以固体形式析出，造成垢下腐蚀。注氨是调节pH值减缓腐蚀的重要措施。石化公司目前都是注氨水，国外用有机胺代替氨水收到更好的效果，因为有机胺的露点高，可避免在水冷凝区发生露点腐蚀，并且能与HCl一起冷凝，有利于中和。

HCl, which is conducive to neutralization.

(4) Corrosion inhibitor injection

There are many types of corrosion inhibitors, and they should be selected appropriately. The corrosion inhibitor can form a protective film on the metal surface.

(5) Water injection

When the mixture gas of oil and water enters the volatilization line from the top of the tower, the temperature is generally above the dew point of water (water is the gas phase), and the corrosion is very slight. When the temperature gradually decreases and reaches the dew point, the water vapor begins to condense into liquid water. During the initial stages of condensation, a small amount of liquid droplets contact with a large amount of hydrogen chloride gas. The concentration of hydrogen chloride in the liquid is very high and the pH value is very low, so its corrosivity is extremely strong. With the increase of condensed water, the concentration of hydrogen chloride in liquid water gradually decreases, while the pH value gradually increases, and the corrosion also decreases. Therefore, the corrosion of tower top system is most serious in the phase change part, followed by the liquid phase part and very light in the gas phase part.

The phase change process is generally occurs at the inlet of the air cooler. The wall of the air cooler is very thin and easy to be corroded and penetrated. Moreover, the structure of air cooler is complex and expensive, so people want to move the phase change part with the most serious corrosion to the volatile line part with simple structure and thick wall. This will not only prolong the service life of the air cooler, but also make it cheaper to replace the pipes of the volatilization line. The method adopted is to inject alkaline water into the volatilization line. After the water is

(4)注缓蚀剂

缓蚀剂种类很多，应适当选择。缓蚀剂能在金属表面形成一层保护膜。

(5)注水

油水混合气体从塔顶进入挥发线时，温度一般在水的露点以上(水为气相)，腐蚀极为轻微。当温度逐渐降低，达到露点时，水汽即开始凝结成液体水。凝结之初，少量的液滴与多量的HCl气体接触，液体中的HCl浓度很高，pH值很低，因而它的腐蚀性极为强烈。随着凝结水量增加，液体水中HCl浓度逐渐降低，pH值则逐渐升高，此时腐蚀也随着减小。故塔顶系统腐蚀以相变部位最为严重，液相部位次之，气相部位很轻。

相变部位一般在空气冷却器(以下简称“空冷器”)入口处，空冷器壁很薄，容易腐蚀穿透。空冷器结构复杂，价格昂贵，因此将腐蚀最严重的相变部位移至结构简单且壁厚的挥发线部位。这样既可延长空冷器寿命，而且更换挥发线管道也比较便宜。采用的方法是在挥发线注碱性水，在挥发线注

injected into the volatilization line, the dew point will move from the air cooler to the volatilization line, thus reducing the corrosion of the air cooler. A large amount of alkaline water injected into the volatilization line can also dissolve the deposited ammonium chloride to prevent ammonium chloride from blocking. In Hdditionally, the alkaline water will neutralizes hydrogen chloride present in the system; On the other hand, diluting the concentration of hydrogen chloride in the condensate in the phase change zone can reduce the corrosion of the medium.

水后，露点部位从空冷器内移至挥发线，从而使空冷器腐蚀减轻。挥发线注入大量的碱性水，还可以溶解沉积的NH_4Cl，防止NH_4Cl堵塞；另外大量的碱性水，一方面中和HCl；另一方面冲稀相变区冷凝水中HCl浓度，可以减轻介质腐蚀。

(1) Selection of corrosion-resistant materials [4]

(1)选用耐蚀材料[4]

① The material selection of the "three roof" condensate cooling system is mainly carbon steel, which mainly depends on the improvement of process anti-corrosion measures to prevent corrosion. If the equipment is seriously corroded, the use of dual-phase steel, titanium and Monel and other high-grade materials can be considered, but the equipment material should be upgraded as a whole to avoid weak links, and attention should be paid to the galvanic corrosion caused by the contact of dissimilar metals. Non-metallic coating can also be brushed on the cooler tube bundle and air cooler tube bundle to prevent corrosion.

①"三顶"冷凝冷却系统的选材以碳钢为主，主要靠完善工艺防腐措施来抑制腐蚀，若设备腐蚀严重，可考虑使用双相钢、钛材和Monel等高等级材料，但设备材质应整体升级，避免出现薄弱环节，同时应注意异种金属接触引起的电偶腐蚀。亦可在冷却器管束、空冷器管束刷涂非金属涂层进行防腐蚀。

② The material selection for processing high temperature parts of high sulfur and low acid crude oil is mainly based on the side line sulfur content, side line temperature and the material to be used. The selected material must have a theoretical corrosion less than 0.25 mm/a. At the same time, the provisions on materials in SH/T 3129—2012 Material selection guideline for equipment and piping in units processing acid crude oils, SH/T 3096—2012 Material selection guideline for design of equipment and piping in units processing

②加工高硫低酸原油高温部位的选材，主要依据侧线硫含量、侧线温度和欲用材质，采用McConomy曲线计算理论腐蚀速率，所选材料计算的理论腐蚀速率应小于0.25mm/a。同时应考虑SH/T 3129—2012《高酸原油加工装置设备和管道设计选材导则》、SH/T 3096—2012《高硫原油加工装置设备和管道

sulfur crude oils, and the Guidelines for the Selection of Materials for Equipment and Piping used in Units for Processing High-sulfur Crude Oil should also be taken into account. Finally, the material selection experience of enterprises processing high-sulfur and low-acid crude oil and the corrosion cases on site should be considered. Based on the above aspects, the reasonable materials for processing high temperature parts of high sulfur and low acid crude oil are given.

设计选材导则》和《加工高含硫原油部分装置在用设备及管道选材指导意见(试行)》,(石化股份炼〔2022〕89号)中有关材质的规定,最后应考虑加工高硫低酸原油企业的选材经验以及现场的腐蚀案例。综上所述,给出加工高硫低酸原油高温部位的合理用材。

③ The selection of materials for processing and high-temperature parts of high-sulfur and high-acid crude oil is mainly based on the side-line acid value, side-line temperature and materials to be used, and the theoretical corrosion rate calculated by API581 is adopted. The theoretical corrosion rate calculated by the selected materials should be less than 0.25 mm/a. In addition, the provisions on materials in SH/T 3129—2012 and SH/T 3096—2012 should be considered. Finally, the material selection experience of enterprises processing high-sulfur and low-acid crude oil and on-site corrosion cases should be considered. Based on these fartors appropriate materials for the high temperature parts of high sulfur and low acid crude oil can be determined.

③加工和高硫高酸原油高温部位的选材,主要依据侧线酸值、侧线温度和欲用材质,采用API581计算理论腐蚀速率,所选材料计算的理论腐蚀速率应小于0.25mm/a。同时应考虑SH/T 3129—2012和SH/T 3096—2012中有关材质的规定,最后应考虑加工高硫低酸原油企业的选材经验以及现场的腐蚀案例。综上所述,给出加工高硫低酸原油高温部位的合理用材。

④ For crude oil that is low in sulfur buthigh in acid, the primany concern is naphthenic acid corrosion at high temperature, and the equipment material selection is basically consistent with high sulfur and high acid value.

④对于低硫高酸原油,其在高温下主要表现为环烷酸腐蚀,设备选材与高硫高酸值基本一致。

(2) Other protection methods

(2)其他防护方法

For crude oil processing of atmospheric and vacuum distillation unit, high sulfur and high acid value crude oil and low sulfur and low acid value crude oil can be mixed

常减压蒸馏装置原油加工,可采用高硫高酸值和低硫低酸值原油混炼,以降低介质

to reduce medium content and reduce corrosion.

Change the equipment structure to make the gas-liquid load evenly distributed, reduce erosion and reduce the flow rate; The pipeline and container should be able to drain the liquid, reduce dead corners and cecum, and reduce gaps.

At present, the refinery has adopted some measures to reduce corrosion at high temperature and vulnerable areas, such as reducing the diameter of low-speed oil transfer line, expanding the bending radius of high-speed oil transfer line, and changing the connection type of high-speed and low-speed oil transfer line.

8.3 Corrosion and protection of catalytic cracking units

8.3.1 Corrosion in catalytic cracking units

1. Influencing factors of high temperature gas corrosion

(1) Effect of oxygen

Oxygen is the most common high-temperature oxidant. The oxidation reaction of the metal material and the air contact surface is faster under high temperature conditions, and forms a layer of oxidation corrosion products on the metal surface.

含量减轻腐蚀。

改变设备结构，使气液负荷分布均匀，减少冲蚀，降低流速；管线和容器要能排净液体不能存水，减少死角和盲肠及减少缝隙等。

目前，炼油厂在高温易受腐蚀部位采用一些措施都有利于减轻腐蚀，如减压低速转油线扩径、高速转油线扩大弯曲半径，改变高速低速线的连接形式等。

8.3 催化裂化装置的腐蚀与防护

8.3.1 催化裂化装置的腐蚀

1.高温气体腐蚀的影响因素

（1）氧的影响

氧是最普通的高温氧化剂。金属材料与空气接触面的氧化反应，在高温条件下氧化速度更快，在金属表面形成一层氧化腐蚀产物。

催化裂化装置
相关知识

(2) Impact of SO_2

The corrosion of sulfur dioxide to metals is second only to the oxidation of air, but its destruction is sometimes more serious than oxidation. When sulfur dioxide is contained in the high-temperature gas, due to the relatively low co-melting point of iron oxide and sulfide, the molten mixture can penetrate into the grain to cause metal corrosion, thus increasing the oxidation rate.

（2）SO_2的影响

SO_2对金属的腐蚀仅次于空气的氧化作用，但其破坏性有时比氧化更严重。当高温气体中含有SO_2时，由于铁的氧化物和硫化物具有较低的共熔点，熔融后的混合物能渗入晶粒间造成金属腐蚀，因此会增加氧化速度。

(3) Effect of high temperature steam

High-temperature cooling steam of the reverse-return system is essential, and its oxidation effect is stronger than that of the air at the same temperature. This is because the loose scale layer formed in the high-temperature steam is more likely to fall off than the scale layer gap formed in the air.

（3）高温蒸汽的影响

反–再系统高温冷却蒸汽是必不可少的，其氧化作用比同温度下的空气强，这是因为在高温蒸汽中生成的疏松垢层，比在空气中形成的垢层空隙更多，更容易脱落。

(4) Effect of high temperature carbon dioxide

The carbon dioxide in the high-temperature gas of the anti-regeneration system can also accelerate the corrosion of the equipment due to the following reactions:

（4）高温CO_2的影响

反–再系统高温气体中的CO_2也能加速设备腐蚀，这是由于存在以下反应：

$$2Fe+3CO_2 \longrightarrow Fe_2O_3+3CO \quad (8\text{–}20)$$

2. Wear-resistant lining for both vessels

The operating conditions of the reaction settler and regenerator of the catalytic cracking unit are harsh. The lining of the reaction-regeneration system should not only bear high temperature (600~750 ℃), but also resist the wear of high-speed catalyst. Generally, the lining of the two devices adopts the wear-resistant and heat-insulating double-layer lining structure supported by tortoiseshell mesh. In actual production, there are often bulges, fractures and lining cracking and falling off in

2.反应沉降器和再生器的耐磨衬里

催化裂化装置的反应沉降器、再生器操作条件苛刻，反–再系统衬里不仅要承受高温（600~750℃），而且还要抵抗高速催化剂的磨损。一般情况下，两器的衬里采用龟甲网支撑的耐磨隔热双层衬里结构。在实际生产中，反–再系统双层带龟甲网衬里经常出现

different degrees in the double-layer with tortoiseshell mesh lining of the anti-regeneration system.

(1) The material selection of tortoiseshell mesh is improper. Some refineries use 18-8 steel. Because of its high coefficient of thermal expansion, bulging often occurs; Some use 1Cr13. Because of its high carbon content and poor weldability, the welded junction is easy to crack, and even the connecting claw of the tortoiseshell mesh is broken.

(2) The arrangement of insulation nails is too sparse, and the specification requires that the arrangement of insulation nails is $16/m^2$, but the intersecting line of the barrel section opening should be properly densified to reduce the occurrence of bulging and cracking.

(3) The small welding area leads to the disconnection of the cylindrical insulation nail and the end plate, and the end plate falls off.

(4) The welding quality of the insulation nail and the wall is poor, which causes the insulation nail at the bulge of the tortoiseshell mesh to be pulled out. During construction, the welding quality of the insulation nail should be checked.

3. Corrosion damage of the expansion joint

(1) Effect of purging steam

Although the expansion joint for purging steam can prevent the catalyst in the flue gas from entering the bellows and keep the bellows at a low operating temperature, the existence of water vapor increases the corrosion conditions of the bellows. Cl^- and SO_2 in flue gas combine with water vapor to form highly corrosive substances. In addition, the partial pressure of the blowback steam is almost equal to the total pressure

不同程度的鼓包、断裂和衬里开裂脱落现象。

（1）龟甲网选材不当，有的炼油厂采用18-8钢，因其热膨胀系数高，经常产生鼓包现象；有的采用1Cr13，由于其含碳量高，可焊性较差，焊口易开裂，甚至龟甲网的连接爪子也发生断裂。

（2）保温钉布置太稀，规范要求布置保温钉为16个/m^2，但在筒节开口相贯线部位则应适当加密，以减少鼓包开裂的发生。

（3）焊接面积少导致柱形保温钉和端板断开、端板脱落。

（4）保温钉与器壁的焊接质量差，造成龟甲网鼓包处保温钉被连根拔起，施工时应检查保温钉的焊接质量。

3.膨胀节的腐蚀破坏

（1）吹扫蒸汽的影响

设置吹扫蒸汽的膨胀节虽然可防止烟气中催化剂进入波纹管，保持波纹管有较低的工作温度，但由于水蒸气的存在，增加了波纹管产生腐蚀的条件。烟气中的Cl^-、SO_2等与水蒸气结合而成为腐蚀性极强的物质。另外，反吹蒸汽分压

of the flue gas, and its dew point temperature also rises, which is easy to produce condensate on the inner surface of the bellows and cause corrosion damage.

几乎和烟气总压相等，其露点温度也随之上升，很容易在波纹管内表面产生冷凝液而导致腐蚀破坏。

(2) Influence of bellows thickness

（2）波纹管厚度的影响

The thickness of the single layer of corrugated pipe is 0.5 mm. Due to its thin thickness, the water and corrosive medium infiltrated after the single layer of corrosion perforation will expand under heat, which is very easy to cause bulging deformation or corrosion perforation.

波纹管单层厚度为0.5 mm，由于太薄，单层腐蚀穿孔后渗入的水分及腐蚀介质受热膨胀，极易产生鼓包变形或腐蚀穿孔。

(3) Impact of startup and shutdown and operation fluctuation

（3）开、停工及操作波动的影响

In most cases, the damage of the corrugated expansion joint is caused during startup and shutdown. Due to the low temperature of flue gas in this room, it is very easy to produce condensate and dissolve corrosive media. Once formed, condensate is easy to concentrate, making the concentration of corrosive media multiply. According to the analysis of the liquid discharged from the condensate vent valve of the expansion joint in the horizontal section, the pH value is 4~6, and the concentration of Cl^- generally exceeds the critical stress corrosion concentration of 18-8 steel.

多数情况下，波形膨胀节损坏是在开、停工期间产生的。由于烟气温度低，极易产生冷凝液并溶解腐蚀性介质，冷凝液一旦形成则易浓缩，使腐蚀介质浓度成倍增加。从水平段膨胀节排凝放空阀放出的液体分析，pH值为4~6，Cl^-浓度一般都超过18-8钢的临界应力腐蚀浓度。

The fluctuation of startup and shutdown operation will cause expansion deformation of expansion joint. This kind of deformation sometimes changes alternately, so it will cause fatigue failure of expansion joint caused by alternating stress.

开、停工操作波动会使膨胀节产生伸缩变形。这种变形有时是交替变化的，因此会产生由交变应力而引起的膨胀节疲劳破坏。

(4) Impact of manufacturing quality

（4）制造质量的影响

The forming of bellows is a plastic process, and there is a certain residual stress after forming. The

波纹管成型是塑性加工过程，成型后存在一定的残余应

service life of the expansion joint will be affected if the welding method and welding process are improper or the weld has defects.

4. Wear of single and double acting slide valves

In recent years, the circulation, regeneration, standby slide valve and double-acting slide valve valve body of the catalytic cracking reverse-return system have been changed to the cold-wall type, and the wear-resistant layers are generally JA-95 and TA-218. The wear-resistance and strength are greatly improved, and the wear is slight after several cycles of use. However, the valve channel, valve rod, valve plate, seat ring and other parts of the slide valve are seriously worn.

力。如果焊接方式和焊接工艺不当，或焊缝存在缺陷，都将影响膨胀节的使用寿命。

4.单双动滑阀的磨损

催化裂化反－再系统的循环、再生、待生滑阀和大烟道的双动滑阀阀体近年都改为冷壁式，其耐磨层一般都选用JA-95和TA-218等，耐磨性和强度都大为提高，使用几个周期后磨损轻微。但是滑阀的阀道、阀杆、阀板、座圈等部位磨损十分严重。

8.3.2 Protection of catalytic cracking units

1. Protection against high temperature gas corrosion

In order to ensure sufficient strength and rigidity under high temperature conditions, the internals of the reaction-regeneration system must also have sufficient strength to resist the erosion and wear of the catalyst at high flow rate. Therefore, the feed nozzle, emergency steam pipe and settler in the riser and the feed leg, tie bar, wing valve and pressure measuring pipe in the regenerator are made of austenitic stainless steel. Low-alloy strength steel such as 12CrMo, Cr5Mo, etc. can be used for the lining retainer of the opening nozzle in the two reactors and the air distribution pipe in the coking tank.

2. Wear-resistant lining corrosion protection

(1) 0Cr13 material with low carbon content, good

8.3.2 催化裂化装置的防护

1.防止高温气体腐蚀

反－再系统内构件为保证在高温条件下有足够的强度和刚性，同时也必须有足够的强度以抵抗高流速催化剂的冲蚀和磨损，因此在提升管内的进料喷嘴、事故蒸汽管和沉降器以及再生器内的旋分器料腿、拉筋、翼阀、测压管等均采用奥氏体不锈钢材料。两器内开口接管的衬里护圈以及烧焦罐内的空气分布管可采用低合金强度钢，如12CrMo、Cr5Mo等。

2.耐磨衬里腐蚀防护

（1）龟甲网选材时可采用

toughness and good weldability can be used for the selection of tortoiseshell.

(2) During welding, it is necessary to ensure the welding quality, reduce the weld defects, appropriately increase the welding area, and conduct timely inspection.

3. Corrosion protection of expansion joints

① Select appropriate corrosion resistant materials. In recent years, many manufacturers have selected 316L as expansion joint in the reverse-reactor system. Its corrosion resistance is greatly improved compared with 18-8 steel. Recently, B-315 steel with better corrosion resistance has been developed in China and has been applied in the catalytic reverse-reactor system.

② Select appropriate bellows structure. Multi-layer structure is used for expansion joints of flue gas machine or other parts. If the thickness is too thin, it is often easy to corrode perforation or bulge, so it is generally appropriate to use δ= 3 layers of 1.0mm steel plate to reduce perforation and bulging damage.

③ During the startup and shutdown of the expansion joint it must be equipped with purging steam, the vent and condensate drain valve at the bottom of the expansion joint should be opened. During the startup and temperature rise of the device, do not rush to open the purging steam. Before opening the valve, drain the condensate first, and pay attention not to ventilate too fast to avoid the swelling deformation of the expansion joint.

④ The expansion joint at the inlet of the flue gas turbine should be externally insulated to increase the temperature of the outer wall of the bellows and prevent the generation of condensate.

含碳量低、韧性好、可焊性好的0Cr13材料。

（2）焊接时需要保证焊接质量，减少焊缝缺陷，适当增加焊接面积，及时进行检查。

3.膨胀节的腐蚀防护

①选择合适的耐蚀材料。近年来很多厂家在反-再系统中选择316L作膨胀节，其抗蚀性能较18-8钢大大提高，最近国内又研制出抗蚀性能更好的B-315钢，已在催化反-再系统中得到应用。

②选择合适的波纹管结构。用于烟机或其他部位的膨胀节选用多层结构。如果厚度太薄，容易腐蚀穿孔或鼓包，一般宜用δ=1.0mm钢板3层，以减少穿孔和鼓包损坏。

③必须设置吹扫蒸汽的膨胀节在开、停工期间，膨胀节底部的放空排凝阀要打开，装置开工升温过程中，不要急于打开吹扫蒸汽，开阀前要先排凝，并注意不要通气过快，以免膨胀节鼓包变形。

④对烟机入口的膨胀节宜进行外部保温以提高波纹管外壁温度，防止冷凝液产生。

8.4 Corrosion and protection of delayed coking units

8.4 延迟焦化装置的腐蚀与防护

Figure 8-3 shows the process flow of a typical delayed coking unit with one furnace and two towers.

典型的一炉两塔延迟焦化装置的工艺流程如图8-3所示。

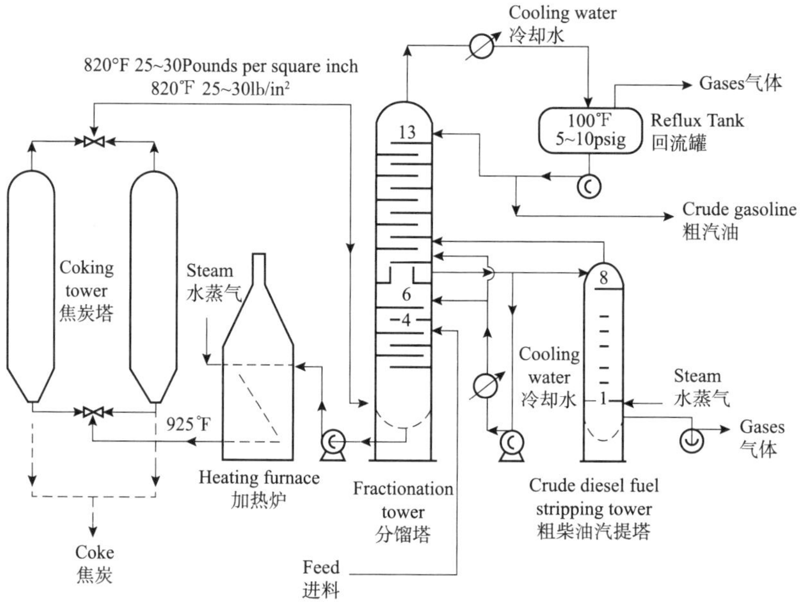

Figure 8-3 Process flow of a typical delayed coking unit with one furnace and two towers[6]

图8-3 典型的一炉两塔延迟焦化装置的工艺流程[6]

8.4.1 Corrosion of delayed coking units

8.4.1 延迟焦化装置的腐蚀

1. Corrosion analysis of raw materials

1.原料腐蚀性分析

As the raw material of coking processing is mainly residual oil, which has the characteristics of high viscosity and complex composition, and it contains a large number of inorganic salts, metal ions, asphaltenes and macromolecular non-hydrocarbon compounds. Heavy oil molecules undergo cracking and condensation reactions at high temperatures. The coke deposited in the furnace tube of coking heating furnace comes from the condensation reaction of asphaltene in heavy

由于焦化加工原料主要是渣油，它具有黏度大和组成复杂的特点，其中含有大量的无机盐、金属离子、沥青质和大分子的非烃化合物。重油分子在高温条件下发生裂解和缩合反应。焦化加热炉炉管所沉积的焦炭来自重油中沥青质的缩合反应。渣油是由饱和烃、芳

oil. Residual oil is composed of four components of saturated hydrocarbon, aromatic hydrocarbon, colloid and asphaltene, as well as non-hydrocarbon compounds such as sulfur, nitrogen, oxygen and metal. It is a colloid dispersion system, whichis a micelle formed by taking asphaltene as the core and adsorbing colloid. In the heating furnace, the heavy oil keeps heating up, and the content of colloid and asphaltene in the heavy oil is relatively high. With the occurrence of the vaporization and cracking reaction of the oil, part of the condensation reaction occurs at the same time. They are easy to be deposited on the surface of hot metal, and the generated coke core deposits on the inner surface of the furnace tube to form a scale layer, and gradually dehydrogenates and condenses to form coke. The content of S, N and other impurity atoms in the raw material is high, which is easy to decompose to produce free radicals under high temperature conditions, Thus, chain reaction is initiated and polymer is gradually formed. In addition, metal ions and dirty metal surface in raw materials play a certain catalytic role in the polymerization reaction.

2. Corrosion parts of the device [6]

(1) Fractionation tower top corrosion and salt formation

The content of chlorine and nitrogen in the feedstock vacuum residue is high. Under high temperature reaction conditions, organic and inorganic chlorides and nitrides undergo high temperature hydrolysis reaction to generate HCl and NH_3. Under appropriate conditions, part of HCl and NH_3 generate NH_4Cl (gas), and NH_4Cl is dissolved in water to form NH_4Cl aqueous solution. The aqueous solution flows back with the liquid phase in the tower

烃、胶质和沥青质4个组分以及硫、氮、氧和金属等非烃化合物组成的，是一种胶质分散体系，其分散体系是以沥青质为核心并吸附胶质形成的胶束。在加热炉中重质油不断升温，重质油中的胶质、沥青质含量较高，伴随着油品汽化和裂解反应的发生，同时发生部分缩合反应，它们容易在热金属表面沉积，生成的焦核沉积在炉管内表面形成垢层，并逐渐脱氢缩合形成焦炭，而原料中的S、N等杂质原子含量较高，在高温条件下易分解产生自由基，从而引发链反应，逐渐形成高分子聚合物。另外，原料中的金属离子和不光洁金属表面对聚合反应起到一定的催化作用。

2.装置腐蚀部位[6]

（1）分馏塔顶腐蚀及结盐

原料减压渣油中的Cl、N含量较高，在高温反应条件下，有机氯化物、无机氯化物和氮化物发生高温水解反应生成HCl和NH_3。在适当的条件下，部分HCl 和 NH_3生成NH_4Cl（气体），NH_4Cl 溶于水形成 NH_4Cl水溶液。该水溶液

to the next tray. As the temperature of the tray in the fractionator gradually rises from top to bottom, the water in the liquid phase decreases, and the salt will become saturated and crystallize, so that NH_4Cl will gradually precipitate and stay in the tray, downcomer, receiving tank, etc., forming part of the tray and downcomer to form salt. In addition to the inside of the tower, NH_4Cl crystals will also deposit on the equipment and pipeline through which the liquid phase oil flows in serious cases, resulting in salt deposition corrosion of the top circulating pump. Low overhead temperature, large cold reflux, low overhead recycle tower temperature, and large amount of water vapor in the tower entering the fractionator are all important reasons for corrosion and salt deposition in the overhead recycle section of the fractionator. At low temperature and low flow rate, it will not only be blocked by the crystallization of NH_4Cl, but also cause the corrosion under the scale and wear corrosion to intensify, resulting in the corrosion of tower top internals and subsequent equipment pipelines, heat exchangers, etc..

(2) Coke tower system corrosion

Each operation cycle of the coke tower will undergo a cold and hot cycle of the coke tower and its accessories. Each cold and hot cycle of the coke tower will cause large material and heat losses. If the top of the coke tower is poorly insulated, the oil and gas containing H_2S, HCl, NH_3 and H_2O corrosive substances in the tower will become condensate at the dew point temperature, and the swirl will wash the inner wall to cause thinning and pitting; The coke tower skirt is prone to crack due to the combined effect of fatigue stress and welding residual stress caused by long-term repeated thermal shock.

随塔内的液相回流，流到下一层塔盘，由于分馏塔内自上而下层塔盘的温度逐渐升高，液相内水分减少，盐会变得饱和并结晶析出，使NH_4Cl逐渐析出停留在塔盘、降液管、受液槽等，造成部分塔盘及降液管结盐。除塔内部外，严重时NH_4Cl晶体也会在液相油流经的设备及管道上沉积，导致顶循泵发生结盐腐蚀。塔顶温度低、冷回流大、顶循返塔温度低、进分馏塔的水蒸气量大等都是分馏塔顶循段腐蚀结盐的重要原因。在低温度、低流速部位不仅会因 NH_4Cl结晶造成堵塞，而且会造成垢下腐蚀及磨损腐蚀的加剧，造成塔顶内部件以及后续设备管线、换热器等的腐蚀。

（2）焦炭塔系统腐蚀

焦炭塔操作中每一个操作周期都要经历焦炭塔及其附件的一次冷热循环过程，每一次焦炭塔冷－热态周期性切换，物料和热量损失较大。如焦炭塔塔顶保温不好，塔内含 H_2S、HCl、NH_3、H_2O 腐蚀质的油气在露点温度变成冷凝液，旋流冲刷内壁造成减薄与坑蚀；焦炭塔长期反复热冲击的疲劳应力与焊接残余应力综合作用，焦炭塔裙座易产生裂纹。

(3) The presence of wet hydrogen sulphide corrosion parts of the device

Wet hydrogen sulfide corrosion of the unit mainly occurs in vessels, heat exchangers and pipelines with operating temperature≤120℃. Hydrogen sulfide corrosion occurred in the upper part of fractionator (T-112), separation tank, absorption tower (T-311) and other parts of gas phase recovery of coking unit. The unit cooler, air cooling inlet and outlet pipeline, air cooling and oil-water separator to fractionator overhead, compressor, absorption and stabilization system, cooler, air cooling and pipeline are all the key parts of various corrosion.

（3）装置湿硫化氢腐蚀存在部位

装置湿硫化氢腐蚀主要发生在操作温度≤120℃的容器、换热器及管线等部位。焦化装置的气相回收，分馏塔（T-112）上部、分离罐、吸收塔（T-311）等部位发生硫化氢腐蚀。装置冷却器、空冷器进出口管线、空冷器及至分馏塔塔顶油水分离器、压缩机、吸收稳定系统、冷却器、空冷器及管线均为发生各类腐蚀的重点部位。

8.4.2 Protection of delayed coking units

8.4.2 延迟焦化装置的防护

1. Corrosion control of raw materials

The sulfur mass fraction of the designed vacuum residue of the delayed coking unit is not more than 1%, and the sulfur mass fraction of the vacuum residue is about 0.98%. The monitoring of the sulfur content, chlorine content, acid value, etc. of the vacuum residue should be strengthened. When mixing catalytic oil slurry, control the solid content of oil slurry to be no more than 6g/L, and control the mixing ratio to be no more than 8%.

2. Fractionation tower top temperature and reflux control

The top temperature of fractionator shall be strictly controlled at 90~115℃, the top reflux temperature at the top shall be higher than 90 ℃, and the internal operating temperature at the top shall be controlled at 28 ℃ higher than the water dew point temperature. Pay attention to

1.原料腐蚀性控制

延迟焦化装置设计减压渣油设防值含硫质量分数不大于1%，减压渣油现含硫质量分数为0.98%左右，加强对减压渣油硫含量、氯含量、酸值等的监控。掺炼催化油浆时，控制油浆固含量不大于6g/L，控制掺炼比不大于8%。

2.分馏塔顶温度及回流控制

严格控制分馏塔顶温度在90～115℃，塔顶回流温度高于90℃，控制塔顶内部操作温度高于水露点温度28℃。注意分馏塔顶积盐及腐蚀问题，确

the problem of salt accumulation and corrosion on the top of the fractionator, ensure that the corrosion inhibitor injection on the top of the fractionator is normal and the acid water washing is normal, strengthen the monitoring of low temperature corrosion on the top of the fractionator, regular corrosion monitoring of equipment and pipelines, and the inspection of high temperature parts, and strengthen the corrosion monitoring of weak parts and the supervision of the implementation effect of industrial anti-corrosion measures.

保分馏塔顶缓蚀剂注入正常及酸性水水洗正常，加强分馏塔顶部低温腐蚀监控及设备管线的定期腐蚀监测和高温部位检查，同时加强薄弱部位的腐蚀监测和对工艺防腐蚀措施实施效果的监督。

3. heating furnace operation [7]

Strengthen the monitoring of the sulfur content of high-pressure gas. The sulfur concentration of high-pressure gas fuel gas should be less than 100 mg/m^3, and strengthen the daily gas dehydration. Optimize the furnace fire adjustment, ensure that the furnace fire is "multi-nozzle, homogeneous flame, and short flame", adjust the gas injection volume, ensure that the heating furnace heat load is uniform, timely adjust the furnace oxygen content and negative pressure, the furnace tube skin temperature does not exceed the design value, increase the gas injection volume appropriately, and reduce the material residence time in the furnace tube. The furnace tube of the heating furnace shall be 15CrMo, the furnace tube skin temperature shall not be greater than 650 ℃, and the furnace temperature shall not be greater than 780 ℃. Control the flue gas temperature of heating furnace at 150~170 ℃ to ensure that the flue gas temperature is higher than the flue gas dew point temperature. The air preheating system regularly drains the accumulated liquid to slow down the flue gas dew point corrosion.

3.加热炉操作 [7]

加强对高压瓦斯硫含量监控，高压瓦斯燃料气含硫质量浓度应小于100mg/m^3，日常加强瓦斯脱液。优化加热炉炉火调整，确保炉火“多火嘴、齐火苗、短火焰”，调整注气量，保证加热炉热负荷均匀，及时调整炉膛氧含量及负压，炉管表皮温度不超设计值，适量增大注气量，减少物料在炉管内停留时间。加热炉炉管使用15CrMo，加热炉炉管表皮温度不大于650℃，炉膛温度不大于780℃。控制加热炉排烟温度在150~170℃，确保排烟温度高于烟气露点温度，空气预热系统定期排积液，减缓烟气露点腐蚀。

8.5 References

[1] WANG J. Exploration of chemical equipment anti-corrosion strategy [J]. Contemporary chemical research, 2021(17):157-158.

[2] HU Z Y. Research progress of HCl-H_2S-H_2O anti-corrosion at the top of tower of atmospheric pressure reduction unit [J]. Guangzhou Chemical Industry, 2022, 50(2):15-17.

[3] ZHU L J, LIU S M. Corrosion of equipment and pipeline for processing high-sulfur and high-acid crude oil in atmospheric and vacuum distillation unit [J]. Plant Maintenance Engineering. 2022(16): 57-58.

[4] QU Y. Study on corrosion and protection of atmospheric pressure tower top air cooling in atmospheric pressure reducing unit [J]. Shandong Chemical Industry, 2020, 49(20):105-106, 108.

[5] DING Y. Damage forms of catalytic cracking unit liners and their repair methods [J]. Green environmental protection Building Materials, 2019(8):193, 195.

[6] ZHANG S, TIAN X D. Typical corrosion and protection of delayed coking units [J]. Petroleum Refining and Chemical Industry, 2021, 52(7):76-81.

[7] ZHAO Z X, CHEN Y J, DING S W. Main corrosion types and protective measures of delayed coking units [J]. Petrochemical Corrosion and Protection, 2020, 37(1):33-36.

8.5 参考文献

[1] 王婕．化工设备防腐策略探析［J］．当代化工研究，2021（17）：157–158.

[2] 胡芝悦．常减压装置塔顶 HCl–H_2S–H_2O 防腐研究进展［J］．广州化工，2022，50（2）：15–17.

[3] 朱立军，刘尚铭．常减压装置加工高硫高酸原油设备及管道腐蚀［J］．设备管理与维修，2022（16）：57–58.

[4] 曲云．常减压装置常压塔顶空冷腐蚀与防护研究［J］．山东化工，2020，49（20）：105–106，108.

[5] 丁义．催化裂化装置衬里的损坏形式及其修补方法［J］．绿色环保建材，2019（8）：193，195.

[6] 张塞，田晓冬．延迟焦化装置的典型腐蚀与防护［J］．石油炼制与化工，2021，52（7）：76–81.

[7] 赵振新，陈泳健，丁书文．延迟焦化装置的主要腐蚀类型及防护措施［J］．石油化工腐蚀与防护，2020，37（1）：33–36.